Industrial Engineering
in Systems Design

This book focuses on and promotes the applications of the diverse tools and techniques of industrial engineering to the design and operation of systems in the industry, business, the government, and the military. Industrial engineering is growing rapidly as an educational option and is a practice favorite in Asia, South America, and many parts of Europe. This book will meet the needs of those growth markets.

Industrial Engineering in Systems Design: Guidelines, Practical Examples, Tools, and Techniques offers a wide range of engineering tools from checklists to in-depth analysis guidelines for systems design and operation. The book discusses the integration of industrial and systems engineering from both qualitative and quantitative techniques for systems design. In addition, guidelines for operational resiliency for the industry in the case of disruptions, such as a pandemic, are covered, and the book provides case examples for industries in developing and under-developed nations. The inclusion of practical examples of where industrial engineering has contributed to the advancement and survival of industries makes this book a very interesting and useful resource.

This is a practical guide for professional engineers and consultants involved in the design and operation of systems, particularly manufacturing, production, and supply chain systems, and can also be used as a reference for students.

Systems Innovation Book Series

Series Editor: Adedeji Badiru

Systems Innovation refers to all aspects of developing and deploying new technology, methodology, techniques, and best practices in advancing industrial production and economic development. This entails such topics as product design and development, entrepreneurship, global trade, environmental consciousness, operations and logistics, introduction and management of technology, collaborative system design, and product commercialization. Industrial innovation suggests breaking away from the traditional approaches to industrial production. It encourages the marriage of systems science, management principles, and technology implementation. Particular focus will be the impact of modern technology on industrial development and industrialization approaches, particularly for developing economics. The series will also cover how emerging technologies and entrepreneurship are essential for economic development and society advancement.

Data Analytics
Handbook of Formulas and Techniques
Adedeji B. Badiru

Conveyors
Application, Selection, and Integration
Patrick M McGuire

Innovation Fundamentals
Quantitative and Qualitative Techniques
Adedeji B. Badiru and Gary Lamont

Global Supply Chain
Using Systems Engineering Strategies to Respond to Disruptions
Adedeji B. Badiru

Systems Engineering Using the DEJI Systems Model®
Evaluation, Justification, Integration with Case Studies and Applications
Adedeji B. Badiru

Handbook of Scholarly Publications from the Air Force Institute of Technology (AFIT), Volume 1, 2000–2020
Edited by Adedeji B. Badiru, Frank Ciarallo, and Eric Mbonimpa

Project Management for Scholarly Researchers
Systems, Innovation, and Technologies
Adedeji B. Badiru

Industrial Engineering in Systems Design
Guidelines, Practical Examples, Tools, and Techniques
Brian Peacock and Adedeji B. Badiru

Industrial Engineering in Systems Design

Guidelines, Practical Examples, Tools, and Techniques

Brian Peacock

Adedeji B. Badiru

CRC Press
Taylor & Francis Group
Boca Raton London New York

CRC Press is an imprint of the
Taylor & Francis Group, an **Informa** business

First edition published 2023
by CRC Press
6000 Broken Sound Parkway NW, Suite 300, Boca Raton, FL 33487-2742

and by CRC Press
4 Park Square, Milton Park, Abingdon, Oxon, OX14 4RN

CRC Press is an imprint of Taylor & Francis Group, LLC

© 2023 Brian Peacock and Adedeji B. Badiru

Library of Congress Cataloging-in-Publication Data

Names: Peacock, J. Brian, 1938- author. | Badiru, Adedeji Bodunde, 1952- author.
Title: Industrial engineering in systems design : guidelines, practical examples, tools, and techniques / Brian Peacock, Adedeji B. Badiru.
Description: First edition. | Boca Raton : CRC Press, [2023] | Series: Systems innovation book series | Includes bibliographical references and index.
Identifiers: LCCN 2022058525 (print) | LCCN 2022058526 (ebook) |
ISBN 9781032356907 (hbk) | ISBN 9781032357638 (pbk) | ISBN 9781003328445 (ebk)
Subjects: LCSH: Systems engineering.
Classification: LCC TA168 .P35 2023 (print) | LCC TA168 (ebook) |
DDC 620.001/1--dc23/eng/20230111
LC record available at https://lccn.loc.gov/2022058525
LC ebook record available at https://lccn.loc.gov/2022058526

ISBN: 978-1-032-35690-7 (hbk)
ISBN: 978-1-032-35763-8 (pbk)
ISBN: 978-1-003-32844-5 (ebk)

DOI: 10.1201/9781003328445

Typeset in Times
by KnowledgeWorks Global Ltd.

Dedicated to human spirit in the work environment

Contents

Preface

This book represents a continuation of the authors' decades long commitment to promoting and disseminating the application of the diverse tools and techniques of industrial engineering to a variety of operational challenges in business, industry, government, the military, and so on. In this book, the focus is on the role of industrial engineering in systems design. The book presents guidelines, practical examples, models, tools, and techniques of industrial engineering. The diverse sub-fields of industrial engineering are covered in the book to differing extents based on topical relevance and context.

Systems view of work requires a systematic approach to managing work. This is exactly what this book provides. The discipline of industrial engineering, using systems thinking, is very versatile for designing work. The ongoing era of COVID-19 makes it imperative that new processes of work be embraced. The premise of this book is to design work in a way that is adaptive and responsive to the realities of work in a digital era. Work is the basis for accomplishing goals and objectives. Work permeates everything we do. Everyday activities, such as writing, producing, gardening, building, cleaning, cooking, and so on, are around us every day. Not all work is justified, not all work is properly designed, not all work is evaluated accurately, and not all work is integrated. A systems model, such as that presented in this proposed book, will make work more achievable through better management. Work is defined as a process of performing a defined task or activity, such as research, development, operations, maintenance, repair, assembly, production, administration, sales, software development, inspection, data collection, data analysis, teaching, and so on. In essence, work defines everything we do. In spite of its being ubiquitous, a very little explicit guide is available in the literature on how to design, evaluate, justify, and integrate work. A unique aspect of this book is the inclusion of the cognitive aspects of work performance. Through a comprehensive systems approach, this book facilitates a better understanding of work for the purpose of making it more effective, efficient, and rewarding. Topics covered include the definition of work, the hierarchy of needs of workers vis-à-vis the organization's hierarchy of needs, work performance measurement, conceptual framework for work management, analytic tools for work performance assessment, work design, work evaluation, work justification, work integration, and work control. This is exactly why using an industrial engineering approach to design work makes sense.

Brian Peacock

Adedeji Badiru

Author Biographies

Brian Peacock has had a long career in Ergonomics with an undergraduate degree from Loughborough University and a PhD from Birmingham University. His career included faculty positions at the University of Oklahoma, Dalhousie University, Embry Riddle University, and the National University of Singapore; industry (General Motors), and NASA. He has published widely on ergonomics topics.

Adedeji Badiru is a professor of Systems Engineering at the Air Force Institute of Technology (AFIT). He is a registered professional engineer and a fellow of the Institute of Industrial Engineers as well as a Fellow of the Nigerian Academy of Engineering. He has a BS degree in Industrial Engineering, MS in Mathematics, and MS in Industrial Engineering from Tennessee University, and a PhD in Industrial Engineering from the University of Central Florida. He is the author of several books and technical journal articles and has received several awards and recognitions for his accomplishments. His special skills, experience, and interests center on research mentoring of faculty and graduate students.

1 Industrial Engineering in Systems Design

INTRODUCTION

Everything we do is a system. Thus, it is imperative that we commit substantive efforts to the design of systems. Industrial engineering (IE) offers the best avenue for achieving sustainable efficiency, effectiveness, and productivity in systems in the work environment (Peacock, 2019, 2020, 2021). This is even more imperative when systems design is involved because of the multiplicity of factors involved in the interplay of people, machines, and tools. This is why the subdisciplines of human factors and ergonomics always feature prominently in any presentation of IE. That expectation is not different in this book.

As industrial engineers, we specialize in fitting the people, the process, and the tools together, using systems principles and techniques. Figure 1.1 aptly conveys this versatility of the discipline of IE in finding a better way to accomplish operational goals and objectives.

IE cares not only about making widgets but also about what widgets are made of, how they are made, where they are made, when they are made, and why they are made. IE optimizes the pathways toward products, services, and results. The framework used throughout the rest of this book capitalizes on the versatile framework presented in Figure 1.1. Badiru (2014, 2019) and all the references therein present the tools, techniques, and methods of IE that are applicable to this context. As conveyed in Figure 1.1, human factors and ergonomics are foundational to the ways of IE, as a people-oriented discipline.

On a daily basis, business, industry, academia, and government face challenges in human work and performance. Not only must work be done, it must be done productively, safely, profitably, and consistently, which is what IE facilitates, from a systems perspective. Thus, the premise of this book is to use engineering systems methodology for enhancing human work and performance. The engineering foundation for the book is based on the various tools, models, and techniques previously developed and implemented by Dr. Brian Peacock. In this book, the Peacock Tools and Techniques are incorporated into the trademarked DEJI Systems Model, providing a structured engineering process for Design, Evaluation, Justification, and Integration.

To understand the traction provided by IE, we have to recap the respective definitions of a system and the profession of IE.

A system is defined as a *collection of interrelated elements, whose collective output (in unison) is higher than the mere sum of individual outputs of the elements.*

Industrial Engineering – A profession that is concerned with the design, installation, and improvement of integrated systems of people, materials, information, equipment, and energy by drawing upon specialized knowledge and skills in the mathematical, physical, and social sciences, together with the principles and

DOI: 10.1201/9781003328445-1

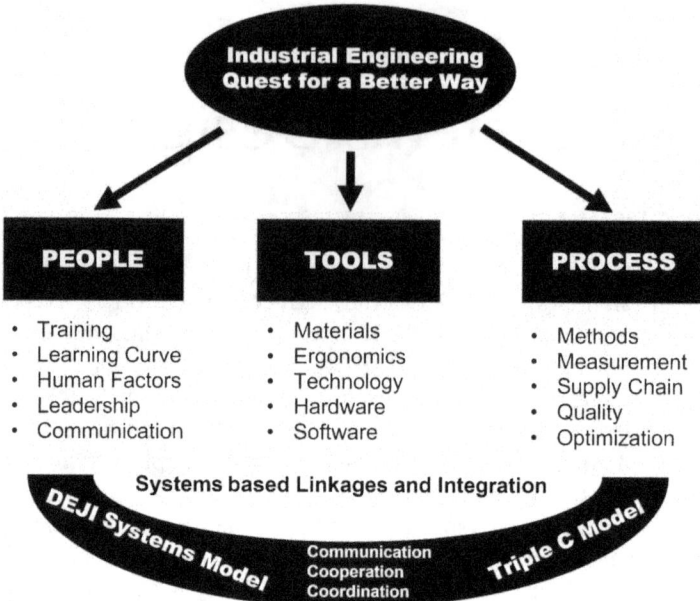

FIGURE 1.1   Industrial engineering and the quest for a better way.

methods of engineering analysis and design to specify, predict, and evaluate the results to be obtained from such systems.

Using systems-thinking approaches, IE makes direct contributions to teamwork, structural communication, critical thinking, smart design, impact assessment, operational literacy, and social responsibility. The above definition embodies the various aspects of what an industrial engineer does. As can be seen, the profession is very versatile, flexible, and diverse. It can also be seen from the definition that a systems orientation permeates the work of industrial engineers. Some of the major functions of industrial engineers involve the following:

- Design integrated systems of people, technology, process, and methods.
- Develop performance modeling, measurement, and evaluation for systems.
- Develop and maintain quality standards for industry and business.
- Apply production principles to pursue improvements in service organizations.
- Incorporate technology effectively into work processes.
- Develop cost mitigation, avoidance, or containment strategies.
- Improve overall productivity of integrated systems of people, materials, and processes.
- Recognize and incorporate factors affecting performance of a composite system.
- Plan, organize, schedule, and control production and service projects.
- Organize teams to improve efficiency and effectiveness of an organization.
- Install technology to facilitate workflow.

- Enhance information flow to facilitate smooth operations of systems.
- Coordinate materials and equipment for effective systems performance.

IE makes systems function better together with less waste, better quality, and fewer resources. The goal of every organization is to eliminate waste. Thus, the above definition is aptly relevant for everyone. IE can be described as the practical application of the combination of engineering fields together with the principles of scientific management. It is the engineering of work processes and the application of engineering methods, practices, and knowledge to production and service enterprises. IE places a strong emphasis on understanding of workers and their needs in order to increase and improve production and service activities. IE activities and techniques include the following:

1. Designing jobs (determining the most economic way to perform work).
2. Setting performance standards and benchmarks for quality, quantity, and cost.
3. Designing and installing facilities.

IE FOUNDATION IN THE INDUSTRIAL REVOLUTION

IE has a proud heritage with a link that can be traced back to the *industrial revolution*. Although the practice of IE has been in existence for centuries, the work of Frederick Taylor in the early twentieth century was the first formal emergence of the profession. It has been referred to with different names and connotations. Scientific management was one of the original names used to describe what industrial engineers do. *Industry*, the root of the profession's name, clearly explains what the profession is about. The dictionary defines industry generally as the ability to produce and deliver goods and services. The "industry" in IE can be viewed as the application of skills and cleverness to achieve work objectives. This relates to how human effort is harnessed innovatively to carry out work. Thus, any activity can be defined as "industry" because it generates a product; be it a service or physical product. A systems view of IE encompasses all the details and aspects necessary for applying skills and cleverness to produce work efficiently. However, the academic curriculum of IE must change, evolve, and adapt to the changing systems environment of the profession.

It is widely recognized that the occupational discipline that has contributed the most to the development of modern society is *engineering*, through its various segments of focus. Engineers design and build infrastructures that sustain the society. These include roads, residential and commercial buildings, bridges, canals, tunnels, communication systems, healthcare facilities, schools, habitats, transportation systems, and factories. Across all of these, the IE process of systems integration facilitates the success of the infrastructures. In this sense, the scope of industrial and systems engineering steps through the levels of activity, task, job, project, program, process, system, enterprise, and society. This handbook of industrial and systems engineering presents essential tools for the levels embodied by this hierarchy of functions. From the age of horse-drawn carriages and steam engines to the present age of intelligent automobiles and aircraft, the impacts of industrial and systems engineering (ISE) cannot be mistaken, even though the contributions may not be recognized in the context of the ISE disciplinary identification.

It is essential to recognize the alliance between "industry" and IE as the core basis for the profession. The profession has gone off on too many different tangents over the years. Hence, it has witnessed the emergence of IE professionals who claim sole allegiance to some narrow line of practice, focus, or specialization rather than the core profession itself. The industry is the original basis of IE and it should be preserved as the core focus, which should be supported by the different areas of specialization. While it is essential that we extend the tentacles of IE to other domains, it should be realized that over-divergence of practice will not sustain the profession. The continuing fragmentation of IE is a major reason to compile a handbook such as this. A fragmented profession cannot survive for long. The incorporation of systems can help to bind everything together.

ENGINEERING ORIGIN OF THE TYPEWRITER

Notable industrial developments that fall under the purview of the practice of IE range from the invention of the typewriter to the invention of the automobile. Ergonomics and human factors provide the functional adeptness of the typewriter, from which we can discern the basis for the design of many modern products. Writing is a basic means of communicating and preserving records. It is one of the most basic accomplishments of the society. The course of history might have taken a different path if early writing instruments had not been invented at the time that they were. The chronological technical evolution of the typewriter is as follows:

1714: Henry Mill obtained British patent for a writing machine.
1833: Xavier Progin created a machine that uses separate levers for each letter.
1843: American inventor, Charles Grover Thurber, developed a machine that moves paper horizontally to produce spacing between lines.
1873: E. Remington & Sons of Ilion, NY, manufacturers of rifles and sewing machines, developed a typewriter patented by Carlos Glidden, Samuel W. Soule, and Christopher Latham Sholes, who designed the modern keyboard. This class of typewriters wrote in only uppercase letters but contained most of the characters on modern machines.
1912: Portable typewriters were first introduced.
1925: Electric typewriters became popular. This made typeface to be more uniform. International Business Machines Corporation (IBM) was a major distributor of this product.

In each case of product development, engineers demonstrate the ability to design, develop, manufacture, implement, and improve integrated systems that include people, materials, information, equipment, energy, and other resources. Thus, product development must include an in-depth understanding of appropriate analytical, computational, experimental, implementation, and management processes.

OPERATIONAL EFFICIENCY

Lee and Johnson (2014) present a comprehensive treatment of operational efficiency, which is the basis for every systems design. The fields of engineering and management associate *efficiency* with how well a relevant action is performed, i.e., "doing things right", and *effectiveness* with selecting the best action, i.e., "doing the right thing". Thus, a firm is *effective* if it identifies appropriate strategic goals, and *efficient* if it achieves them with minimal resources. This chapter focuses on *operational efficiency,* or the ability to deliver products and services cost effectively without sacrificing quality. In this chapter, we investigate a firm's operational efficiency with both queueing models and productivity and efficiency analysis methods that identify maximum productivity and measure efficiency as a ratio of observed productivity to maximum productivity. The maximum productivity level serves as a benchmark for desired performance. The methods for analysis will vary depending on the level of analysis. For example, at the microlevel, we measure operational efficiency at points (machine, workstation, laborer, etc.) on the shop floor, whereas the macrolevel might be at the firm, industry, or national level. We begin by evaluating performance at the operational level, and then apply productivity and efficiency analysis to aggregate performance at higher levels. The analysis of productivity and efficiency is associated with production economics which focuses on assessment and uses an aggregate description of technology to answer questions such as (Hackman, 2008):

- How efficient is the firm in utilizing its inputs to produce its outputs?
- Is the firm using the right mix of inputs or producing the right mix of outputs given prevailing prices?
- How will the firm respond to a price hike in a critical input?
- How efficient is the firm in scaling its operations?
- Has the firm improved its productive capability over time?
- How does the firm compare to its competitors?

The *strategic level* includes long-term planning issues such as make-or-buy decisions. The *tactical level* describes midterm actions that are done perhaps on a weekly or monthly basis, while the *operational level* emphasizes daily scheduling and shop floor control. PEA (Productivity and Efficiency Analysis) supports tactical-level decisions and is part of mid-term production planning. PEA provides performance benchmarking and production guidance. It can also provide ex post analysis to quantify efficiency for complex production processes that use multiple inputs to generate multiple outputs, or ex ante analysis to suggest guidelines for resource allocation.

ABSOLUTE OPERATIONAL EFFICIENCY

Ideal benchmarks to measure efficiency are usually developed in a design laboratory under perfect operating conditions. However, it is not easy to identify the sources of efficiency loss between ideal performance and the best observed performance. For instance, in a manufacturing process operating in perfect conditions, one machine's ideal throughput is 100 units per hour, yet the actual throughput is 80 units per hour due to operator's skill, scheduling, etc. We can estimate an absolute operational efficiency (AOE) as:

$$AOE = \frac{actual\ throughput}{ideal\ throughput} = \frac{80}{100} = 0.8$$

Note that ideal benchmarks can be observed at the machine or process level, but are almost never observed at the firm level. Thus alternative metrics are beneficial in the cases when ideal benchmarks are not observable.

RELATIVE OPERATIONAL EFFICIENCY

Relative operational efficiency (ROE) is the ratio of actual throughput compared to best observed throughput. Relative benchmarks are often used to measure efficiency because a similar comparable machine, process, firm, etc., are often easily identifiable. We estimate ROE by identifying the best observed performance in a data set of multiple operations performing the same task, for instance, a data set of multiple machines performing the same manufacturing process. We find that the best observed throughput is 90 units per hour, but machine A produces 80 units per hour. We can estimate the relative operational efficiency (ROE) of machine A as:

$$ROE = \frac{actual\ throughput}{best\ observed\ throughput} = \frac{80}{90} = 0.88$$

Best observed throughput is often determined by using historical performance data under the assumption, if all conditions are unchanged, actual throughput should be equal to/or close to the historically best performance.

In the real world, a firm's resources are always limited. When a firm would like to provide a product or service, it must consume input resources to generate the output level. In this setting, operational efficiency is determined by the outputs produced as well as the input resources or costs consumed. Thus, we can define productivity and efficiency as:

$$productivity = \frac{output}{input}$$

$$efficiency = \frac{productivity}{productivity\ of\ best\ practice}$$

In other words, productivity is the ratio of the output level to the input level and efficiency is the ratio of the current productivity level to the best practice productivity level. The best practice is defined as the largest productivity achievable.

The relationship between the output levels produced as input levels change is the production function.

EFFICIENCY EVALUATION AND PERFORMANCE INDICES

At the shop-floor level, queueing models provide a method for evaluating machine performance. In the model below, we use the notation M/M/1 to describe the inter-arrive process, the service process for a single-server queueing system. The first M indicates customer arrivals following a Poisson (Markovian) Process and the inter-arrival time is the exponential distribution. The second M indicates the service time follows an exponential distribution. The 1 indicates there is a single server. We use two parameters to describe the M/M/1 queueing system. Let λ be the arrival rate and μ be the service rate. For example, if $\lambda = 2.5$ customers per hour, it means on average 2.5 individuals arrive every hour. Thus, $1/\lambda$ is the mean inter-arrival time and $1/\mu$ is the mean service time. The condition $\lambda < \mu$ is necessary for the system to be stable, i.e., for the queue to be finite in length. $\rho = \lambda/\mu$ is the probability the server is busy and p_0 is the probability the server is idle. p_i is the probability of the server with i customers. We use the following set of algebraic equations to analyze the queue's performance.

In the beginning, we want to know the stable probability p_0. To characterize a queueing system with the state transition between 0 and 1, a rate-balance equation between the arrival rate and service rate can be shown as:

$$\lambda p_0 = \mu p_1 \rightarrow p_1 = \left(\frac{\lambda}{\mu}\right) p_0 = \rho p_0.$$

Intuitively, an empty system needs 1 arrival to become state 1; a system with 1 customer needs 1 departure to become state 0. This idea is the foundation of the rate-balance equation.

Similarly, we can derive the rate-balance equation for state 1 associated with state 0 and state 2.

$$(\lambda + \mu) p_1 = \lambda p_0 + \mu p_2 \rightarrow p_2 = (1 + \rho) p_1 - \rho p_0 = \rho^2 p_0$$

We can also derive a general formula, $p_n = \rho^n p_0$, for the probability that there are n customers in the system (p_n).

We obtain p_0 since the sum of all probability p_n for $n = 1, \ldots, \infty$ must be equal to 1:

$$\sum_{n=0}^{\infty} p_n = p_0 \sum_{n=1}^{\infty} \rho^n = \frac{p_0}{1-\rho} = 1 \rightarrow p_0 = 1 - \rho.$$

Thus, we derive the steady-state probability:

$$P[\text{server idle}] = p_0 = 1 - \rho$$

$$P[\text{server busy}] = 1 - p_0 = \rho = \lambda/\mu \text{ (also called the "utilization")}$$

$$P[\text{n customers in the system}] = p_n = \rho^n (1 - \rho)$$

To derive the probability of n or more customers in the system:

$$\sum_{m=n}^{\infty} p_m = (1-\rho)\sum_{m=n}^{\infty} \rho^m = (1-\rho)\sum_{k=0}^{\infty} \rho^{n+k} = \rho^n(1-\rho)\sum_{k=0}^{\infty} \rho^k = \rho^n(1-\rho)\frac{1}{1-\rho} = \rho^n$$

$$P[\text{n or more customers in the system}] = \rho^n$$

$$P[\text{less than n customers in the system}] = 1 - \rho^n$$

So far the probability distribution of steady state is derived for a single-server queueing system. We can construct two indices to evaluate the queueing system's performance by asking:

What is the expected number of customers in the system/in the queue?
What is the expected time of a customer staying in the system/in the queue?

Let L be the expected number of customers in the system.

$$L = \sum_{n=0}^{\infty} n p_n = \sum_{n=0}^{\infty} n\rho^n(1-\rho) = \sum_{n=0}^{\infty} n(\rho^n - \rho^{n+1}) = 1(\rho^1 - \rho^2) + 2(\rho^2 - \rho^3) + 3(\rho^3 - \rho^4) + \cdots$$

$$= \rho + \rho^2 + \rho^3 + \rho^4 + \cdots = \rho(1 + \rho + \rho^2 + \rho^3 + \cdots) = \frac{\rho}{1-\rho} = \frac{\lambda}{\mu - \lambda}$$

The expected number of customers in the queue, L_q, can be derived similarly. Note that we assume the customer being served is not in the queue, so n customers in the system means the queue length is $n-1$.

$$L_q = \sum_{n=1}^{\infty} (n-1)p_n = \sum_{n=1}^{\infty} n p_n - \sum_{n=1}^{\infty} p_n = L - (1-p_0) = \frac{\rho}{1-\rho} - (1-(1-\rho)) = \frac{\rho}{1-\rho} - \rho$$

$$L_q = L - \rho = L\rho$$

Let W be the expected time spent in the system by a customer. Intuitively, it is equal to the expected number of customers in the system divided by arrival rate λ. The equation is:

$$L = \lambda W$$

This equation, or Little's Law, defines the relationship between L and W. Similarly, $L_q = \lambda W_q$, where W_q denotes the expected time spent in the queue by a customer:

$$W = \frac{L}{\lambda} = \frac{L\rho}{\lambda\rho} = \frac{L_q/\lambda}{\rho} = \frac{W_q}{\rho} \rightarrow W_q = \rho W$$

$$W = \frac{L}{\lambda} = \frac{L_q + \rho}{\lambda} = \frac{\mu L_q + \lambda}{\lambda\mu} = \frac{L_q}{\lambda} + \frac{1}{\mu} = W_q + \frac{1}{\mu}$$

The relationship between W and W_q results because the expected time spent in the system is equal to the expected time spent in the queue plus the mean service time.

Above, Little's Law is defined as a general queueing system. In a manufacturing system, Little's Law is interpreted as the relationship among work-in-process (WIP), throughput (TH), and cycle time (CT):

$$WIP = TH \times CT$$

WIP is the number of unfinished units in the production system, TH is the number of finished products manufactured per unit of time, and CT is the amount of time the units remain in the production system. Given a fixed WIP, an inverse relationship characterizes TH and CT, i.e., an increase in TH will decrease CT.

Little's Law is useful because it applies to a wide variety of production systems. Given a fixed TH, WIP and CT will maintain an almost linear relationship until the capacity limit is approached, but if WIP continues to increase, CT will deteriorate rapidly. When utilization approaches 100%, the increase of arrival rate λ will deteriorate WIP or CT. Thus, $\lambda > \mu$ implies the workstation is no longer stable. The typical performance metrics for queueing systems are utilization and throughput. We calculate utilization as:

$$Utilization = \frac{\lambda}{\mu} = \frac{actual\ throughput}{theoretical\ (ideal)\ throughput}$$

Given CT and the level of WIP, we use Little's Law to calculate the M/M/1 system's productivity by dividing TH by 1. More complicated network analyses are possible with multiple processors linked in a network (Gautam, 2012). Queueing theory can be used to calculate throughput, and productivity can be estimated by dividing throughput by the number of processors. However, all processors may not be identical and throughput will clearly be impacted by the underlying network structure. Further, the human component of operating machines adds additional complications and uncertainty that are difficult to capture in queueing models. Thus production functions are useful for estimating complex systems or firm-level performance.

PRODUCTION FUNCTION AT THE DESIGN LEVEL

A production function $f(x)$ is the maximum output that can be achieved using input vector $x = (x_1, \ldots, x_N)$ (Hackman, 2008). Outputs are units a firm generates and inputs are the factors of production or the commodities used in production. In economics, there are at least five types of factors of production: capital, labor, land, energy, and raw materials. We can analyze the performance of a firm's production system by using either the long-run production function or the short-run production function. In the short run, the factors can be divided into fixed factors and variable factors. Fixed factors are the factors that cannot be changed in the short run such as building and land, and variable factors are the factors that can be changed in the short run such as temporary workers. In the long run, all of the production factors are variable.

Theoretically, four properties characterize a production function (Chambers, 1988; Coelli et al., 2005):

Nonnegativity: the production output is a finite, nonnegative, real number.

Weak essentiality: the production output cannot be generated without the use of at least one input.

Monotonicity: additional units of input will not decrease output; also called nondecreasing in x.

Concavity: any linear combination of the vectors x^0 and x^1 will produce an output that is no less than the same linear combination of $f(x^0)$ and $f(x^1)$. That is, $f(\lambda x^0 + (1-\lambda)x^1) \geq \lambda f(x^0) + (1-\lambda)f(x^1)$. This property implies the "law of diminishing marginal returns".

These properties can be relaxed to model specific production behaviors. For example, monotonicity is relaxed to model input congestion (Färe et al., 1985, 1994) and concavity is relaxed to characterize an S-shaped production function (Frisch, 1964; Henderson and Quandt, 1980).

SHORT-RUN PRODUCTION FUNCTION

Because of the fixed factors in the short run, the production function is characterized by monotonically increasing levels and diminishing returns, i.e., increasing one variable factor of production will increase output levels at a decreasing rate while holding all others constant. The fixed factors limit the growth of the output. This is also called the law of diminishing marginal returns (product). Three concepts of production characterize a short-run production function:

Total product (TP): the total amount of output generated from the production system, $TP = y = f(x)$. Average product (AP): the average amount of output per unit input, $AP = \dfrac{f(x)}{x}$.

Marginal product (MP): the marginal change while adding one more unit of input, $MP = \dfrac{df(x)}{dx}$.

As a firm increases its input levels, the output levels also increase. The firm reaches point A, an inflection point, i.e., where the maximal marginal product is achieved. As inputs continue to increase, the single-input and single-output production function show diminishing marginal product as it reaches the most productive scale size (MPSS). MPSS is the point on the production function that maximizes the average product (or productivity). Finally, input and output levels continue to increase until point B, beyond which input congestion occurs due to the fixed factors and negative marginal product.

LONG-RUN PRODUCTION FUNCTION

All of the factors of production are variable in the long run. Consider production using multiple inputs. It is common practice to plot the relationship between two of the variables while holding all others fixed. Graphical analysis can show the relationship between the inputs x_n and x_m while holding the output fixed at the value y^0 and holding all other inputs fixed. The resulting curve is the input isoquant, which gives all combinations of x_n and x_m capable of producing the same output level y^0. It is convex toward the origin if it satisfies all properties of the production function. For different output levels $y^2 > y^1 > y^0$, these isoquants form nonintersecting functions. The slopes of the isoquants are the marginal rate of technical substitution (MRTS) which measures the rate of using x_n to substitute x_m while holding the output level constant:

$$MRTS_{nm} = -\frac{\partial x_m (x_1,\ldots,x_{m-1},x_{m+1},\ldots,x_N)}{\partial x_n} = \frac{MP_n}{MP_m}$$

$$MP_n\, \partial x_n + MP_m\, \partial x_m = 0$$

where $x_m (x_1,\ldots,x_{m-1},x_{m+1},\ldots,x_N)$ is an implicit function indicating how much x_m is needed to produce the same output level given fixed levels of $x_1,\ldots,x_{m-1},x_{m+1},\ldots,x_N$. Thus, the rate of substitution of input m for input n along the isoquant is equal to the ratio of the marginal productivity of n relative to the marginal productivity of m. To remove the unit of measurement, the direct elasticity of substitution (DES) is the percentage change in the input ratio relative to the percentage change in the MRTS, and quantifies the curvature of the isoquant:

$$DES_{nm} = \frac{d(x_m/x_n)}{d(MP_n/MP_m)} \times \frac{MP_n/MP_m}{x_m/x_n}$$

Next, we describe three typical production functions for a two-input case. Leontief production functions or fixed proportions functions describe production that occurs in fixed proportions, e.g., cars that require wheels (x_n) and bodies (x_m). The mathematical form is $y = \min\{\beta_n x_n, \beta_m x_m\}$ and $\beta_n, \beta_m > 0$. The horizontal part of the isoquant indicates that an increase in x_n does not contribute to the output (y), and $MP_n = 0$ and $MRTS_{nm} = 0$, and that the vertical part of the isoquant indicates that an increase in x_m does not contribute to the output (y), and $MP_m = 0$ and $MRTS_{nm} = \infty$. $MRTS_{nm}$ is not defined at the corner. Therefore, a Leontief production function is used to model production where there is no substitution between x_n and x_m, i.e., $DES_{nm} = 0$.

LINEAR PRODUCTION FUNCTION

A linear production function assumes that inputs are substituted at a constant rate regardless of the level of either input or output. The mathematical form is $y = \beta_n x_n + \beta_m x_m$ and $\beta_n, \beta_m > 0$; The production function implies a constant rate of substitution, $MRTS_{nm} = \frac{MP_n}{MP_m} = \frac{\beta_n}{\beta_m}$, and also imposes perfect substitution between x_n and x_m, i.e., $DES_{nm} = \infty$.

COBB-DOUGLAS PRODUCTION FUNCTION

A Cobb-Douglas production function assumes that inputs are substitutable. However, consistent with the law of diminishing marginal productivity, additional inputs are needed to maintain the same output level as the mix of inputs becomes more skewed. The mathematical form is $y = \alpha x_n^{\beta_n} x_m^{\beta_m}$ and $\alpha, \beta_n, \beta_m > 0$; Computational analysis shows that the production function is a smooth curve and convex toward the origin and that $\mathrm{MRTS}_{nm} = \dfrac{\mathrm{MP}_n}{\mathrm{MP}_m} = \dfrac{\alpha \beta_n x_n^{\beta_n - 1} x_m^{\beta_m}}{\alpha \beta_m x_n^{\beta_n} x_m^{\beta_m - 1}} = \dfrac{\beta_n x_m}{\beta_m x_n}$ decreases with respect to x_n. Thus, substitution exists in this production function and $0 < \mathrm{DES}_{nm} < \infty$.

PROPERTIES OF PRODUCTION FUNCTION

Production functions are convex toward the origin because the absolute value of the slope of the isoquant decreases while increasing x_n, thus, MRTS_{nm} also decreases. This is called the law of diminishing marginal rate of technical substitution. The mathematical representation is $\dfrac{\partial}{\partial x_n} \mathrm{MRTS}_{nm} < 0$.

In addition, if a proportionate increase in all inputs results in a less than proportionate increase in output, we say that the production function exhibits decreasing returns to scale (DRS). Alternatively, if increasing all inputs results in the same proportional increase in output, we say that it exhibits constant returns to scale (CRS). Finally, if the increase of all inputs results in a more than proportionate increase in output, we say that the production function exhibits increasing returns to scale (IRS). There are many reasons why firms may exhibit different returns to scale. For example, a firm may exhibit IRS if hiring more personnel allows specialization of labor, but the firm may eventually exhibit DRS if the firm becomes so large that management is no longer able to control operations. Firms that can replicate all aspects of their operations exhibit CRS. Operating at decreasing returns to scale would indicate decentralization or downsizing might be appropriate whereas operating at increasing returns to scale would indicate mergers, acquisitions, or other changes in organizational structure might be appropriate.

PERFORMANCE AND EFFICIENCY ESTIMATION

IE analysts design the production function to define a benchmark to measure how efficiently production processes use inputs to generate outputs. Given the same level of input resources, inefficiency is indicated by lower levels of output. In a competitive market, if a firm is far from the production function and operates inefficiently, it needs to increase its productivity to avoid going out of business.

Production theory provides a useful framework to estimate the production function and efficiency levels of a firm in three ways: (1) using parametric functional forms in regression-based methods, e.g., SFA (Aigner et al., 1977; Meeusen and van den Broeck, 1977); (2) using nonparametric linear programming methods, e.g., DEA (Charnes et al., 1978; Banker et al., 1984); or (3) integrating regression and programming methods, e.g., StoNED (Kuosmanen and Johnson, 2010; Kuosmanen and Kortelainen, 2012).

STOCHASTIC FRONTIER ANALYSIS

Aigner and Chu (1968) use the logarithmic form of the Cobb-Douglas production function to estimate a deterministic frontier:

$$\ln y_k = x_k'\beta - u_k$$

where $k = 1,\ldots,K$ and y_k indicates the single output of the firm k; x_k is a $I \times 1$ vector with the elements of logarithm inputs; β is a vector of unknown parameters; and u_k is a nonnegative random variable associated with technical inefficiency. Several methods can be used to estimate the parameter β, such as maximum likelihood estimation (MLE) or ordinary least squares (OLS) (Richmond, 1974). However, the Aigner and Chu method neglects statistical noise and assumes that all deviations from the frontier are a result of technical inefficiency. Therefore, Aigner et al. (1977) and Meeusen and van den Broeck (1977) proposed the stochastic frontier production function and introduced the random variable representing statistical noise as:

$$\ln y_k = x_k'\beta + v_k - u_k$$

where v_k models the statistical noise using a symmetric random error. The function is bounded from above due to the stochastic variable $\exp(x_k'\beta + v_k)$. To illustrate, we use a Cobb-Douglas stochastic frontier model with a single input variable:

$$\ln y_k = \beta_0 + \beta_1 \ln x_k + v_k - u_k$$

$$y_k = \exp(\beta_0 + \beta_1 \ln x_k) \times \exp(v_k) \times \exp(-u_k)$$

In this functional form, $\exp(\beta_0 + \beta_1 \ln x_k)$ is the deterministic component, $\exp(v_k)$ is the statistical noise, and $\exp(-u_k)$ is the inefficiency component. Lee and Johnson (2014) illustrate the deterministic frontier $y_k = \exp(\beta_0 + \beta_1 \ln x_k)$, the noise effect, and the inefficiency effect of firm A and firm B. Firm A has a negative random noise component, whereas firm B has a positive noise random noise component. The observed output level $y_k = \exp(\beta_0 + \beta_1 \ln x_k + v_k - u_k)$ and the frontier output level (i.e., without the inefficiency effect) is $y_k^* = \exp(\beta_0 + \beta_1 \ln x_k + v_k)$. The observed output of firm B lies below the deterministic part of the frontier because the sum of the noise and inefficiency is negative. Since SFA estimates the inefficiency effects, we can define the output-oriented measure of technical efficiency (TE) by using the observed output over the frontier output:

$$TE_k = \frac{y_k}{\exp(x_k'\beta + v_k)} = \frac{\exp(x_k'\beta + v_k - u_k)}{\exp(x_k'\beta + v_k)} = \exp(-u_k)$$

This TE_k estimate shows the measure of the observed output of firm k relative to the frontier output of an efficient firm given the same input vector. This benchmarking with best practice provides the estimation of technical inefficiency.

We need to estimate the parameter vector β before calculating TE. Note that the model is complicated by the two random terms, v_i and u_i, where v_i is usually a symmetric error and u_i is a nonnegative term. The parameter β is estimated under the following assumptions:

$$E(v_k u_l) = 0, \ \forall k, l: \text{ uncorrelated}$$

$$E(v_k) = 0: \text{ zero mean}$$

$$E\left(v_k^2\right) = \sigma_v^2: \text{ homoskedastic}$$

$$E(v_k v_l) = 0, \ \forall k \neq l: \text{ uncorrelated}$$

$$E\left(u_k^2\right) = \text{constant: homoskedastic}$$

$$E(u_k u_l) = 0, \ \forall k \neq l: \text{ uncorrelated}$$

Further, v_k and u_k are uncorrelated with the explanatory variables x_k. Note that $E(u_k) \neq 0$ since $u_k \geq 0$.

To estimate β, Aigner et al. (1977) assume $v_k \sim N\left(0, \sigma_v^2\right)$ and $u_k \sim N^+\left(0, \sigma_u^2\right)$, where v_k follows the independently and identically distributed (iid) normal distribution with zero mean and variance σ_v^2, and u_k follows the iid half-normal distribution which is a truncated normal distribution with zero mean and variance σ_u^2. This is called the "half-normal model" in SFA. Under these assumptions, the OLS estimator will provide consistent estimators of slope in β but a downward-biased intercept coefficient since $E(u_k) \neq 0$. Therefore, we use the maximum likelihood estimator (MLE) on the log-likelihood function with $\sigma^2 = \sigma_v^2 + \sigma_u^2$ and $\xi^2 = \sigma_u^2 / \sigma_v^2$:

$$\ln L(y|\beta, \sigma, \lambda) = -\frac{K}{2} \ln\left(\frac{\pi \sigma^2}{2}\right) + \sum_{k=1}^{K} \ln \Phi\left(-\frac{\varepsilon_k \xi}{\sigma}\right) - \frac{1}{2\sigma^2} \sum_{k=1}^{K} \varepsilon_k^2$$

where y is a vector of log-outputs, $v_k - u_k = \ln y_k - x_k' \beta$ defines a composite error term ε_k i, and Φ is a cumulative distribution function of the stand normal random variable. Finally, we use the iterative optimization procedure to estimate the coefficient β (Judge et al., 1985).

DATA ENVELOPMENT ANALYSIS

DEA is an optimization-based approach that imposes the axiomatic assumptions of monotonicity and convexity and the minimum extrapolation principle (MEP) (Banker et al., 1984). MEP identifies the smallest set that satisfies the imposed production assumptions and envelops all the data. Thus, DEA estimates a piecewise linear production function based on the observed data points. The producer on the SDH can decrease the input level without reducing the output level or decrease the output level without changing the input level. We measure the slack in inputs or outputs along the dashed line segments distinguishing the Farrell efficiency measure (Debreu, 1951;

Farrell, 1957) and Koopmans efficiency measure (Koopmans, 1951). The Farrell measure defines technical efficiency as the maximum radial reduction in all inputs consistent with the equivalent production of output. The Koopmans measure states that it is impossible for a firm to increase any output without simultaneously reducing another output (or increasing any input). Note that after all inputs have been radially reduced, additional slack may still exist in some but not all inputs. Thus, a Farrell-efficient firm may not be Koopmans efficient.

Here, we present the widely used Farrell measure. First, we introduce the linear programming technique to estimate the production function and production possibility set. Let $x \in R_+^I$ denote the inputs and $y \in R_+^J$ denote the outputs of the production system. We define the production possibility set as $T \equiv \{(x,y): x \text{ can produce } y\}$. X_{ik} is the ith input resource, Y_{jk} is the amount of the jth production output, and λ_k is the multiplier for the kth firm. The following model defines the feasible region of the production possibility set $\tilde{T}$. This is called the variable return to scale (VRS) DEA model (Banker et al., 1984) because decreasing marginal product is observed along the frontier:

$$\tilde{T} = \left\{ (x,y): \sum_k \lambda_k Y_{jk} \geq Y_j, \forall j; \sum_k \lambda_k X_{ik} \leq X_i, \forall i; \sum_k \lambda_k = 1; \lambda_k \geq 0, \forall k \right\}$$

We use the DEA estimator to measure the efficiency. We describe the input-oriented technical efficiency (ITE) as measured using the distance function $D_x(x,y) = \inf\{\theta \mid (\theta x, y) \in \tilde{T}\}$.

Input-oriented DEA efficiency model

$$\min_{\theta} \left\{ \theta \mid \sum_k \lambda_k Y_{jk} \geq Y_j, \forall j; \sum_k \lambda_k X_{ik} \leq \theta X_i, \forall i; \sum_k \lambda_k = 1; \lambda_k \geq 0, \forall k \right\}$$

Output-oriented DEA efficiency model

$$\max_{\omega} \left\{ \omega \mid \sum_k \lambda_k Y_{jk} \geq Y_j \omega, \forall j; \sum_k \lambda_k X_{ik} \leq X_i, \forall i; \sum_k \lambda_k = 1; \lambda_k \geq 0, \forall k \right\}$$

We calculate $\theta = 1/\omega$ from the output-oriented DEA efficiency model i to get an output-oriented technical efficiency (OTE), θ. $\theta = 1$ implies an efficient firm and $\theta < 1$ implies an inefficient firm. As an illustrative example (see Lee and Johnson, 2014), three firms, A, B, and C are located in an input space constructed by holding the output level constant at $y = \bar{y}$. The solid line is the piecewise linear efficient frontier estimated by DEA. Firms B and C are located on the frontier, but firm A is on the interior of the estimated PPS, $\tilde{T}$. Using the Farrell measure to estimate the technical efficiency shows that the inputs of firm A can be reduced radially. Point D is the intersection of the line segments $\overline{OA}$ and $\overline{BC}$. In fact, point D is a convex combination of firms B and C. We estimate Firm A's technical efficiency as:

$$TE_A = \theta = D_x(x_A, y_A) = \frac{\overline{OD}}{\overline{OA}}$$

The benefits of both SFA (stochastic frontier analysis) and DEA can be achieved using the nonparametric regression approach, StoNED. The first stage of StoNED uses convex nonparametric least squares (CNLS) proposed by Hildreth (1954) and extended by Hanson and Pledger (1976) to estimate a function satisfying continuity, monotonicity, and global concavity – the standard regularity conditions for a production function. To include both random noise and technical inefficiency, Kuosmanen and Kortelainen (2012) combine the CNLS piecewise linear production function with the composite disturbance term concept from SFA. Let $x_k \in R_+^I$ be an input vector, $y_k \in R_+$ be an output, and f be an unknown frontier production function satisfying continuity, monotonicity, and concavity. The regression model is:

$$y_k = f(x_k) + \varepsilon_k \qquad \forall k = 1,\ldots,K$$

where ε_k is a disturbance term with $E(\varepsilon_k) = 0 \; \forall k$, $Var(\varepsilon_k) = \sigma^2 < \infty \forall i$ and $Cov(\varepsilon_k \varepsilon_j) = 0 \; \forall k \neq j$. We formulate the CNLS problem as the quadratic program:

$$\min_{\alpha,\beta,\varepsilon} \sum_k \varepsilon_k^2$$

s.t.
$$\varepsilon_k = y_k - (\alpha_k + x_k' \beta_k) \qquad \forall k = 1,\ldots,K$$
$$\alpha_k + x_k' \beta \leq \alpha_h + x_k' \beta_h \qquad \forall h, k = 1,\ldots,K$$
$$\beta_k \geq 0 \qquad \forall k = 1,\ldots,K$$

where α_k and β_k are the coefficients characterizing the hyperplanes of the frontier production function f. Note that α_k and β_k are specific to each firm k. The objective function minimizes the sum of squared disturbance terms. The equality constraint defines the disturbance term as the difference between an observed output and an estimated output. The inequality constraints comprise a system of Afriat inequalities (Afriat, 1972), imposing the underlying frontier production function to be continuous and concave. The last constraints enforce monotonicity. Unlike DEA, CNLS uses all of the data points to estimate a production function, making it more robust to outliers.

The CNLS estimator of the production function, $\hat{f}(x)$, is generally not unique, but the fitted output values at observed inputs, $\hat{f}(x_k)$, are unique (Kuosmanen, 2008). In fact, given the fitted output values, it is possible to derive the tightest lower bound of the frontier production function as the explicit lower bound representor function:

$$\hat{f}_{min}(x) = \min_{\alpha,\beta} \left\{ \alpha + x_k' \beta \mid \alpha + x_k' \beta \geq \hat{y}_k \forall k = 1,\ldots,K \right\}$$

where $\hat{y}_k = \hat{f}(x_k)$ is the fitted output value. Since the tightest lower bound $\hat{f}_{min}$ is a piecewise linear function satisfying continuity, monotonicity, and concavity, we can use it as the unique CNLS estimator of the frontier production function f.

StoNED uses a similar approach to SFA for modeling inefficiency and noise terms. Consider the composite disturbance term:

$$\varepsilon_k = v_k - u_k \qquad \forall k = 1,\ldots,K$$

where the same properties for v_k and u_k are assumed.

The composite disturbance term violates the Gauss-Markov property that $E(\varepsilon_k) = E(-u_k) = -\mu < 0$; therefore, we modify the composite disturbance term as:

$$y_k = \left[f(x_k) - \mu\right] + \left[\varepsilon_k + \mu\right] = g(x_k) + \vartheta_k \qquad \forall k = 1, \ldots, K$$

where $\vartheta_k = \varepsilon_k + \mu$ is a modified composite disturbance with $E(\vartheta_k) = E(\varepsilon_k + \mu) = 0$ and $g(x_k) = f(x_k) - \mu$ is an average production function. Since g inherits the continuity, monotonicity, and concavity, the CNLS method can find the estimator of the average production function g. We formulate the composite disturbance CNLS problem as:

$$\min_{\alpha, \beta, \vartheta} \sum_k \vartheta_k^2$$

s.t.
$$\vartheta_k = y_k - (\alpha_k + x_k'\beta_k) \qquad \forall k = 1, \ldots, K$$
$$\alpha_k + x_k'\beta_k \le \alpha_h + x_k'\beta_h \qquad \forall k, h = 1, \ldots, K$$
$$\beta_k \ge 0 \qquad \forall k = 1, \ldots, K$$

where α_k and β_k are the coefficients that characterize the hyperplanes of the average frontier production function g. Note that the composite disturbance CNLS problem only differs from the CNLS problem the sum of squared modified composite disturbances is minimized.

To illustrate the StoNED estimator, 100 observations of a single-input single-output Cobb-Douglas production function are generated, $y = x^{0.6} + v - u$. The observations, x, were randomly sampled from a Uniform [1,10] distribution, v was drawn from a normal distribution with a standard deviation of 0.5, and u was drawn from a half-normal distribution with a standard deviation of 0.7.

The second stage of StoNED uses the modified composite residuals, $\hat{\vartheta}_k \ \forall k$, to separate the technical inefficiencies and random noises by applying the method of moments (Aigner et al., 1977; Kuosmanen and Kortelainen, 2012). Assuming that technical inefficiency has a half-normal distribution, $u_k \sim \left|N(0, \sigma_u^2)\right|$, and that random noise has a normal distribution, $v_k \sim N(0, \sigma_v^2)$, the estimated standard deviation of technical inefficiency and random noise is:

$$\hat{\sigma}_u = \sqrt[3]{\frac{\hat{M}_3}{\left(\frac{2}{\pi}\right)\left(1 - \frac{4}{\pi}\right)}}$$

$$\hat{\sigma}_v = \sqrt{\hat{M}_2 - \left(\frac{\pi - 2}{\pi}\right)\hat{\sigma}_u^2}$$

where $\hat{M}_2 = \frac{1}{n}\sum_k \left(\hat{\vartheta}_k - \hat{E}(\vartheta_k)\right)^2$ and $\hat{M}_3 = \frac{1}{n}\sum_k \left(\hat{\vartheta}_k - \hat{E}(\vartheta_k)\right)^3$ are the second and third sample central moments of the modified composite residuals. Moreover, $\hat{M}_3$ should be negative so that $\hat{\sigma}_u$ is positive. Intuitively, the composite residuals should have negative skewness reflecting the presence of technical inefficiency. We calculate the expected technical inefficiency by:

$$\hat{\mu} = \hat{\sigma}_u \sqrt{2/\pi}.$$

Given $\left(\hat{\alpha}_k, \hat{\beta}_k\right)$ from the CNLS problem, we write the unique StoNED estimator of the frontier production function as:

$$\hat{f}_{min}(x) = \min_{\alpha,\beta}\left\{\alpha + x_k'\beta \mid \alpha + x_k'\beta \geq \hat{y}_k \; \forall k = 1,\ldots,K\right\} + \hat{\mu}$$

where $\hat{y}_k = \min_{h\in\{1,\ldots,n\}}\left\{\hat{\alpha}_h + x_k'\hat{\beta}_h\right\}$. We obtain the unique CNLS estimator of the average frontier production function, $\hat{g}_{min}$, by using the tightest lower bound representor function with the fitted output values, $\hat{y}_k$. Recall that $\hat{y}_k$ is calculated from the representor function and $\left(\hat{\alpha}_k, \hat{\beta}_k\right)$. Therefore, we obtain the frontier production function by additively shifting the unique CNLS estimator of the average frontier production function upward by the expected value of technical inefficiency.

Given $\hat{\sigma}_u$ and $\hat{\sigma}_v$, the method introduced by Jondrow et al. (1982) can estimate firm-specific inefficiency. Specifically:

$$\hat{E}\left(u_k|\hat{\varepsilon}_k\right) = -\frac{\hat{\varepsilon}_k\hat{\sigma}_u^2}{\hat{\sigma}_u^2 + \hat{\sigma}_v^2} + \frac{\hat{\sigma}_u^2\hat{\sigma}_v^2}{\hat{\sigma}_u^2 + \hat{\sigma}_v^2}\left[\frac{\phi\left(\hat{\varepsilon}_k/\hat{\sigma}_v^2\right)}{1 - \Phi\left(\hat{\varepsilon}_k/\hat{\sigma}_v^2\right)}\right]$$

where $\hat{\varepsilon}_k = \hat{\vartheta}_k - \hat{\mu}$, ϕ is the standard normal density function, and Φ is the standard normal cumulative distribution.

CONCEPT OF EFFICIENCY IMPROVEMENT

Overall equipment effectiveness (OEE) is a time-based metric to assess productivity and efficiency, particularly for the semiconductor manufacturing industry (Gilliland et al., 1995; SEMI, 2000, 2001; de Ron and Rooda, 2005). The traditional single index metrics of productivity, throughput, and utilization do not allow easy identification of the root cause for reduced productivity. The OEE definition describes six standard equipment states:

Nonscheduled state: equipment is not scheduled to be used in production, such as unworked shifts, weekends, or holidays (including startup and shutdown).
Unscheduled down state: equipment is not in a condition to perform its intended function due to unplanned downtime events, e.g., maintenance delay, repair, change of consumables or chemicals, and out-of-spec input.
Scheduled down state: equipment is not available to perform its intended function due to planned downtime events, e.g., production test, preventive maintenance, and setup.
Engineering state: equipment is in a condition to perform its intended function but is operated to conduct engineering experiments, e.g., process engineering, equipment engineering, and software engineering.

Standby state: equipment is in a condition to perform its intended function but is not operated; the standby state includes no operator available (including breaks, lunches, and meetings), no items available (including no items due to lack of available support equipment), and no support tools.

Productive state: equipment is performing its intended functions, e.g., regular production (including loading and unloading of units), work for third parties, rework, and engineering runs done in conjunction with production units.

We define OEE as:

$$OEE = \frac{\text{theoretical production time for effective units}}{\text{total time}}$$

We decompose OEE into the following subcomponents: availability efficiency (AE), operational efficiency (OE), rate efficiency (RE), and quality efficiency (QE) (de Ron and Rooda, 2005):

$$OEE = AE \cdot (OE \cdot RE) \cdot QE = \text{Availability} \cdot \text{Performance} \cdot \text{Quality}$$

where

$$\text{Availability} = AE = \frac{\text{equipment uptime}}{\text{total time}}$$

$$\text{Performance} = OE \cdot RE$$

$$OE = \frac{\text{production time}}{\text{equipment uptime}}$$

$$RE = \frac{\text{theoretical production time for actual units}}{\text{production time}}$$

$$\text{Quality} = QE = \frac{\text{theoretical production time for effective units}}{\text{theoretical production time for actual units}}$$

The availability captures the difference between machine breakdown and processing. Performance characterizes the production time and throughput. The quality is described by the yield metric which is typically driven by scrap, rework, defects, and reject types. In other words, OEE is a metric to estimate the efficiency of theoretical production time for effective units. In particular, the theoretical production time means the production time without efficiency losses. In addition, two popular indices can be integrated into the OEE framework: mean time between failure (MTBF) or the average time a machine operates before it fails, and mean time to repair (MTTR) or the average time required to repair a failed component and return the machine to operation:

$$AE = \frac{\text{equipment uptime}}{\text{total time}} = \frac{\text{equipment uptime}}{\text{equipment uptime} \times \dfrac{(MTTR + MTBF)}{MTBF}} = \frac{MTBF}{MTTR + MTBF}$$

OEE has two practical benefits. First, we can use its subcomponents to identify bottlenecks and improve productivity. In general, machines with high utilization are typically the bottlenecks. Because bottlenecks can shift depending on the product mix, it is important for engineers to identify and release bottlenecks quickly to maintain high throughput levels. Note that the utilization is a necessary condition for bottleneck identification, but does not mean that all high-utilization machines are bottlenecks. If the processing time of each product is the same and the variation in the production line is low, a machine may have high utilization without affecting throughput. Second, we can use OEE to separate the machine's status into regular operating conditions and down. The availability level quantifies the time used for production. A lower throughput is sometimes the result of low availability rather than poor performance. Thus, OEE decomposition helps with machine diagnosis and productivity improvement.

LEAN THINKING AND MANUFACTURING

Lean manufacturing has its roots in the manufacturing processes developed by Henry Ford in the 1920s. The Ford Motor Company increased its revenue during the post-World War I depression by developing assembly line methods and eliminating activities that were either unnecessary or did not add value to the cars produced. Toyota coined the name and the concept of lean manufacturing in its production system in the 1980s, and also developed additional supporting methods and concepts such as the Just-in-Time (JIT) system (Ohno, 1988a, 1988b). We call a production system "lean" if it produces the required output levels with minimal buffering costs.

In fact, the only time a machine adds value is when it processes a part. Industrial analysts a Gantt chart to visualize processing time, transportation time, and wait time. Note that loading products into tools is handling, not processing, and thus a nonvalue-adding activity. Most of the processing time of a product is waiting and nonvalue-adding activities. Smith (1998) proposed a manufacturing performance index called manufacturing cycle efficiency:

$$\text{Manufacturing cycle efficiency} = \frac{\text{Value} - \text{adding time}}{\text{Total cycle time}}$$

He pointed out that this index is often less than 1% in practice, meaning that firms usually waste resources performing nonvalue-adding activities. The basic philosophy of lean manufacturing is to eliminate the waste by buy(ing) only enough material to fit the immediate needs of the production plan considering the transportation resource.

In the following, we describe the three main principles of lean manufacturing:

Waste elimination
Continuous flow
Pull production system

BENEFITS OF LEAN IMPLEMENTATION

Waste elimination reduces all forms of waste in the manufacturing process. Continuous flow smoothes and balances the production flow. Pull production system, or "make-to-order production" allows a firm to produce units only when it receives an order. There are four steps to implementing lean manufacturing:

Eliminate waste: seven types of waste are identified and need to be eliminated.
Use buffers: build up, adjust, and swap buffers to manage for variability.
Continuous improvement: a commitment to productivity improvement.
Reduce variability: identify and reduce internal and external causes.

A firm can allocate resources dynamically and switch buffers to manage internal or external variability. Internal variability results from uncertain processing times, setups, machine breakdown, yield loss, rework, engineering change orders, etc., and external variability results from demand fluctuation, customer change orders, supplier uncertain delivery, etc. Lean manufacturing uses three buffers: inventory, capacity, and time. Inventory hedges against uncertain demand. Capacity is somewhat flexible due to hiring/layoffs of temporary workers, adjusting overtime, or outsourcing some activities. The time coordinates supply chain or manufacturing activities.

The benefits of lean manufacturing include:

Productivity improvement
Total manufacturing time saved
Less scrap
Lower inventory
Quality improvement
Plant space saved
Better labor utilization
Lower production cost, higher profits, and wages
Shorter cycle time: make-to-order vs make-to-stock
Safety of operations

WASTE ELIMINATION

Womack and Jones (2003) describe seven types of "muda", or waste, in a production system:

Transportation: move products or materials that are not being processed between workstations, or between supplier and customer.
Inventory: hold excess inventory of raw materials, WIP, or finished units.
Motion: worker or tools move more than necessary, such as picking, sorting, stacking, or storing parts.
Waiting: wait for upcoming tools, materials, parts, or for the next production step.
Overproduction: generate excess products beyond the demand level.
Overprocessing: working more than is necessary because of poor tool or product design.
Defects: cost of poor quality such as rework, scrap, inspection, and repair.

In general, all seven types of waste described above belong to the category of nonvalue-adding activities.

CONTINUOUS FLOW

Continuous flow, or the series of continuous and smooth processes, is the second principle. Each production step performs only the jobs necessary for the next step. Workstations do not hold unnecessary WIP and materials that block incoming and downstream flows.

SINGLE MINUTE EXCHANGE OF DIE (SMED)

Single Minute Exchange of Die (SMED), or "Shingo", can significantly reduce setup time and improve productivity. Long setup time leads to a small number of setups, larger batch sizes, larger WIP inventories, and poor process flow. SMED divides the setup time into internal and external activities. An internal activity is one that can only be done when the machine is stopped such as multi-chambers adjustments; an external activity is anything that can be performed before or after the setup without stopping the machine, such as pre-heating of raw material. To achieve a quick setup and changeover of dies, SMED recommends reducing internal setup time or converting internal activities to external activities.

PRODUCTION LINE BALANCING

Line balancing is a typical problem of the assembly system design in IE (Nof et al., 1997). To compensate for demand fluctuations, the goal is to organize tasks into different groups with each group taking the same amount of time. The line balancing problem is an NP-hard problem (Garey and Johnson, 1979); thus, heuristic methods are usually applied to provide good solutions. Helgeson and Birnie (1961) proposed a heuristic method called the ranked positional weight technique. This heuristic is a task-oriented technique considering the combination of precedence relationships and task processing time. Three steps are applied in this algorithm. Calculate the positional weight (PW) of each task using the processing time (PT) of the task plus the processing time of all tasks having this task as a predecessor.

RANK ORDERING OF TASKS

Assigns tasks to workstations sequentially in the ranked order, given the precedence relationships and CT (cycle time) constraint. If the CT is 10 minutes for each workstation, we calculate the minimal number of workstations according to the sum of the 8 task times over the CT, i.e., 38/10 = 3.8 and round up to 4. However, this minimum number does not consider the precedence constraints. Thus, we use the ranked PW technique for line balancing. We find that the required number of workstation is 5 and the total idle time is 12, both of which tend to increase at downstream stations. The smoothing can be done by product type or by volume; both are quite efficient and can bring substantial efficiencies and savings. Note that a smoothed and continuous flow can be reviewed from a firm's internal production or its supply chain. The benefits include:

- Enhance flexibility by reducing the batch size to accommodate changes in product mix or demand fluctuation.

- Reduce material, WIP and inventory levels since there is no severe over or under production.
- No bottlenecks because of similar burdens for each workstation.
- Enhance loyalty and commitment to the firm, i.e., a stable workforce without temporary labor.
- Shorten changeover and setup times to reduce machine idleness.

PULL PRODUCTION SYSTEM

Push systems release work without consideration of system status and hence do not have an inherent limitation on WIP. The work is released based on a schedule of demand and controlled release rates, typically referred to as a due-date-driven production system. A pull system developed by Toyota releases work based on the status of the systems and has an inherent WIP limitation. The system authorizes work releases based on system status and controls WIP level. It is an order-driven production system (Hopp and Spearman, 2004). There are two techniques in the customer-pull production system: just-in-time and kanban. Just-in-time attempts to reduce inventory, holding costs, and WIP using small lot size or even single unit processing. A "kanban" is a signboard or visual for realizing just-in-time and often leads to significant quality improvement. The advantages of using a pull production system include:

- Reduce WIP and CT: limit releases into the production line.
- Improve quality: short queues allow errors to be identified quickly and shut down the production line to correct the problems.
- Reduce cost: switch the control from release rate to WIP level and reduce WIP progressively.
- Logistical benefits: less congestion, easier control, and WIP cap control.
- Kanban provides for efficient lot tracking and predetermines WIP level by the number of kanban.

In fact, based on Little's Law, $WIP = CT \times TH$, given the same rate of throughput, reducing the WIP level will lead to a reduction in CT. Thus, a pull production system reduces CT by controlling the WIP level. For further study of the pull system, see Ohno (1988a, 1988b), Liker (2004), and Nahmias (2009).

CONCLUSIONS

Operational efficiency can be measured and improved using the approaches described in this chapter. Today, many manufacturing firms define a metric for efficiency and concentrate on operational improvement activities to increase it. The specific approaches developed to identify best practice performance or to determine if a particular activity adds value are often product or industry specific. However, the evolution of new – and global – industries will require more sophisticated efficiency analysis techniques and metrics.

REFERENCES

Afriat, S. N. (1972). Efficiency Estimation of Production Functions, *International Economic Review* 13(3): 568–598.

Aigner, D. J., and S. F. Chu (1968). On Estimating the Industry Production Function, *American Economic Review* 58: 826–839.

Aigner, D., C. A. K. Lovell, and P. Schmidt (1977). Formulation and Estimation of Stochastic Frontier Production Function Models, *Journal of Econometrics* 6: 21–37.

Badiru, A. B. (2014), editor; *Handbook of Industrial & Systems Engineering* (2nd ed.). Taylor & Francis CRC Press, Boca Raton, FL.

Badiru, A. B. (2019). *Systems Engineering Models: Theory, Methods, and Applications.* Taylor & Francis/CRC Press, Boca Raton, FL.

Banker, R. D., A. Charnes, and W. W. Cooper (1984). Some Models for Estimating Technical and Scale Inefficiencies in Data Envelopment Analysis, *Management Science* 30(9): 1078–1092.

Chambers, R. G. (1988). *Applied Production Analysis: A Dual Approach.* Cambridge University Press, New York, NY.

Charnes, A., W. W. Cooper, and E. Rhodes (1978). Measuring the Efficiency of Decision Making Units, *European Journal of Operational Research* 2(6): 429–444.

Coelli, T. J., D. S. Prasada Rao, C. J. O'Donnell, and G. E. Battese (2005). *An Introduction to Efficiency and Productivity Analysis* (2nd ed.). Springer, New York, NY.

de Ron, A. J., and J. E. Rooda (2005). Equipment Effectiveness: OEE Revisited, *IEEE Transactions on Semiconductor Manufacturing* 18(1): 190–196.

Debreu, G. (1951). The Coefficient of Resource Utilization, *Econometrica* 19: 273–292.

Färe, R. S., S. Grosskopf, and C. A. K. Lovell (1985). Technical Efficiency of Philippine Agriculture, *Applied Economics* 17: 205–214.

Färe, R. S., S. Grosskopf, and C. A. K. Lovell (1994). *Production Frontiers.* Cambridge University Press, Cambridge.

Farrell, M. J. (1957). The Measurement of Productive Efficiency, *Journal of the Royal Statistical Society, Series A* 120(3): 253–281.

Frisch, R. (1964). *Theory of Production.* Rand McNally & Company, Chicago, IL.

Garey, M. R., and D. S. Johnson (1979). *Computers and Intractability: A Guide to the Theory of NP-Completeness.* W. H. Freeman, New York, NY.

Gautam, N. (2012). *Analysis of Queues: Methods and Applications.* CRC Press (Taylor and Francis), Boca Raton, FL.

Gilliland, J., J. Konopka, K. Barber, R. Schnabl, and V.A. Ames (1995). Semiconductor manufacturing productivity: Overall equipment effectiveness (OEE) guidelines. Technology transfer 950327443 A-GEN, Revision 1.0. Sematech.

Hackman, S. T. (2008). *Production Economics: Integrating the Microeconomic and Engineering Perspectives.* Springer-Verlag, Heidelberg.

Hanson, D. L., and G. Pledger (1976). Consistency in Concave Regression, *Annals of Statistics* 4(6): 1038–1050.

Helgeson, W. P., and D. P. Birnie (1961). Assembly Line Balancing Using the Ranked Positional Weight Technique, *Journal of Industrial Engineering* 12: 394–398.

Henderson, J. M., and R. E. Quandt (1980). *Microeconomic Theory: A Mathematical Approach.* (3rd ed.). McGraw-Hill, New York, NY.

Hildreth, C. (1954). Point Estimates of Ordinates of Concave Functions, *Journal of the American Statistical Association* 49(267): 598–619.

Hopp, W. J., and M. L. Spearman (2004). To Pull or Not to Pull: What Is the Question? *Manufacturing and Service Operations Management* 6(2): 133–148.

Jondrow, J., C. A. K. Lovell, I. S. Materov, and P. Schmidt (1982). On the Estimation of Technical Inefficiency in the Stochastic Frontier Production Function Model, *Journal of Econometrics* 19(2–3): 233–238.

Judge, G. G., W. E. Griffiths, R. C. Hill, H. Lutkepohl, and T.-C. Lee (1985). *Introduction to the Theory and Practice of Econometrics*. John Wiley & Sons, Inc, New York, NY.

Koopmans, T. (1951). An Analysis of Production as an Efficient Combination of Activities. In: Koopmans, T.C. (Ed.), *Activity Analysis of Production and Allocation*. Cowles Commission for Research in Economics, Monograph No. 13. John Wiley & Sons, Inc, New York, NY.

Kuosmanen, T. (2008). Representation Theorem for Convex Nonparametric Least Squares, *Econometrics Journal* 11: 308–325.

Kuosmanen, T., and A. L. Johnson (2010). Data Envelopment Analysis as Nonparametric Least Squares Regression, *Operations Research* 58(1): 149–160.

Kuosmanen, T., and M. Kortelainen (2012). Stochastic non-Smooth Envelopment of Data: Semi-Parametric Frontier Estimation Subject to Shape Constraints. *Journal of Productivity Analysis*, 20128(1), 11–28.

Lee, Chia-Yen, and Andrew L. Johnson (2014), "Operational Efficiency," Chapter Two. In: Adedeji Badiru (Ed.), *Handbook of Industrial and Systems Engineering*, pp. 17–44. Taylor and Francis/CRC Press, Boca Raton, FL.

Liker, J. K. (2004). *The Toyota Way: 14 Management Principles from the World's Greatest Manufacturer*. McGraw-Hill, New York, NY.

Meeusen, W., and J. van den Broeck (1977). Efficiency Estimation from Cobb-Douglas Production Functions with Composed Error, *International Economic Review* 18(2): 435–444.

Nahmias, S. (2009). *Production and Operations Analysis*. (6th ed.). McGraw-Hill, New York, NY.

Nof, S. Y., W. E. Wilhelm, and H.-J. Warnecke (1997). *Industrial Assembly*. Chapman & Hall, Boca Raton, FL.

Ohno, T. (1988a). *Toyota Production System: Beyond Large-Scale Production*. Productivity Press Inc, New York, NY.

Ohno, T. (1988b). *Just-In-Time for Today and Tomorrow*. Productivity Press Inc, New York, NY.

Peacock, B. (2019). *Human Systems Integration*, Self-published manuscript, Fernandina Beach, FL.

Peacock, B. (2020). *How Ergonomics Works*, Self-published manuscript, Fernandina Beach, FL.

Peacock, B. (2021). *Ergonomics Tools and Applications*, Self-published manuscript, Fernandina Beach, FL.

Richmond, J. (1974). Estimating the Efficiency of Production, *International Economic Review* 15: 515–521.

Smith, W. (1998). *Time Out: Using Visible Pull Systems to Drive Process Improvements*. John Wiley & Sons, New York, NY.

Standard for Definition and Measurement of Equipment Productivity, Semiconductor Equipment and Material International (SEMI) E79-0200, 2000.

Standard for Definition and Measurement of Equipment Reliability, Availability, and Maintainability, SEMI E10-0701, 2001.

Womack, J. P., and D. T. Jones (2003). *Lean Thinking: Banish Waste and Create Wealth in Your Corporation* (2nd ed.). Free Press Simon & Schuster, New York, NY.

2 Human Systems Integration

INTRODUCTION

Everything we do as humans involves enmeshing human requirements with technical requirements (Peacock, 2019, 2020, 2021). Inasmuch as industrial engineering is viewed as a people-oriented profession, we must recognize the integral role of humans in systems design. For example, our transportation vehicles, including boats, trains, cars, and space vehicles require humans in the loop of systems design.

People make systems. People are the key components of any system. Systems-oriented humans form the foundation for the success of programs in business and industry.

This chapter addresses some opportunities for human factors to become involved early in the design cycle. Brief descriptions are given of five case studies in which different human factors approaches and systems engineering tools were applied successfully. The case studies include a mass transit railway, supertankers, cars and car manufacturing, and, finally, space vehicles. All design processes start with requirements – the customer wants the device or service to fulfill some function. Of course, there are conditions associated with the requirements that are articulated in criteria such as quickly and safely and inexpensively. In all cases, the measurement and communication of human factors information benefits from the design of a common communication currency.

When we discuss human factors, it is common to describe a machine, a user, an interface, and an environment, in the broadest sense of the words. In systems design, we focus on the interfaces with the equipment, tools, environments, and organizational structures. Human factors usually faces complex "human-machine systems". For example, an operator in the Space Shuttle may be controlling a robotic arm on the end of which is perched another astronaut. A third, tethered astronaut is outside assisting with the placement of a multimillion-dollar piece of hardware on a space telescope. The environment is characterized by minimal gravity, zero air pressure, alternating hot, cold, light, and dark. The interface has to deal with the control of six- or seven–degrees of freedom "robotic arm" using quite limited camera views and communication facilities while the whole world is watching. Whatever the context the rules are always the same – we must design the tools and tasks so that the users can be successful and safe.

SYSTEMS VISIONS AND MISSIONS

All good ideas start with visions. "Beam me up Scotty". Visions are free, but when we translate a vision into a mission we need funding. We also need specific mission requirements that can be measured so that we may know whether or not the mission has been accomplished. Missions usually have constraints – "put a man on the moon before the end of this decade" or "put a group of people on Mars". Next, come the details – the

DOI: 10.1201/9781003328445-2

specifications of how, what, when, and how much? There may also be a "where" thrown in although usually, the "whys" are beyond question. Most complex missions have a way of costing more than the first estimates because as the plans unfold, problems arise due to a lack of information or opportunistic subcontractors holding the main organization to ransom. The more successful missions have clear requirements and accurate specifications very early in the design process.

In the 1970s, Hong Kong wanted to solve its traffic problem – 5 million people all wanting to go to work at the same time. Rickshaws and old buses could not handle the demand, so a mission to build a mass transit railway was born. The mission had requirements – move 1 million people from A to B quickly, safely, comfortably, and cheaply. And "oh by the way expect to move 2 million people from A to B to C in a few years' time". The plans involved digging a channel up the main street, putting in a tube, and then returning the main street to its former condition a couple of years later. In the 1980s, the oil crisis directed attention to reserves in Arctic Canada, but it was deemed to be unwise to have a ship full of LNG collide with an iceberg or run aground. After considerable human factors, economic and engineering analysis, the plans proved to be untenable. In the 1980s, General Motors was feeling the heat from overseas competition and so developed a mission to improve the quality and productivity of their automobiles and reduce the costs. They embraced many concepts of systems engineering and learned all about "the voice of the customer". They found that there are many external and internal customers and they learned the language of systems engineering. By 1990, the U.S. unions decided that "ergonomics" would help to reduce the effects of the increasing levels of repetition needed to improve productivity. Work-related cumulative trauma disorders began to increase epidemic proportions. The plans included the creation of the General Motors Manufacturing Ergonomics Laboratory and the development of a plant-based "reactive ergonomics program".

The mission to Mars turned out to be more formidable than the mission to the moon in the 1960s, despite the enormous advances in technology. It may be tolerable to lose a robotic mission, but a manned mission had to have more guarantees. So NASA refocused its attention to Low Earth Orbit and the International Space Station (ISS) in a hope to answer some of the questions associated with long-duration manned space flight. Many people still have a vision of going to Mars and many have laid out elaborate plans but at the present time there is no mission. All of these case studies confirmed the importance of clear operational definitions in human factors and systems engineering. Unfortunately, the evolution of this subject area has resulted in ambiguities. The next few pages outline some definitions that may help to improve the reliability of communication as human factors engineering interacts with the system design process.

SYSTEMS AND PROCESSES

A key concept lies in the definition of what is a "system" and what does a "system" do? A convenient operational description of a "system" is any hardware, software, and naturally occurring or human entity that, by themselves, have no functions. When two or more systems interact, in a physical and organizational context, to achieve an objective then this interaction is called a "process". Usually, the objective or outcome of a "process" is a change in the characteristics of one or more contributing systems. In the case of human factors, one of the contributing "systems" is a human "system". For example, a person may have the characteristic of "being at home". Only when he/she interacts with a car and a roadway does he change his location characteristic to "being at work". During this journey, the human subsystem may interact with other systems – such as a coffee cup, a cell phone, a frosty road, and another human-vehicle system to engage in a process that results in an accident – characterized by a change in the shape of the vehicle and the owner's wallet.

Design processes create various human, hardware, information, and organizational systems with the purpose of producing a new entity or service. There are multiple purposes of such processes. First, the product must meet with customer expectations – this in its broadest sense is called product quality. The next process objective is to be efficient or productive; i.e., it must achieve its quality objectives with minimal use of consumable resources (systems) such as people, money, materials, or time. This last resource "time" often stands out as a key aspect of process design. The customer would like the elapsed time between his want being expressed and fulfilled to be as short as possible. "Time to market" is a key objective of most processes. One way of achieving this objective is through the practice of "concurrent engineering" in which phases of the process are implemented in parallel rather than in sequence so that, e.g., the demands of manufacturing can be addressed during the product design phase. These process objectives are of particular interest to the eventual paying customer, management, and shareholders. However, the unique nature of the human system elements is that they may have their own agendas and objectives. For example, employees would like to maximize their own salaries and minimize the risk of accidents, both of which may conflict with other process purposes – such as productivity. A more detailed look at the design process identifies multiple overlapping stages. The term "concurrent" is somewhat optimistic in practice.

Given the vision of putting a man on Mars, there are distinct but interdependent phases that must be addressed. The first phase is a function identification – launching, navigating, eating – each with its own purposes that are characterized by "quality", "productivity", "safety", etc. Next comes the realization of these functions through the design and construction of the appropriate hardware, software, "humanware", and "organizationware". The process integration phase focuses on interactions, interdependencies, and interfaces. Of course, the advantages of concurrent engineering are particularly evident here as, e.g., the human and hardware systems must be compatible. The penultimate phase of operations design really addresses the time element. In the Mars mission example, it is critical that various supplies (food, water, oxygen, shelter, etc.) would be on the planet before the humans arrive. Another good example is to be found in automobile production – it is one thing to design and build a car, but to produce 1000 cars a day presents altogether new operational challenges, not the least of which is just-in-time materials delivery. Finally, there is operations implementation, which has its own local objectives and its contribution to the next mission through "lessons learned".

THE GRAMMAR OF DESIGN

The grammar of design offers a discipline for communication that increases the effectiveness and efficiency of the design process. The first concept is that "processes have requirements" and that requirements relate to the adverbs associated with the process functions (verbs). The process "verb" may be "transporting" a vehicle and human systems in some context or environment. The purposes or objectives of "transporting" may include speed and safety. They will certainly include "quality" – the payload should arrive at the correct destination. Thus "transporting" may be measured in terms of how "quickly", safely", and "accurately" – adverbs. It is important to emphasize that quantification of these adverbs is important if the process requirements are to be reliably assessed.

The achievement of these process requirements will depend on the characteristics of the contributing systems. In the above example, if speed were emphasized, then a vehicle with a big engine and a driver with a heavy foot would assure the desired objective. Again, for precise system design, it is necessary to quantify the adjectives – "big" and heavy" associated with the system nouns – engine and foot. Otherwise, the engineer cannot design the system and the human factors engineer cannot evaluate the quantitative relationships between the system specifications and the process requirements. Give the engineer a number!

VERIFICATION AND VALIDATION

Once the system is built (or modeled) and the process implemented (or modeled), then the human factors engineer is faced with the important task of evaluation. This consists of two sub-processes – verification and validation. If the system characteristics have been specified precisely and quantitatively, then verification of adherence to these specifications is simply a matter of measurement. On the other hand, evaluation of process requirements implies the process of validation, which in turn implies the performance of the interacting systems in a real-world context. The key challenge to validation is the inevitable presence of user, context, and temporal variability. A precisely specified and a constructed car may not "perform" adequately with an inebriated driver in thick fog. Development of the validation process begs the question of "humanware" design – who is the expected user and possible misuser. Validation also may exclude certain contextual conditions such as fog or ice or 100 mph. The contribution of human factors engineering lies in a clear description of usage requirements, user capabilities and limitations, design specifications, and evaluation conditions.

PERFORMANCE, BEHAVIOR AND PREFERENCE

Human factors measures of process requirements may be classified at three levels – performance, behavior, and preference. Performance can usually be measured in terms of time and accuracy, given the context. For example, running a mile on level ground will differ from the time taken to run the same distance uphill. The accuracy (quantitative deviation from the objective) needed to thread a needle is different from that needed to park a car. Behavior relates to how a task is performed. In cricket, it is possible for the bowler to achieve his objective by swinging or spinning. Behaviors can be categorized and counted. Using the cricket example again, three bouncers get the bowler suspended; in baseball, one beamer gets the pitcher mugged. However, these "bench-clearing brawls" (ungentlemanly behaviors) can be modified (despite the preaching of Skinner) by negative feedback – fines. The most elusive measure associated with humans is that of preference. Preferences may be stated and counted but may not affect behavior or performance. However, extraction of the "voice of the customer" or the mechanisms of "usability studies" often resort to the assessment of preferences or subjective judgments of differences. We must be ever vigilant that we do not put too much store in observations elicited by improper application of one of our most widespread techniques – psychophysics. Unfortunately, this is often the only technique we have available.

Working in space involves many processes such as staying in one place, moving, eating, and assembling. The more complex activity of assembling involves the interaction between multiple human systems, components, robotic arms, and communications facilities. The context of microgravity, the vacuum of space, radiation, and very high cost present unique constraints on the process requirements. The interacting tasks of controlling the robotic arm whilst perching exemplify the challenges, especially when the higher level task of assembling an expensive component may take all day. The adverbs related to this task include carefully, slowly, and comfortably. Slowly can be and is defined precisely and carefully is described in terms of deviation from a prescribed, tight trajectory. Comfortably is one of those unfortunate human factors challenges that defy reliable quantification, although it is well known that uncomfortable perching may create a distraction that, in turn, may compromise carefully.

REQUIREMENTS AND SPECIFICATIONS

These process objectives and requirements can only be realized through the precise specification of the contributing system characteristics. What kind of restraint design results in comfortable perching? What kind of joystick design contributes to the activity of careful control of a heavy payload (crew colleagues, components, and tools)? What kind of organization of multiple pairs of eyes and brains is conducive to reliable communicating? How much light should be provided, given that the daylight only lasts 45 minutes up there? How much oxygen should be provided for the strenuous tasks of fighting a pressurized space suit? How long should the workday be? Give the engineer a number.

The development of process requirements and design specifications is not simple. It is rarely possible to simply translate empirical data into a number that can be applied to reliable validation. At a simple level, if asked why the height of a door opening is 7 feet, we may waffle about percentiles and allowances for shoes and hats. Similarly, we may look at accident statistics on the freeway and determine that 100 mph (160 kph) is acceptable for 95% of journeys. Or we may state unequivocally that 15 minutes of arc is the design specification for font height on a computer screen. If we don't give the engineer a number, he can't design or verify. But we all know that the number includes a policy overlay and will usually be modulated by domain experience. Consequently, it is essential that requirements and specifications be developed by consensus, with management (or the law) imposing policy, our scientists providing the logic and the data, and our engineers – the eventual users of the standard – providing the domain experience. Of course, all standards (requirements and specifications) should be subject to iterative evaluation and an effective technical memory should lead us to convergence. Unfortunately, those policymakers often change their minds when faced with tradeoffs.

THE HONG KONG MASS TRANSIT RAILWAY

The main performance requirements of this transportation process were to maximize the safe throughput of passengers, given the constraints of size, speed, and the need to show a profit. Throughput is constrained by spatial capacity and passenger behavior, which in turn is affected by spatial arrangements. The seat design and grab rail specifications were based on anthropometry and human behavior. The approaches used to generate the anthropometry and behavior evidence involved the human factors literature, surveys, analysis, and evaluation of physical mockups. The use of adjustable physical mockups of both the passenger compartment and the operator's cab proved to be very instructive. In fact, there were substantial discrepancies between the simple application of anthropometric accommodation principles and the actual behaviors of representative samples of subjects in the physical mockups. An ironical twist in the seating systems development, based on input from the Hong Kong Fire Department, resulted in the adoption of stainless steel bench seats rather than the scalloped aluminum that was first proposed. This resulted in an adaptive rather than constrained seating arrangement. The vertical poles and horizontal grab rails were positioned to allow optimum accessibility, stability, and motion, given the wide range in anthropometric characteristics of the expected user population. A horizontal bar reachable by a 5th-percentile female would hit a 95th-percentile male on the chin! Compromise!

The operator's seat design followed by a task analysis that indicated that the operator would have to get in and out of the train every 90 seconds as he checked that the platform was clear prior to starting the train. This resulted in the design of a seat that could be folded back (allowing easy egress in an emergency) or in the down position for the longer between station transits. The seat also had a padded front edge to accommodate the preferred lean sitting posture. In these examples, human factors was applied in a somewhat ad hoc way in the very early design stages. The principal "tool" was a physical mock-up evaluation.

LIQUEFIED NATURAL GAS TRANSPORT

In the early 1980s, serious consideration was given to the exploitation of the vast oil and gas reserves in Arctic Canada. The two transportation options were pipelines and large double-hulled ice-breaking tankers. Given the cargo and the context, there were substantial safety concerns – a collision or a grounding could result in a cloud of escaped gas descending on a town and then exploding. The preferred analytic approach was to use fault tree analysis – both for the mechanical and electrical systems and the human systems. It should be noted that, unlike military vessels, commercial vessels are designed to be operated with very small crews – thus, reducing the human redundancy in case of error. The approach to the assessment of human error was based on the human reliability assessment techniques developed for the nuclear power plant industry at Sandia National Laboratories. A massive (paper) fault tree was developed and an assessment of the performance-shaping factors indicated that an incapacitated crewmember would be a likely cause of catastrophic failure. Six years later, the Exxon Valdez confirmed the findings. Fortunately, the LNG project was abandoned for a combination of environmental, engineering, economic, and human factors reasons.

CAR AND TRUCK DESIGN

Car and truck design is a fashion business. Some say function before form, whereas others say form before function. This can be translated into preference before a performance. There are, however, basic functional requirements of capacity, operating, maintaining, etc., before aesthetics takes over. Human factors contributions in product design covers the full spectrum of customer needs – from basic physical issues, through sensory and information processing to their requirements for alternative features and styling. Different vehicle types attract different customers and have different uses. Much of this evidence is elicited early in the design process through competitive review, clinics, and more precise laboratory investigations, involving simulators of various levels of sophistication. As the design process progresses, various iterations of prototypes are assessed using modeling, checklists, and "drives" on closed courses and the open road. The formal process is iterative and involves concept evaluation, selection, and refinement, through processes of analysis, testing, and board review.

Quality Function Deployment is a technique that has been applied widely in the automotive industry to translate the voice of the customer into design specifications and on down through the manufacturing, production, and distribution processes. Early uses of the technique resulted in very large and unwieldy matrices that became increasingly less than useful. However, the principles are sound and lend themselves well to the discipline of requirements and specifications development. Unfortunately, the user (customer) does not always adhere to these grammatical rules of design. The dutiful customer should ask for a vehicle that "goes fast", "is easy to maintain", "enhances his social image", and "protects him in the event of an accident." Instead, the customer may stipulate engineering nouns and adjectives such as: 300 horsepower, maintenance-free, red vehicle with side air bags. This lack of discipline occurs among the many internal and external customers.

An example of QFD in product design would be to address the operation or driving of the vehicle. One adverb might be top speed and the range of top speeds of competitive vehicles might be available from market research. The engineer would recognize that top speed would be accomplished by, among other things, engine size, which would be described in engineering units of cu ins. The adverb "fast" (speed) is, of course, affected by more factors than engine size – there is the mass of the car, the gearing ratio, the aerodynamics and the type of fuel, etc. Similarly, the "safety" requirement might conflict with the "fast" requirement. Thus the task of the human factors engineer becomes more complex in the optimization of conflicting requirements.

The human factors engineer is a surrogate for the end user. He should identify the populations of interest on the dimensions of interest. He should communicate clearly with the design engineer by relating the associations between levels of design specifications (independent variables) and performance outcomes (dependent variables). The relationships are affected by human variability, which can only be reduced by curtailing the "expected user population" by selection or training. There will be many occasions where the relationships are affected by the prevailing conditions and by interactions with other variables; there may also be multiple, sometimes conflicting outcomes. Eventually, the designer will have to settle for a single value on each dimension, unless he can design an adjustable feature. The task of the human factors engineer is to communicate the acceptable ranges for each independent variable, given the percent accommodation policy, and to articulate the likely sources of interaction.

DESIGN FOR MANUFACTURING AND ASSEMBLY

Contemporary vehicles have many more features and components, especially on the engines, than they had a few decades ago. However, the drivers have not changed in stature and so it is not possible to increase the height of the hood. Thus more things have to be compressed into a smaller space, which produces challenges for packaging, assembling, and maintenance. Contemporary "design for assembly" approaches use mock ups and computer models to assess these manufacturing challenges. There are also certain well defined, though not always feasible, ground rules for design – such as layered assembly and upward and outward facing fasteners. Given the best possible design, with manufacturability in mind, the ergonomist is next faced with materials delivery and presentation, tools, workplaces, and task content. This last challenge of "line balance" attempts to maximize the utilization of every second of the assembly operation. The ergonomist looks for ways of increasing physical (and mental) job variety through team structure, job enlargement, and rotation but may be constrained by seniority agreements and quality concerns.

One fundamental challenge of manufacturing ergonomics lies in the difficulties associated with measuring people in their working environment. These conditions do not lend themselves to the rigorous demands of the experimental laboratory for accuracy, precision, reliability, and even sometimes validity. The sample size is also usually restricted. Consequently, the thrust of manufacturing ergonomics should be the assessment of the workplace using population data while allowing the eventual individual operator(s) to fine-tune the arrangements to suit their particular needs. Systems approach to workplace and task evaluation using various levels of analytic tools should therefore limit themselves to population data.

Manufacturing ergonomics assessments are applied at all stages of the manufacturing system design and implementation process. These assessments take the form of computer modeling, "wall reviews", "prototype reviews", and "slow build" reviews in which each motion is evaluated in great detail. This upfront assessment leads to much improved designs of production systems. The practicing manufacturing ergonomist has made available a wide variety of analysis tools that range from checklists through integrated analysis methods to digital simulations.

Ultimately any design, design change, or operational intervention will be based on a risk cost – benefit assessment. The solution may be a change in the component (the product engineer's responsibility), a change in the tool or workplace (the manufacturing engineer's responsibility), a change in the amount of work in the job cycle or the line rate (the industrial engineer's responsibility), and a change in who does the job (the supervisor's responsibility with due regard to seniority) or the method by which the job is performed (the operator's or trainer's responsibility).

OPPORTUNITIES FOR INTERVENTION

The application of systems engineering and human factors in car and truck manufacturing addresses the opportunities for change that are presented during each of the product, manufacturing process, production, and operations phases. By way of example, rather than use a specialized manufacturing process, one can consider the baggage handling processes at an airport. The first design phase is the product – an item of baggage. There are restrictions on shape, size, weight, materials, and content. The handling process design includes consideration of all sub-processes that occur between the parking lot and the aircraft's hold and back to the parking lot. These sub-processes include mechanical handling and information processing devices, human handling and information processing activities, the design of interfaces, and due consideration of the environment. The production system design element takes the problem from the handling of a single item to that of millions of items a year. It requires the coordinated activities of sufficient handling devices, sufficient information processing capacity, sufficient people (with appropriate training), and sufficient numbers of interfaces. Finally, operations management involves a full complement of baggage handlers, maintainers, customer service agents, second-level problem solvers, and managers all being appropriately selected, trained, and assigned to achieve a desired level of customer service. As the overall process moves toward operations, there will be increasingly greater levels of scrutiny. Hopefully, the baggage and handling systems design issues have been dealt with early in the design process. Think of seasonal demands to handle golf clubs, skis, bicycles, and fish.

WALLS AND THE ERGO COP

One way of modeling each phase of the process is through the development of "walls" that contain an array of standardized details of the systems and processes as they mature toward implementation. These "walls" provide the media for multi-disciplinary teams to comprehensively evaluate each stage of the process so that late developments don't interfere with the critical path and late changes don't result in excessive costs.

Manufacturing ergonomics has its own elements on the "walls". Assuming that the product design issues have been addressed on an earlier "wall", the manufacturing ergonomics wall will contain questions that address workplace design issues of fit, reach, targets, and task content. At the production level, the physical and temporal aspects of an operation will be amalgamated to assure an acceptable job cycle work-load. Later the operations wall will address individual job and team assignment questions. Finally, the operations output wall will document quality, productivity, and health and safety issues associated with each operation.

The practice of manufacturing ergonomics provides important lessons for many other practice domains. The traditional practice of imprecise process requirements and unrealistic design specifications leads to inappropriate designs to be addressed by the "ergo cop". Eventually, battles ensue in the "review boards" often result in requests for waivers and either a loss of face or an inflated ego of the ergonomist. This process is both inefficient for the company and unhealthy for the profession. The ergonomist should participate with engineering, management, and the operations/user community in the establishment of clear performance requirements and suf-ficiently precise design specifications and design implementation. In this way, there will be no surprises at the board (performance) reviews.

SPACE VEHICLES

The design of space vehicles differs from high-volume manufacturing in product cost and product life cycle. The environmental challenges, power requirements, and human interactions are unique. The complexity and remoteness of the operations lead to massive information management challenges and costs. The space program is deliberately very visible – the whole world is watching. Finally, because of these things, there is relatively limited opportunity for the program to capture sufficient "lessons learned". Much of the evidence that cannot be based on analysis, must be based on small samples of empirical evidence. Human factors specialists become acutely aware of the challenges of human variability, given the relatively small num-ber of experienced astronauts. Manned exploration missions, e.g., to Mars, present even greater challenges of evidence from robotic missions.

Over the past two decades NASA has developed extensive statements regarding the human factors issues of manned space flight. These statements are in addition to the extensive medical requirements. The NASA Standard 3000 – the Man-Systems Integration Standard (MSIS) is a compilation of evidence from both the profession of human factors (and other sources) and domain knowledge. Military standards such as Mil-Std. 1472 were particularly influential in MSIS development. The basic MSIS standard has been adapted to program-specific statements for Space Transportations System (Shuttle) and the ISS. These basic and derived standards have, like the pro-grams to which they refer, been subject to hostile attacks in the requests for waivers from engineers, programs, and contractors. As the space program matures working groups, tiger teams, and review boards all contribute evidence on which the next generation of standards will be based.

A general challenge to human factors is exemplified by the NASA review pro-cesses. Almost all human experiences are dogged by individual, contextual, and temporal variability. For example, an ideal thermal environment is affected by activity, clothing, individual acclimatization/tolerance, and duration of expo-sure. The requirements for strenuous exercise are different from those of reading. Consequently, an overly specific statement like 72°F will inevitably be inappropriate much of the time. The challenge to a human factors review panel is to address all the, possibly interacting, criteria in coming up with a decision that the engineer can design. Clearly, adjustability is required, but how much adjustability? Further complications arise because of constraints on design or change. For example, a particular intervention may be too costly or not feasible in the time scale of the overall project. The final complication of the review process is that the judges (usually experienced managers) overlay their own experience/prejudice on the decision. The task of the human factors engineer is to apply his/her own principles to the conditions surrounding the review process. It is the responsibility of the human factors engineer to be "user friendly" in his own practice. The response: "come back in a year when I have done a comprehensive study" is only occasionally warranted. Similarly, an answer that says 72°F, because the textbook says so, is equally naïve.

The human factors community at NASA makes extensive use of digital modeling in the design, evaluation, and real-time mission support phases. The primary contractor – Boeing – made extensive use of anthropomorphic modeling during the early design phases of the ISS. Currently, the Interior Volume Control Working Group uses models of the ISS interior, together with anthropomorphic models to evaluate additions and changes such as sleep quarters, the galley, exercise equipment, and protruding racks that interfere spatially and temporally with routine and emergency activities. The application of digital human modeling in the evaluation of the conditions of work in an assembly task was shown to be more precise, faster, and far less expensive than the alternative of a full-blown trial in the Neutral Buoyancy Labor. Modeling was particularly useful in the iterative design and analysis cycle of the crew quarters rack which provides facilities for sleep, computer workstation storage, and privacy.

Lighting models, using the Lawrence Berkeley Laboratories "Radiance" software, are critical to operations, given the changes from extreme brightness to complete darkness every 45 minutes. Differing viewing points for crewmembers and cameras, shadows and glare compound the difficulties. Just-in-time modeling and prediction of lighting conditions are invaluable to many Extra Vehicular Activity operations. Exterior robotic operations in the rapidly changing day/night cycles make use of both human vision and camera vision for both training and real-time activities. These models are particularly useful to aid decision-making when contingencies change the timeline and hence the lighting conditions for particular activities.

Although there are very few crewmembers, the multiple demands on their time and the many resource constraints, such as equipment, power, materials, and lighting, make activity scheduling a very difficult challenge. The difficulty is compounded by sparse and imprecise evidence regarding the duration of human activities in the microgravity environment and the ever-present challenge of human variability. The problem is being addressed by enhanced data collection approaches and a range of complex and simple scheduling models. The crew work day on the ISS is broken up into three main categories – work (including scheduled science investigations, assembly, maintenance, planning, and communications), sustaining activities (sleep, exercise, eating, and personal time), and responding to contingencies (such as caution and warning signals). The considerable spatial restrictions of the ISS complicate work activities through stowage constraints and spatio-temporal interference. The personal preferences of individual crew members in the highly congested conditions sometimes result in excessive time being spent in finding tools and materials. The crew workday is categorized in detail, however, the variability of times of activities within categories is not well understood. Consequently, steps are being taken to collect better data and develop simulations of activities on a daily, weekly, and mission basis. These models show not only the occasions when the schedule is overbooked but also how different priorities of activities can be used to accommodate this overbooking – such as sacrificing sleep or personal time.

GLOBAL INTEGRATION

Human Factors and Systems Engineering are essential to an effective and efficient design. However, all designs of processes and contributing systems are complicated by change and human, situational and temporal variability. A major thrust over the past decade has been attention to common processes of both the design activity itself and the resulting product and manufacturing processes. Unfortunately, times and best practices change so the processes must be flexible to assess and accommodate these changes. These challenges are particularly evident in international operations, where economies of common processes often conflict with different national practices that have been established over many years. Of particular value in the human factors area is the establishment of a common communication currency that enables comparisons to be made between widely differing alternatives and conditions and which resolves the ubiquitous importance weighting problems.

Global integration efforts are often the source of conflict between the efficiencies of common processes and the perception of what are best practices. Not invented here is often an underlying motive. The challenges of competition in the automotive industry, coupled with the explosion of computing and telecommunication facilities have combined to fuel the fires of globalization. Manufacturing organizations seek out high-quality, but lower cost labor markets. Also, it is not efficient to have an engineering design center in every country – why develop essentially the same product separately in multiple markets? But engineers worldwide are conservative and resistant to imposed change. In these cases, there is no substitute (other than dictatorship) for extensive face-to-face interactions among the design teams, including the human factors engineers.

The ISS faces similar challenges. The program has very important political underpinnings, the costs are extremely high and national identities need to be clear. And there are other constraints – only three crew members at a time, relatively few modules and only occasional opportunities to visit. The management of such a program is not limited to the handful of astronauts and cosmonauts; there are very large support staff in operations, engineering, medicine, and science management. The ISS is at once a miracle of systems and safety engineering and at the same time a management nightmare.

SYSTEM CHANGES

The mechanisms of dealing with change are described in different organizations as a request for waivers, change requests, or engineering change orders. In many cases, these requests are appropriate, albeit due sometimes to poor planning or unclear specifications earlier in the process. Often, however, they are seen as frivolous – made only to accommodate the failure of a supplier to be able to deliver on earlier agreements. Where waivers are processed on an individual basis, they may not comprehend the implications on other aspects of the process or the trickle-down effects to other subsystems. Cost is a common reason for the change and the systems engineer and the human factors engineer must work with the managerial accounting community to establish a common basis for the rational processing of waiver requests.

Human Factors cannot be practiced without engineering – the people who design the systems – and management – the people who make the policies. Sometimes policies are imposed from elsewhere – through technical standards, government regulations, or labor agreements. Some human factors practitioners feel that it is their duty to convey policy, especially where engineering and management do not have the appropriate information to decide on policy. On occasion, human factors specialists substitute dogma for policy. The notorious 5th percentile is a prime example. It is a useful concept, often with good rationale, but it is widely misunderstood and often inappropriately applied.

A major problem is that someone who represents the 5th percentile on one measure is unlikely to hold that relative position on another; furthermore, when accommodation is based on multiple dimensions then it may be difficult to define who or what is a 5th percentile. Monte Carlo simulation methods may be applied to somewhat relieve this problem. Another difficulty is that the implications of a design decision may be more or less important. Thus in the case of a highly sensitive design decision, it may be appropriate to accommodate the 1st percentile. On other occasions, design for the average may be an adequate approach, given that the dimension in question is not related to an important outcome. An example of a highly sensitive design decision would be the walking speed of old people crossing a busy road.

OUTCOME AND DESIGN SCALES – COMMON CURRENCIES

Human factors engineers are made aware of the processes and theoretical underpinnings of scaling methods from their earliest training in the statistical methods applied in the broad context of human variability. Scaling systems abound – percentages and Yes/No are separated by Lickert-type scales of varying degrees of resolution. A prerequisite of any scale is the establishment of anchors for both endpoints and intermediate thresholds. The fuzzy classification reflects reality but is often a practical inconvenience. Appropriate resolution is always needed. The zero to ten scale is probably the most universally familiar one and has stood the test of time. It usually has adequate resolution and can easily be linked to a response categorization.

Given this ten-point scale, it is relatively easy to visualize a nonlinear mapping function that covers the full range of outcomes from ideal to unacceptable. Examples include lighting, noise, temperature, spatial, and force scales of design specification ranges. It is also possible to comprehend single and complex variables, although, for engineering design purposes the evaluation will ultimately have to identify individual variables for change. Where the relationships are not monotonic, as in the choice of an optimal temperature, then it is convenient to use two scales – one for hot, the other for cold. There may also be multiple outcomes – some of which might be conflicting. For example, a spatial scale related to controls, such as vehicle pedals, may have movement time and inadvertent actuation conflicts. Such conflicts point toward the importance of consensus processes in the establishment of mapping statements and cut-offs. The reality of human variability is such that a single mapping function will never be precisely "right". Again the inclusion of human population accommodation policy in the consensus decision is essential to assure "buy-in" of all concerned as the design process develops.

SYSTEMS CONSENSUS

The development of human factors design standards is best pursued through a consensus process, using the common currency described earlier. The credo that standards should be data driven is over simplistic. Policy, scientific logic, technical feasibility, and experience must all contribute to the establishment of a standard. It is also essential that representatives of the customers – internal or external – who will have to apply or be affected by the standard should be involved. In this way, up front agreement in both the principles and the values related to the standard will help to assure "buy-in" and less demand for waivers. Human factors standards are also iterative in that they should be verified, validated, and evaluated as the project or program evolves. An excellent example lies in the establishment of speed limits for different road conditions.

Design tradeoffs should be made with a full global view of all the relevant information, preferably with a common communication currency. The choice of cut-off points on individual variables can be used to assign explicit weightings. A broad range would imply wide tolerance (less importance) and a narrow range of greater importance. The common currency outcome prediction scale facilitates the process of amalgamation. At the simplest level, a count of the number of variables in each of the outcome ranges produces an index or profile that reflects the general nature of the problem. The count can also be used as a decision aid – e.g., decision policies could be "no reds" or not more than "5 yellows".

Addition (as opposed to counting) is rarely justified, although this is the preferred method for some checklists. However, the case for multiplication in the amalgamation process may be justified where interactions are likely. Such situations are best handled by two-dimensional matrices using the same common currency described here. The special case of interactions between basic variables and time may also be approached in this way. This big picture decision aid can be used to indicate before and after change situations, comparisons between alternatives or progress of a project through the design process. The common currency scale can be viewed as an estimate of the "probability of failure on a single transaction". For example, the probability of "failure to accommodate" of a 12-inch wide seat would be of the order of 0.7. Similarly, the probability of failure of a 2-mm high font, given an elderly reader population could be 0.99. At the other extreme, the probability of failure of a 24-inch diameter escape hatch might be 0.05.

SYSTEMS DESIGN DECISIONS

The decision process must also involve to benefit of the transaction, the number of transactions per unit time, the number of people affected, and the various costs of failure. To complete the assessment an evaluation must be made of the costs, benefits, and probabilities of alternatives. Where possible, objective evidence (including data) should contribute to the probability and costs/benefits and exposure estimations. Where individual, personal decisions are made then subjective probability and cost estimates may be sufficient – this is, of course, the basis of most naturalistic decision-making.

The hypothetical example of our choices of transport to work – car, bus, or tank – can be used to illustrate the quantum decision process. Assuming a decision horizon – a day, year, or project lifetime – all measures can be reduced to base ten arithmetic. In the case of choice of transportation mode based on cost and safety then the analysis shows that we should ride the bus. However, if we are the president of a country in political turmoil then we may wish to revise our probability, exposure, cost, and outcome estimates and at least buy an armored car. This quantum arithmetic approach is appropriate for a cursory analysis and exploration of the sensitivity of the different elements of the decision to changes. Greater resolution may be obtained by including multipliers and decimal components, while still adhering to the basic decision logic. However, in this case, it may be appropriate to use some computational aid.

SYSTEMS DESIGN OF PRODUCTION LINES

1. **Ergonomics and production lines**
   a. During the first part of the last century Henry Ford, Frederick Taylor, and Lillian Gilbreth were among the leaders of the process of scientific management. Their focus was the measurement, simplification, and standardization of manual work, particularly in product assembly. They demonstrated unequivocally that these processes can greatly improve product quality and productivity. Toward the end of that century, the Toyota Production System continued this prescriptive approach to job design but added the principles of continuous improvement and participatory quality teams. The production line can now be found in most industries around the world such as automotive, electronics, textiles, consumer products, call centers, and food processing. Unfortunately, these tremendous gains in productivity and product quality come with a human cost and some of this cost is due to the misapplication or narrow application of physical ergonomics by a focus on the twin issues of force and posture/movement and the removal of "nonvalue-added work". These ergonomics applications certainly contributed to continuous improvement in productivity and product quality through greater work intensity and repetition. But these gains were offset by increases of work-related musculoskeletal disorders and many less tangible cognitive and social detriments, such as vigilance decrement and boredom. These issues may appear less important where labor is readily available, cheap, mobile, and dispensable, but ironically they become more important as the employee pool becomes more stable.
   b. There are fortunately a number of variations on the theme of the production line. The first engineering approach is through mechanization and robotics for repetitive and forceful work, although these processes

often require human operators to complete the task cycle through such activities as materials input, operation initiation, and parts removal. The second, administrative, approach is through job structuring, rotation, and enlargement, perhaps involving a sequence of work cells within the production line. These human-centered approaches can create a more knowledgeable, flexible, and stable workforce, especially where the job assignments include such tasks as inspection, materials, maintenance, and a share in the supervisory tasks of training, monitoring, health, and safety, etc.

c. A human-centered or sociotechnical systems approach to the design of production lines can maintain these gains in productivity and product quality while providing safe, healthy, and satisfying employment. One key component of this participatory approach is a reversal of Taylor's scientific management view that the worker "should do and not think". There is no doubt that line workers develop considerable task knowledge and that this knowledge can be harnessed for the benefit of all stakeholders. However, where this participative approach is mismanaged sub-optimization and conflict replace optimization and cooperation.

d. These concepts of human-centered manufacturing will be supported by case studies in automobile assembly, textiles manufacturing, bookbinding, and medical claims processing.

2. **Ergonomics research and ergonomics design are two sides of the same coin**

a. This chapter discusses three ergonomics research and three ergonomics design examples to demonstrate the similarities between the two processes. The ergonomics research topics all focus on running and relate to gait biomechanics, the physiology of fatigue and aging, and the effects of terrain on performance. The design topics all focus on transportation – the design of mass transit railways, the design of cars for the elderly driver, and the design of aviation displays.

b. Research and design can both be described by a control model involving inputs, outputs, feedback and feed forward, and adaptation and learning. In research, the hypothesis is an anticipation or feed forward of the relationship between selected technological, environmental, or operational variables and a human response. Design involves the control of the technological (or operational) context to withstand uncertainties in the operational or environmental contexts. Reliable designs maintain their performance levels over the life cycle of the product, given expected and predictable contexts. Resilient designs can withstand unpredicted and sometimes extreme contexts, often by "failing safe".

c. A major challenge in research is the validity of the investigation design and the results. In the gait biomechanics research, the investigation addressed the hell and mid-foot strike variations. However, as these factors are affected by stride length and running speed as well as individual variation and shoe design, the conclusions may not be widely generalizable. Indeed, the major challenges of most human factors investigations are sample selection and size.

d. Similar challenges of validity are found in design. There are the intended users of a product and sometimes but not always predictable misusers. Also, there are the intended conditions of use and possibly more challenging unexpected conditions. For example, a family car may not be able to protect the inexperienced, inebriated, or incompetent driver in heavy traffic. Similarly, the family car may not perform well in off-road conditions or even on a road with ice or flooding.

e. Research into the physiology of fatigue and aging usually relies on selected or pseudo-random samples representing the ranges of interest. The analysis of differences relies on means and variances. However, these are simply reflections of the samples and not necessarily the age variable of interest – it is possible to select a sample of 60-year-old who will run faster than a sample of 20-year-old. One way out of this sampling dilemma is to use age-based records to model the age effect. Such analyses show very clearly the age effect without being contaminated by such things as aptitude, training, or illness. Similar approaches may be used with cognitive variables.

f. The design of mass transit systems presents both physical and operational challenges for passengers. The physical design of features such as seats, grab rails, and exits may be entirely suitable in uncrowded conditions but in rush hour and emergency conditions, the abilities and behaviors of a few passengers can disrupt the intended objectives of stability and mobility.

g. Human fatigue research shows deterioration in physiological and cognitive functions as a function of time and activity. The Weibull distribution is used to model fatigue and wear out of both mechanical systems and human systems. The fatigue curve in long-distance running is very similar to that of a car, albeit with different time scales. Both curves can be changed by maintenance, or training, but only where the demands are limited. Resilient designs rely on operational factors such as alternating periods of activity and rest/maintenance.

CONCLUSIONS

It is important to reiterate that it is the role of management or governments to communicate policy, given human factors evidence of outcome likelihood and effect. The role of consensus in the establishment of design standards was also addressed. In this respect, it is important to address the realities of false consensus and the sometimes inappropriate or overly weighted influence of experts. The opinion of experts should always be weighted heavily, but only in the area of their expertise. A better approach is to use independent and interdependent "voting" processes, with sufficient allowance for discussion as the standard or decision scenario is developed.

> The design of boats, trains, cars, and space vehicles as most other human factors opportunities necessarily involves teams of one kind or another. Generally, the teams consist of an exhaustive and exclusive "set" of people with technical and domain knowledge. Individual team members may have discrepant objectives. Team dynamics create challenges to both the effectiveness (accuracy) and efficiency (speed) of standard, risk, and design decisions. Greater effectiveness and efficiency can be achieved through the application of common currency and clear visual aids for comparison, tradeoff, and decision-making.

Even the decision processes common in contemporary design projects face the challenges of over/under reliance on expertise. On occasion, the efforts to substitute process for expertise may also be counterproductive. Overly enthusiastic attention to "process" can be cumbersome, where simple experience may be sufficient. One hundred years ago craftsmen built outstanding automobiles, slowly. Nowadays, processes result in the high volume production of automobiles. Hospitals used to be run by medical experts but now the processes of HMOs have diverted the purpose to profit rather than a caring motive. Soccer is essentially a game of experts, it has experts that are bound by processes. Music was once the realm of experts, now it is relegated to simplistic marketing processes. We went to the moon on the backs of experts; processes will get us to Mars.

Human factors is alive and well early in many design processes. The appropriate place for human factors is as a branch of engineering, not as an "ergocop" in safety or consumer protection functions. Human factors is rightly a component of systems engineering, it can contribute important knowledge and tools both to the designs themselves and to the design processes. One important contribution is the establishment of a common currency for communication of human factors and implications of design.

REFERENCES

Peacock, B. (2019). *Human Systems Integration*, Self-published manuscript, Fernandina Beach, FL.

Peacock, B. (2020). *How Ergonomics Works*, Self-published manuscript, Fernandina Beach, FL.

Peacock, B. (2021). *Ergonomics Tools and Applications*, Self-published manuscript, Fernandina Beach, FL.

3 DEJI Systems Model in Engineering Design

INTRODUCTION

Industrial engineering encompasses diverse elements of the general practice of engineering. Design is fundamental to what industrial engineering brings to business and industry. Engineering ergonomics is the core of every design effort. In that regard, ergonomics, as it relates to the use of DEJI (Design, Evaluation, Justification, and Integration) Systems Model, is the focus of this chapter. It presents the various considerations in the design of work systems from the physical rather than the mental or cognitive aspects. Topics covered include the design process, product lifecycle evaluation, service justification, and systems integration.

This chapter is based on Peacock (2014). Human factors is the mental or cognitive aspects of a system design while ergonomics is the physical aspect. Design is the process of converting the voice of the customer into some product or process. The problem, however, is which customer's voice do designers listen to. Take the examples of a smartphone or a car or the process of getting money from the bank or the university you choose to attend. All of these have many customers (or stakeholders), each with different requirements. The end users are interested in effectiveness and ease of use of the product or process for their particular needs, which may differ. They will also be interested in the cost. Others may emphasize safety and security. But there are other "customers" perhaps with different requirements. What about the line workers who assemble the product or the employees of the bank or university? There are also the managers and shareholders of the companies that manufacture the products or manage the processes. They are interested in sales and profits. The result is that design is a process of resolving many, sometimes conflicting requirements.

One way of addressing this problem of design is to look at the ergonomics of the design process. There are many stages in this process including: concept development and selection through manufacturing process design and production, to sales, operations, and maintenance. Recently, design for disposal has become a concern for many products. Each stage has different customers. Surprisingly much of this design process takes place around a table, where the different stakeholders communicate the importance of their particular requirements. And this is where the fun starts and the ergonomics of process design can have its opportunity to shine (Peacock, 2004, 2019, 2020, 2021; Peacock and Orr, 2001; ReVelle et al., 1998).

The first step is to agree on the scope of the many requirements. In general, the purposes of all products, processes, and organizations are: Effectiveness, Efficiency, Ease of Use, Elegance, Safety, Security, Sustainability, and Satisfaction (E4S4). These purposes are deliberately general and comprehensive, but they will include more specific and quantitative concepts in particular instances. For example, efficiency includes the optimal use of resources such as money, time, people, and materials; and security includes process failure due to malicious or accidental acts of third parties.

DOI: 10.1201/9781003328445-3

Most organizations will identify growth and stability as primary purposes, but in reality, these are dependent on E4S4.

The next step is to agree on the language of communication. Perhaps the best language is an adaptation of Quality Function Deployment, which has been used in the automobile industry for a couple of decades. It is important to separate the concepts of requirements and specifications. Requirements are what the customers want, and specifications are what the designers and engineers need. Requirements reflect the use of the product or process by a customer in a context. Specifications are those numbers that accompany the lines on the engineering drawing or project plan. Car owners may want vehicles that are comfortable to drive on long journeys or they may require cell phones that can also be used for e-mail and surfing the web. The manufacturing employees want products that are easy to assemble and maintainers like easy access to their tasks.

THE DESIGN PROCESS

Design is the process of converting the voice of the customer into some product or process. The problem, however, is: which customers' voice do designers listen to? Take the examples of a smartphone or a car or the process of getting money from the bank or the university you choose to attend. All of these products and services have many customers (or stakeholders), each with different requirements. The end users, who often have a choice among competitive products or services, are interested in effectiveness and ease of use for their particular needs. Some may also be interested in aesthetics and cost while others may emphasize safety. Examples of processes and considerations for the design and evaluation of products are provided below:

- Transportation: automobile, train
- Communication: Mobile phone
- Education: Curriculum, Course
- Payment: ATM
- Recreation: Game
- Control: TV remote control

There are other "customers" such as the production line workers and service organization employees who have less choice; they will have different requirements, such as ease of access for assembly or to their computer workstation. There are also the managers and shareholders of the companies that manufacture the products or oversee the services; they are interested in quality, productivity, sales, and profits. Design, therefore, is a process of resolving many, sometimes conflicting, requirements.

DESIGN PURPOSES AND OUTCOMES

A general model of product or service design objectives can be summarized as "E4S4", which is explained below:

- E4: Effectiveness, Efficiency, Ease of Use, Elegance
- S4: Safety, Security, Sustainability (Reliability and Resilience), Satisfaction

Effectiveness, or quality, is a measure of how well the product meets customer requirements. Efficiency reflects the use of resources such as time, money, materials, energy, and people. Ease of use indicates the comfort and convenience of various users as they interact with the product or service. Elegance is the esthetic appeal of the product that reflects the emotional attachment with the user. Safety is a description of the hazards associated with product or service interaction, including barriers to failure and mitigation following product or service failure. Security of a product or service is the degree to which malicious or accidental misuse can be prevented. Satisfaction is a composite measure of the various customers' experience with the product or service. Finally, sustainability indicates the reliability of the outcomes of a product or service over time (reliability) and the degree to which the product or service can withstand unexpected and sometimes extreme contexts (resilience).

The effectiveness of a transportation system is a measure of whether the chosen product or service achieves the objectives of a particular "transaction" or journey. For example, the choice among walking, bicycling, driving a car, or flying in an airplane will depend on the distance of the journey – the transaction requirements. Given the selection of an effective system, such as a car, the customer may be concerned with the efficiency of this choice such as vehicle cost or fuel consumption. The ease of use criterion may relate to the vehicle information systems or the ease of parking. The esthetic measure – "elegance" will indicate those subjective impressions of the vehicle's styling or nameplate. The safety objective may reflect the sophistication of the vehicle control features such as antilock brakes or reversing indicators; also the mitigation features such as air bags and "friendly" dashboards may be included in the safety evaluation. Security in the case of a car may include locks and GPS tracking systems. Sustainability is measured by the evaluation of the vehicles' reliability and appeals over time. Resilience reflects the performance of the vehicle in extreme conditions including natural contexts such as off-road routes and floods and human-made challenges such as crashes and fuel shortages. Satisfaction is a composite measure of the total experience of the car user as reflected in owner surveys – customer loyalty to a brand will depend on the weighting of the various evaluation criteria, such as cost, styling or safety features.

LIFE CYCLE

One way of addressing the multiple purposes of design is to look at the ergonomics of the design process. There are many stages in this process, including: concept development and selection through manufacturing process design and production, to sales, operations, and maintenance. Recently design for disposal has become a concern for many products. Each stage has different customers and stakeholders. Many of the design decisions are made around a table, where the different stakeholders communicate the importance of their particular requirements. The interfaces in the production environment can be summarized as follows:

- Design Team:
  - Requirements
  - Concept development
  - Concept selection

- • Specifications
- • Tooling
- • Production
- • Marketing
- • Use and Misuse
- • Maintenance
- • Disposal
- • Customers

In between these, we have feedback, Technical Memory, and Kaizen. Also of note are prediction and anticipation that are needed. The "cradle to grave" description of a product or service encompasses design, production, use, and eventual disposal. In a large organization, the marketing group will interpret the requirements articulated by the end user, sometimes with the help of an ergonomist. The human factors specialists will also work with other customers and stakeholders, such as manufacturing and maintenance, to extend these requirements. The final set of requirements involving various E4S4 criteria will inevitably be a compromise worked out among marketing, design, engineering, manufacturing, human factors, and cost departments with management oversight. Given these requirements, the design team will explore many concepts that conform to the different requirements to various degrees. Convergence on a single concept for development will inevitably be a compromise among the many customer and stakeholder biases.

For example, a major design decision for a mobile phone will be between hard and soft keys which imply both different technologies and different use of the product's interface "real estate" surface. The human factors specialist will evaluate effectiveness, efficiency, and customer satisfaction as reflected, e.g., by the feedback provided to the user in time or light-constrained operations. Given the very compact nature of this product, the ergonomist will assess the design of the production line layout, tools, and the various jigs and fixtures needed to provide access and stability for assembly and inspection.

STRUCTURES, PROCESSES, AND OUTCOMES

Donabedian (1988) described the healthcare processes in terms of structures, processes, and outcomes. These concepts may also be applied to any complex entity such as a university where the structures include buildings, facilities, laboratories, and faculty members. Structures or entities are designed by reference to specifications. The processes include lectures and examinations and various management processes needed to assure that the institution runs smoothly. Generally, processes have requirements. The outcomes include the employment of graduates and publications by faculty members. Generally, the structures are entities described by nouns and designed with reference to adjectives. Processes, on the other hand, are activities or verbs that are measured by reference to adverbs. The outcomes will generally be a change in the status (adjectives) of one or more of the input structures.

The contribution of the ergonomist is to help to translate requirements into specifications. In the case of a car, the requirement "comfortable to drive" is translated into the interior package dimensions, seat adjustment range, contours, and foam density.

Other specifications may relate to the driving environment and perhaps to the driver. Surfing the web can be achieved by specifying a command structure for hard keys or a menu process using icons and large buttons on a touch screen. The manufacturing design requirements of easy access may be achieved by a layering of modules and orientation of fasteners toward the assembler or maintainer. The process of QFD can be used to further analyze the relationships between requirements and specifications (Akao, 1990).

The evaluation or validation of processes will take place in a real-world context or realistic simulations with representative users and transactions. Validation will address process purposes, requirements, and outcomes. For example, an ATM transaction may require multiple layers of security including temporary passwords that may confuse elderly users.

The customer requirements for getting money out of the bank include ease of use of the teller machine and security of the process. Both requirements may be achieved by specifying primary and secondary PIN designs that are easy to remember and hard to steal. This challenge indicates a trade-off between ease of use and security. The efficiency/effectiveness trade-off of a transaction will be reflected in the length of the encrypted password and the availability of "quick cash" options in common amounts. The ATM example is a simplified version of a web page where the number and complexity of choices and navigation through sequences of choices may be accomplished by such interactions as point and click or involve keyboard choices and inputs. User performance (effectiveness and efficiency) and satisfaction with web page transactions will be reflected by objective measures of speed and accuracy and subjective measures that are related to the transaction content.

Product design specifications are driven by process requirements which in turn depend on intended outcomes. Although a requirement may be relative – the speed and accuracy of a text message on one mobile phone should be "better than" those of another – the design specifications must be objective and quantitative. A specification is absolute, perhaps with some tolerance: the font size should be at least 18pts. There may also be trade-offs among multiple requirements. For example, the size of a key-pad and hence the spacing of the keys will give rise to speed – accuracy trade-offs. Furthermore, these performance trade-offs may be preempted by a requirement to minimize the size of the overall device. Similar trade-offs between effectiveness and efficiency will arise with the screen and associated symbol design. The requirements for the Holy Grail of the ideal communication devices – (e.g., iPhone or iPad) – will depend on particular usage and context of use criteria. Consequently, the specifications, such as screen/symbol and key size and spacing will derive from a compromise among performance requirements. It is usually an easy matter to verify that the design specifications have been met. However, where design constraints are encountered such as cost, weight, or technology, there may have to be trade-offs among the specifications with resulting implications for the outcomes.

In all these cases – cars, smartphones, bank account access, and website design – the role of the ergonomist is to work with the customers and the designers on the translation of requirements into specifications. The first task is to ensure that customers talk in terms of functions or activities and not the means of implementing these functions. Designers and engineers, on the other hand, should not try to dictate customer requirements but rather understand what the customer wants, and the context of the use of the product or service. The ergonomist, using a box of investigation, analysis, simulation, and experimentation tools can contribute effectively to the task of translating the voice of the customer into the specifications of the product or service.

SYSTEM PREDICTIONS

The main challenge for design is prediction. Seen as a control problem, design is about the development of structures to contribute to processes with desirable outcomes. Most "designs" have precedents and therefore can benefit from feedback; over time designs are adapted to the context and the designers learn from their successes and failures. But progress involves uncertainty and risk, and technical memory is not always as effective as it should be. Also, the contexts and uncontrollable external forces are not always precisely predictable. Therefore, the designers must design for intended users, use and contexts, and foreseeable misuse and unexpected contexts. The product liability courts are full of cases where system failures resulted in unwanted outcomes. Also, design compromises may result in vulnerabilities to misuse or hostile contexts. A second challenge in prediction is that of user variability. Although the majority of experienced and trained users may use the product or process effectively, efficiently and safely, novice and otherwise degraded users may misuse the product, perhaps with catastrophic results.

The transportation example will be used to illustrate these challenges of prediction in design. The designer of a car only has imperfect knowledge of the potential user. Furthermore, the car buyer may choose which car he or she buys and what journeys to make; they may also have exaggerated opinions regarding their driving abilities. Finally, the unpredictable context of other drivers, traffic, traffic regulations, roads, and weather add to the complexity of prediction, requirements, specifications, and outcomes. For example, one may attempt to develop the design specifications for a car aimed at the older driver. The requirements are derived from predictions regarding intended and unintended contexts and outcomes. Peacock and Karwowski (1993) provide extensive background on requirements and specifications for vehicle design. The minivan has two advantages – ease of access and plenty of room for the grandchildren. Where mass and safety are concerned, the full-size pickup will provide good visibility and good performance in the event of a crash. If the elderly driver has accumulated sufficient wealth and wishes to relive his youth, then the Corvette may be the vehicle of choice, although entry, egress, and visibility may be a concern for the older driver, not to mention the response of the accelerator. The family sedan is perhaps the best compromise, especially if the physical, cognitive, and operational interfaces are geared toward the diminishing sensory and cognitive capacities of the older driver. If the driving environment is away from the busy freeways, the journeys are short, and the weather is fine, then a golf cart may suffice

THE 6 Us AND 2Ms

An alternative way of analyzing product use with the purpose of predicting outcomes and therefore requirements is to use the 6Us and 2Ms template (Peacock and Resnick, 2010) either for descriptive or quantitative purposes. The first question is: "how will the product or service be used?" Further depth to this question asks: "how useful is the product or service?" This second "utility" question may reflect criteria of efficiency and safety over and above the primary question of effectiveness. The next questions are about the intended user and possible misuser. Next, come questions about the method and conditions of "usage" and the frequency of "utilization". Finally, questions are asked about "ease of use and ease of misuse".

A close cousin of the mobile phone – the TV remote control will be used to describe this analysis process. Both devices are handheld and therefore portable, they both have a plethora of frequently and less frequently used functions and buttons and they both may be used with minimal visual feedback, although a wrong number or choice may eventually provide unwanted feedback. The TV remote control is typically used to choose among many broadcast information and entertainment products. Its utility is really in the realm of efficiency – one does not have to move from the chair to search and select the desired program. Because of this effectiveness and efficiency, the device is used by the widest possible range of users – from the very young to the very old. The utilization or frequency of use can range from a few times a day to dozens of times because of this effectiveness and convenience. The usage or conditions of use are where the trouble starts. Typically, the device will be used with minimal environmental lighting. Misuse opportunities abound. A common misuse is to select the wrong device from the four or five that typically adorn coffee tables. The user and usage conditions, coupled with the large number of closely spaced and undifferentiated keys (from the tactile perspective) lead to a high probability of error. This basic shortcoming is compounded by the large number of features and choices, many of which are not used or rarely used by most users. A second design shortcoming is the lack of a simple accessible feature to help the user recover from an inadvertent error.

A CASE STUDY IN DESIGN: HONG KONG MASS TRANSIT RAILWAY

The Hong Kong Mass Transit Railway carries millions of passengers a day. For the passenger compartment, the principal passenger criteria are stability, safety, and mobility. Given the relatively short journeys, comfort is a lesser consideration. Journey time and cost are also key issues. The organization is interested in payload – passengers per square meter, and dwell time at stations – 25 passengers off and 25 passengers on the train in 25 seconds. Other considerations included emergency evacuation, maintenance, and resilience. The primary ergonomics opportunities to fulfill these requirements were in the design of the inward-facing bench seats and the arrangement of the horizontal rails and vertical poles.

The design process involved extensive use of physical mockups and usability trials to supplement the basic principles of anthropometry and biomechanics. Compromises had to be made between reach and fit – could the shorter passengers reach the high-level horizontal bar and would the taller passengers have sufficient head clearance? Given that standing passengers were intended to cluster around the vertical poles, these were located to allow the best possible access and mobility for passengers leaving or entering the train. The design of the train operator compartment ("operator" was preferred to "driver", given the substantial amount of automation) also had important ergonomics criteria such as control effectiveness, comfort, and stability; forward vision was also important. The travel times between stations were generally of the order of a few minutes; at each station, the operator was required to alight and check the status of passenger movement between the platform and the train, with the help of video cameras. Because of the constraints on the size of the operator's compartment, a full static seat was impractical. Consequently, a novel seat was designed that allowed the driver to choose among sitting, leaning on the folded seat, or standing, depending on the segment duration. The control panel included a prominent emergency stop button and a handle that fulfilled acceleration (forward) and braking (backward) functions.

QUALITATIVE AND QUANTITATIVE DESIGN TOOLS

The initial use of an ergonomics design tool is to serve as a checklist to ensure that the full spectrum of relevant questions is asked. Next, the tool must ensure sufficient depth and detail to facilitate the development of relevant requirements and clear specifications. Two such tools are described here – to facilitate the choices among design options at both the general and the detailed level. The first tool lists the general E4S4 purposes with opportunities for more detailed criteria. It should be noted that the criteria may be differentially weighted and evaluated in a way that minor issues (tipping or halo factors) do not dominate the overall decisions.

The use of this worksheet may be either qualitative or quantitative. For convenience, simple cryptic words may be entered to summarize the performance of the alternative products or services on each of the criteria. There is a plethora of scoring rubrics – varying from simple pass/fail, through rankings and ratings to detailed measurements of times and costs, and composite customer scores.

A second worksheet addresses the 6Us and 2Ms, again with the purpose of comparing a set of alternative designs. This investigation method may be supplemented by the 5Ws and a How (Who, What, Where, When, Why, How) and the "5 Whys" in which each use evaluation is queried by asking "Why?" a series of times until a root cause of a problem or requirement is defined.

CONCURRENT ENGINEERING

Concurrent engineering is the process of strategic overlap of design phases (product, manufacturing, production, etc.) to ensure that all customer requirements are considered as early as possible in the design process. Failure to do this can result in costly and time-consuming changes. Successful concurrent engineering results in a reduction of the total time between concept development and production. Also, the process facilitates effective trade-offs among the requirements of different customers.

Car design from scratch can take up to 5 years, by which time the faster competitors have their vehicles on the road and the design may contain aging technology. The concurrent engineering process can greatly reduce this "time to market" challenge. A complementary strategy is to use proven technology, such as engine and chassis modules and fasteners, and differentiate the product by cosmetic changes.

COMMUNICATION IN THE DESIGN PROCESS

Communication is an important issue in the design process. A useful language is an adaptation of Quality Function Deployment, which has been used in the automobile industry for a couple of decades. It is important to separate the concepts of requirements and specifications. It is also important to distinguish among structures, processes, and outcomes. A process is an activity that is described by a verb and quantified or qualified by an adverb. Processes have requirements. On the other hand, structures or entities are things that can be designed by reference to specifications. Specifications are quantitative adjectives. Requirements are what the customers want and specifications are what the designers and engineers need. Requirements reflect the use of the product or service by a customer in a context. Specifications are those numbers that accompany the lines on the engineering drawing or project plan. Outcomes are the result of a process and should reflect the process purposes and requirements on various E4S4 dimensions. However, because of compromise, the actual outcomes of a process may deviate from the requirements. Purposes and requirements will be a strategic compromise among E4S4 criteria. Outcomes may also reflect failure or misuse of a product or service.

The requirements of the manufacturing employee can be communicated upstream to the product design engineer, who already has the requirements of the eventual product user in mind. This communication is facilitated by the use of a common "currency" or language. For example, a designation of "A/green" can indicate that a particular requirement has been met by a design specification. A "B/yellow" will indicate that a particular design dimension may interact with another criterion to create an intolerable situation. A "C/orange" will suggest that the condition will certainly interact with other criteria. Finally, a "D/red" will indicate that the specified design of an individual dimension is intolerable no matter what the other conditions are.

The case of a spare wheel in a large car will be used to illustrate this process. From both the product use and manufacturing points of view, a large wheel placed horizontally in a well at the bottom of the trunk will be both difficult to install and extract – it will receive a "D" for ease of extraction or assembly from the vehicle user and the manufacturing employee, respectively. In practice, the assembly of such a large component may require the use of a mechanical or robotic articulating arm, both of which add time to the assembly process. Placing the wheel in a vertical location at one side of the trunk will reduce the access challenge but the weight may still be intolerable. Replacing the full-sized wheel with a smaller and lighter emergency wheel will improve the situation somewhat and certainly reduce the installation time on a fast-paced production line. However, elimination of the need for a spare by using "run flat" tires will remove the product, manufacturing, and production challenges altogether. These decisions must be made early in the design process as the basic shape of the pressed metal forming the trunk will have implications for other components such as gas tanks, rear axles, brakes, and lighting. In the end, there will probably be trade-offs among product, manufacturing, and production engineering. These trade-offs will be greatly facilitated by a simple metric such as the "A, B, C, D" one described above.

The translation between the operations and manufacturing requirements designations and the weight, location, and frequency specifications may require an in-depth analysis using an ergonomics tool such as the NIOSH Lift Equation. This complex tool assesses adjectives – object weight, vertical and horizontal location, grasping interface, and the temporal dimensions of frequency and shift length. The composite NIOSH Lifting Index 1, 2, 3, etc., will translate easily into the A, B, C, and D "adverbs" of the assembly requirements.

An assembly worksheet can be developed that addresses the postural (location), force, and frequency aspects of all assembly tasks. A similar worksheet using the E4S4 or 6Us and 2Ms criteria could be developed for the vehicle owner and the maintainer. Eventually, a composite picture of the whole product life cycle can be used to show the clusters of satisfactory and unsatisfactory features from many viewpoints. The use of color is a very effective way of showing a "forest full of trees".

SYSTEMS AND PROCESS ENGINEERING

The design of Boats, Trains, and Planes requires coordinated systems and process engineering. This segment addresses some opportunities for human factors to become involved early in the design cycle. Brief descriptions are given of five case studies in which different human factors approaches and systems engineering tools were applied successfully. The case studies include a mass transit railway, supertankers, cars, car manufacturing, and space vehicles. All design processes start with requirements – the customer wants the device or service to fulfill some function. Of course, there are conditions associated with the requirements that are articulated in criteria such as quickly, safely, and inexpensively. In all cases, the measurement and communication of these and other human factors benefit from the design of a common communication currency. The chapter employs frequent but consistent use of "jargon" – requirements, specifications, verbs, adjectives, etc., and examples. The reader is encouraged to develop alternative examples to investigate the concepts.

When we discuss human factors, it is common to describe a machine, a user, an interface, and an environment in the broadest sense of the words. In systems design, we focus on the interfaces with the equipment, tools, environments, and organizational structures with due regard to user (and misuser) capabilities and limitations. Human factors usually faces complex "human-machine systems". For example, an operator in the Space Shuttle may be controlling a robotic arm on the end of which is perched another astronaut. A third, tethered astronaut is outside assisting with the placement of a multimillion-dollar piece of hardware on a space telescope. The environment is characterized by minimal gravity, zero air pressure, and alternating hot, cold, light, and dark. The interface has to deal with the control of six or seven degrees-of-freedom "robotic arm" using quite limited camera views and communication facilities while the whole world is watching. Whatever the context the rules are always the same – we must design the tools and tasks so that the users can be successful and safe.

VISIONS AND MISSIONS

All good ideas start with visions. "Beam me up Scotty". Visions are free, but when we translate a vision into a mission we need funding. We also need specific mission requirements that can be measured so that we may know whether the mission has been accomplished. Missions usually have constraints – "put a man on the moon before the end of this decade" or "put a group of people on Mars". Next, come the details – the specifications of how, what, when, and how much? There may also be a "where" thrown in, although usually the "whys" are beyond question. Most complex missions have a way of costing more than the first estimates because as the plans unfold, problems arise due to a lack of information or opportunistic subcontractors holding the main organization to ransom. The more successful missions have clear requirements and accurate specifications very early in the design process.

In the 1970s, Hong Kong decided to solve its traffic problem – 5 million people all wanting to go to work at the same time. Rickshaws and old buses could not handle the demand, so a mission to build a mass transit railway was born. The mission had requirements – move 1 million people from A to B quickly, safely, comfortably, and cheaply. And "oh by the way expect to move 2 million people from A to B to C in a few years' time". The plans involved digging a channel up the main street, putting in a tube, and then returning the main street to its former condition a couple of years later. In the 1980s, the oil crisis directed attention to reserves in Arctic Canada, but it was deemed to be unwise to have a ship full of LNG collide with an iceberg or run aground. After considerable human factors, economic and engineering analysis, the plans proved to be untenable. In the 1980s, General Motors was feeling the heat from overseas competition and so developed a mission to improve the quality and productivity of their automobiles and reduce the costs. They embraced many concepts of systems engineering and learned all about "the voice of the customer". They found that there are many external and internal customers and they learned the language of systems engineering. By 1990, the U.S. unions decided that "ergonomics" would help to reduce the effects of the increasing levels of repetition needed to improve productivity. Work-related cumulative trauma disorders began to increase to epidemic proportions. The plans included the creation of the General Motors Manufacturing Ergonomics Laboratory and the development of a plant-based "reactive ergonomics program".

The mission to Mars turned out to be more formidable than the mission to the moon in the 1960s, despite the enormous advances in technology. It may be tolerable to lose a robotic mission, but a manned mission had to have more guarantees. So NASA refocused its attention to Low Earth Orbit and the International Space Station (ISS) in a hope to answer some of the questions associated with long-duration manned space flight. Many people still have a vision of going to Mars and many have laid out elaborate plans, but at the present time there is no mission. All of these case studies confirmed the importance of clear operational definitions in human factors and systems engineering. Unfortunately, the evolution of this subject area has resulted in ambiguities. The next few pages outline some definitions that may help to improve the reliability of communication as human factors engineering interacts with the system design process.

SYSTEMS AND PROCESSES

A key concept lies in the definition of what is a "system" and what does a "system" do? A convenient operational description of a "system" is any hardware, software, and naturally occurring or human entity that, by themselves, have no functions.

When two or more systems interact, in a physical and organizational context, to achieve an objective then this interaction is called a "process". Usually, the objective or outcome of a "process" is a change in the characteristics of one or more contributing systems. In the case of human factors, one of the contributing "systems" is a human "system". For example, a person may have the characteristic of "being at home". Only when he/she interacts with a car and a roadway does he change this location characteristic to "being at work". During this journey, the human subsystem may interact with other systems – such as a coffee cup, a cell phone, a frosty road, and another human-vehicle system to engage in a process that results in an accident – characterized by a change in the shape of the vehicle and the owner's wallet.

Design processes create various human, hardware, information, and organizational systems with the purpose of producing a new entity or service. There are multiple purposes of such processes. First, the product must meet with customer expectations – this in its broadest sense is called product quality. The next process objective is to be efficient or productive; i.e., it must achieve its quality objectives with minimal use of consumable resources (systems) such as people, money, materials or time. This last resource "time" often stands out as a key aspect of process design. The customer would like the elapsed time between his want being expressed and fulfilled to be as short as possible. "Time to market" is a key objective of most product design processes. One way of achieving this objective is through the practice of "concurrent engineering" in which phases of the process are implemented in parallel rather than in sequence so that, e.g., the demands of manufacturing can be addressed during the product design phase. These process objectives are of particular interest to the eventual paying customer, management, and shareholders. However, the unique nature of the human system elements is that they may have their own agendas and objectives. For example, employees would like to maximize their own salaries and minimize the risk of accidents, both of which may conflict with other process purposes – such as productivity. A more detailed look at the design process identifies multiple overlapping stages. The term "concurrent" is somewhat optimistic in practice.

Given the vision of putting a man on Mars, there are distinct but interdependent phases that must be addressed. The first phase is a function identification – launching, navigating, and eating – each with its own purposes that are characterized by "quality", "productivity", "safety", etc. Next comes the realization of these functions through the design and construction of the appropriate hardware, software, "humanware", and "organizationware". The process integration phase focuses on interactions, interdependencies, and interfaces. Of course, the advantages of concurrent engineering are particularly evident here as, e.g., the human and hardware systems must be compatible. The penultimate phase of operations design really addresses the time element. In the Mars mission example, it is critical that various supplies (food, water, oxygen, shelter, etc.) would be on the planet before the humans arrive. Another good example is to be found in automobile production – it is one thing to design and build a car, but to produce 1000 cars a day presents altogether new operational challenges, not the least of which is "just-in-time" materials delivery. Finally, there is operations implementation, which has its own local objectives and its contribution to the next mission through "lessons learned" (feedback).

THE GRAMMAR OF DESIGN

The grammar of design offers a discipline for communication that increases the effectiveness and efficiency of the design process. The first concept is that "processes have requirements" and that requirements relate to the adverbs associated with the process functions (verbs). The process "verb" may be "transporting" a vehicle and human systems in some context or environment. The purposes or objectives of "transporting" may include speed and safety. They will certainly include "quality" – the payload should arrive at the correct destination. Thus "transporting" may be measured in terms of how "quickly", safely", and "accurately" – adverbs. It is important to emphasize that quantification of these adverbs is important if the process requirements are to be reliably assessed.

The achievement of these process requirements will depend on the characteristics of the contributing systems. In the above example, if speed were emphasized, then a vehicle with a big engine and a driver with a heavy foot would assure the desired objective. Again, for precise system design, it is necessary to quantify the adjectives – "big" and "heavy" associated with the system nouns – engine and foot. Otherwise, the engineer cannot design the system and the human factors engineer cannot evaluate the quantitative relationships between the system specifications and the process requirements. Give the engineer a number!

VERIFICATION AND VALIDATION

Once the system is built (or modeled) and the process implemented (or modeled), then the human factors engineer is faced with the important task of evaluation. This consists of two sub-processes – verification and validation. If the system characteristics have been specified precisely and quantitatively then verification of adherence to these specifications is simply a matter of measurement. On the other hand evaluation of process requirements implies the process of validation, which in turn implies the performance of the interacting systems in a real-world context. The key challenge to validation is the inevitable presence of user, context, and temporal variability. A precisely specified and a constructed car may not "perform" adequately with an inebriated driver in thick fog. Development of the validation process begs the question of "humanware" design – who are the expected user and possible misuser? Validation also may exclude certain contextual conditions such as inebriation, fog, ice, or 100 mph. The contribution of human factors engineering lies in a clear description of usage requirements, user capabilities and limitations, design specifications, and evaluation conditions.

PERFORMANCE, BEHAVIOR AND PREFERENCE

Human factors measures of process requirements may be classified at three levels – performance, behavior, and preference. Performance can usually be measured in terms of time and accuracy, given the context. For example, running a mile on level ground will differ from the time taken to run the same distance uphill. The accuracy (quantitative deviation from the objective) needed to thread a needle is different from that needed to park a car. Behavior relates to how a task is performed. In cricket, it is possible for the bowler to achieve his objective by swinging or spinning. Behaviors can be categorized and counted. Using the cricket example again, three bouncers gets the bowler suspended; in baseball, one beamer gets the pitcher mugged. However, these "bench-clearing brawls" (ungentlemanly behaviors) can be modified (despite the preaching of Skinner) by negative feedback – fines. The most elusive measure associated with humans is that of preference. Preferences may be stated and counted, but may not affect behavior or performance. However, extraction of the "voice of the customer" or the mechanisms of "usability studies" often resort to the assessment of preferences or subjective judgments of differences. We must be ever vigilant that we do not put too much store in observations elicited by improper application of one of our most widespread techniques – psychophysics. Unfortunately, this is often the only technique we have available.

Working in space involves many processes such as staying in one place, moving, eating and assembling. The more complex activity of assembling involves the interaction between multiple human systems, components, robotic arms, and communications facilities. The context of microgravity, the vacuum of space, radiation, and very high cost present unique constraints on the process requirements. The interacting tasks of controlling the robotic arm while perching exemplify the challenges, especially when the higher level task of assembling an expensive component may take all day. The adverbs related to this task include carefully, slowly, and comfortably. Slowly can be and is defined precisely and carefully is described in terms of deviation from a prescribed, tight trajectory. Comfortably is one of those unfortunate human factors challenges that defy reliable quantification, although it is likely that uncomfortable perching may create a distraction that in turn may compromise carefully.

REQUIREMENTS AND SPECIFICATIONS

These process objectives and requirements can only be realized through the precise specification of the contributing system characteristics. What kind of restraint design results in comfortable perching? What kind of joystick design contributes to the activity of careful control of a heavy payload (crew colleagues, components, and tools)? What kind of organization of multiple pairs of eyes and brains is conducive to reliable communicating? How much light should be provided, given that the daylight only lasts 45 minutes up there? How much oxygen should be provided for the strenuous tasks of fighting the resistance of a pressurized space suit? How long should the workday be? Give the engineer a number!

The development of process requirements and design specifications is not simple. It is rarely possible to simply translate empirical data into a number that can be applied to reliable validation. At a simple level, if asked why the height of a door opening is 7 feet, we may waffle about percentiles and allowances for shoes and hats. Similarly, we may look at accident statistics on the freeway and determine that 100 mph (160 kph) is acceptable for 95% of journeys. Or we may state unequivocally that 15 minutes of arc is the design specification for font height on a computer screen. If we don't give the engineer a number he can't design or verify. But we all know that the number includes a policy overlay and will usually be modulated by domain experience. Consequently, it is essential that requirements and specifications be developed by consensus, with management (or the law) imposing policy, our scientists providing the logic and the data, and the engineers – the eventual users of the standard – providing the domain experience. Of course, all standards (requirements and specifications) should be subject to iterative evaluation and an effective technical memory should lead us to convergence. Unfortunately, those policymakers often change their minds when faced with trade-offs.

THE HONG KONG MASS TRANSIT RAILWAY

The main performance requirements of this transportation process were to maximize the safe throughput of passengers, given the constraints of size, speed, and the need to show a profit. Throughput is constrained by spatial capacity and passenger behavior, which in turn is affected by spatial arrangements. The seat design and grab rail specifications were based on anthropometry and human behavior. The approaches used to generate the anthropometry and behavior evidence involved the human factors literature, surveys, analysis, and evaluation of performance and behavior in physical mock-ups. The use of adjustable physical mockups of both the passenger compartment and the operator's cab proved to be very instructive. In fact, there were substantial discrepancies between the simple application of anthropometric accommodation principles (5th and 95th percentiles) and the actual behaviors of representative samples of subjects in the physical mockups. An ironical twist in the seating systems development, based on emphatic input from the Hong Kong Fire Department, resulted in the adoption of flat stainless steel bench seats rather than the scalloped aluminum that was first proposed. This resulted in an adaptive rather than constrained seating arrangement. The vertical poles and horizontal grab rails were positioned to allow optimum accessibility, stability, and motion, given the wide range in anthropometric characteristics of the expected user population. A horizontal bar reachable by a 5th-percentile female would hit a 95th-percentile male on the chin! Compromise!

The operator's seat design followed a task analysis, which indicated that the operator would have to get in and out of the train every 90 seconds as he checked that the platform was clear prior to starting the train. This resulted in the design of a seat that could be folded back (allowing easy egress in an emergency) or in the down position for longer, between station, transits. The seat also had a padded front edge to accommodate the preferred lean sitting posture. In these examples, human factors was applied in a somewhat ad hoc way in the very early design stages. The principal "tool" was a physical mock-up evaluation.

LIQUEFIED NATURAL GAS TRANSPORT

In the early 1980s, serious consideration was given to the exploitation of the vast oil and gas reserves in Arctic Canada. The two transportation options were pipelines and large double-hulled ice-breaking tankers. Given the cargo and the context, there were substantial safety concerns – a collision or a grounding could result in a cloud of escaped gas descending on a town and then exploding. The preferred analytic approach was to use fault tree analysis – both for the mechanical and electrical systems and the human systems. It should be noted that, unlike military vessels, commercial vessels are designed to be operated with very small crews – thus reducing the human redundancy in case of error. The approach to the assessment of human error was based on the human reliability assessment techniques developed for the nuclear power plant industry at Sandia National Laboratories. A massive (paper) fault tree was developed and an assessment of the performance-shaping factors indicated that an incapacitated crewmember would be a likely cause of catastrophic failure. Six years later, the Exxon Valdez confirmed these findings. Fortunately, the LNG project was abandoned for a combination of environmental, engineering, economic, and human factors reasons.

CAR AND TRUCK DESIGN

Car and truck design is a fashion business. Some quip "function before form", whereas others say "form before function". This can be translated into preference before a performance. There are, however, basic functional requirements of capacity, operating, maintaining, etc., before aesthetics takes over. Human factors contributions in product design cover the full spectrum of customer needs – from basic physical issues, through sensory and information processing to their requirements for alternative features and styling. Different vehicle types attract different customers and have different uses. Much of this evidence is elicited early in the design process through competitive review, clinics, and more precise laboratory investigations, involving simulators of various levels of sophistication. As the design process progresses various iterations of prototypes are assessed using modeling, checklists, and "drives" on closed courses and the open road. The formal process is iterative and involves concept evaluation, selection, and refinement, through processes of analysis, testing, and, eventually, board review.

Quality Function Deployment is a technique that has been applied widely in the automotive industry to translate the voice of the customer into design specifications and on down through the manufacturing, production, and distribution processes. Early uses of the technique resulted in very large and unwieldy matrices that became increasingly less than useful. However, the principles are sound and lend themselves well to the discipline of requirements and specifications development. Unfortunately, the user (customer) does not always adhere to these grammatical rules of design. The dutiful customer should ask for a vehicle that "goes fast", "is easy to maintain", "enhances his social image", and "protects him in the event of an accident". Instead, the customer may stipulate engineering nouns and adjectives such as: 300 horsepower, maintenance-free, red vehicle with side air bags. This lack of discipline (customers stating specifications rather than requirements) occurs among all the many internal and external customers.

An example of QFD in product design would be to address the operation or driving of the vehicle. One adverb might be top speed and the range of top speeds of competitive vehicles might be available from market research. The engineer would recognize that top speed would be accomplished by, among other things, engine size, which would be described in engineering units of cu ins. The adverb "fast" (speed) is, of course, affected by more factors than engine size – there is the mass of the car, the gearing ratio, the aerodynamics, and the type of fuel, etc. Similarly, the "safety" requirement might conflict with the "fast" requirement. Thus the task of the human factors engineer becomes more complex in the optimization of often conflicting requirements.

The human factors engineer is a surrogate for the end user. He or she should identify the populations of interest on the dimensions of interest. He should communicate clearly with the design engineer by relating the associations between levels of design specifications (independent variables) and performance outcomes (dependent variables). The relationships are affected by human variability, which can only be reduced by curtailing the "expected user population" by selection or training. There will be many occasions where the relationships are affected by the prevailing conditions and by interactions with other variables; there may also be multiple, sometimes conflicting outcomes. Eventually, the designer will have to settle for a single value on each dimension, unless he can design an adjustable feature. The task of the human factors engineer is to communicate the acceptable ranges for each independent variable, given the percent accommodation policy, and to articulate the likely sources of interaction.

DESIGN FOR MANUFACTURING AND ASSEMBLY

Contemporary vehicles have many more features and components, especially on the engines, than they had a few decades ago. However, the drivers have not changed in stature and so it is not possible to increase the height of the hood. Thus more things have to be compressed into a smaller space, which produces challenges for packaging, assembling, and maintenance. Contemporary "design for assembly" approaches use mock ups and computer models to assess these manufacturing challenges. There are also certain well defined, though not always feasible, ground rules for design – such as layered assembly and upward and outward facing fasteners. Given the best possible design, with manufacturability in mind, the ergonomist is next faced with materials delivery and presentation, tools, workplaces, and task content. This last challenge of "line balance" attempts to maximize the utilization of every second of the assembly operation. The ergonomist looks for ways of increasing physical (and mental) job variety through team structure, job enlargement, and rotation, but may be constrained by seniority agreements and quality concerns.

One fundamental challenge of manufacturing ergonomics lies in the difficulties associated with measuring people in their working environment. These conditions do not lend themselves to the rigorous demands of the experimental laboratory for accuracy, precision, reliability, and even sometimes validity. The sample size is also usually restricted. Consequently, the thrust of manufacturing ergonomics should be the assessment of the workplace using population data while allowing the eventual individual operator(s) to fine-tune the arrangements to suit their particular needs. Systems approach to workplace and task evaluation using various levels of analytic tools should therefore limit themselves to population data while leaving room for some flexibility.

Manufacturing ergonomics assessments are applied at all stages of the manufacturing system design and implementation process. These assessments take the form of computer modeling, "wall reviews", "prototype reviews", and "slow build" reviews in which each motion is evaluated in great detail. This up-front assessment leads to much improved designs of production systems. The practicing manufacturing ergonomist has available a wide variety of analysis tools that range from checklists through integrated analysis methods to digital simulations.

Ultimately any design, design change, or operational intervention will be based on a risk-cost – benefit assessment. The solution may be a change in the component (the product engineer's responsibility), a change in the tool or workplace (the manufacturing engineer's responsibility), a change in the amount of work in the job cycle or the line rate (the industrial engineer's responsibility), a change in who does the job (the supervisor's responsibility with due regard to seniority), or the method by which the job is performed (the operator's or trainer's responsibility).

OPPORTUNITIES FOR INTERVENTION

The application of systems engineering and human factors in car and truck manufacturing addresses the opportunities for change that are presented during each of the product, manufacturing process, production, and operations phases. By way of example, rather than use a specialized manufacturing process, one can consider the baggage handling processes at an airport. The first design phase is the product – an item of baggage. There are restrictions on shape, size, weight, materials, and content. The handling process design includes consideration of all sub-processes that occur between the parking lot and the aircraft's hold and back to the parking lot. These sub-processes include mechanical handling and information processing devices, human handling and information processing activities, the design of interfaces, and due consideration of the environment. The production system design element takes the problem from the handling of a single item to that of millions of items a year. It requires the coordinated activities of sufficient handling devices, sufficient information processing capacity, sufficient people (with appropriate training), and sufficient numbers of interfaces. Finally, operations management involves a full complement of baggage handlers, maintainers, customer service agents, second-level problem solvers, and managers all being appropriately selected, trained, and assigned to achieve a desired level of customer service. As the overall process moves towards operations there will be increasingly greater levels of scrutiny. Hopefully, the baggage and handling systems design issues have been dealt with early in the design process. However unusual, seasonal demands to handle golf clubs, skis, bicycles, and fish may expose the shortcomings in design accommodation.

WALLS AND THE ERGO COP

One way of modeling each phase of the process is through the development of physical or computer-based "walls" that contain an array of standardized details of the systems and processes as they mature toward implementation. These "walls" provide the media for multidisciplinary teams to comprehensively evaluate each stage of the process so that late developments don't interfere with the critical path and late changes don't result in excessive costs.

Manufacturing ergonomics has its own elements on the "walls" Assuming that the product design issues have been addressed on an earlier "wall" the manufacturing ergonomics wall will contain questions that address workplace design issues of fit, reach, targets, and task content. At the production level, the physical and temporal aspects of an operation will be amalgamated to assure an acceptable job cycle workload. Later the operations wall will address individual job and team assignment questions. Finally, the operations output wall will document quality, productivity, and health and safety issues associated with each operation.

The practice of manufacturing ergonomics provides important lessons for many other practice domains. The traditional practice of imprecise process requirements and unrealistic design specifications leads to inappropriate designs to be addressed by the "ergo cop". Eventually, battles ensue in the "review boards" that often result in requests for waivers and either a loss of face or an inflated ego, of the ergonomist. This process is both inefficient for the company and unhealthy for the profession. The ergonomist should participate with engineering, management, and the operations/user community in the establishment of clear performance requirements and sufficiently precise design specifications and design implementation. In this way, there will be no surprises at the board (performance) reviews.

SPACE VEHICLES

The design of space vehicles differs from high-volume manufacturing in product cost and product life cycle. The environmental challenges, power requirements, and human interactions are unique. The complexity and remoteness of the operations lead to massive information management challenges and costs. The space program is deliberately very visible – the whole world is watching. Finally, because of these things, there is relatively limited opportunity for the program to capture sufficient "lessons learned". Much of the evidence that cannot be based on analysis, must be based on small samples of empirical evidence. Human factors specialists become acutely aware of the challenges of human variability, given the relatively small number of experienced astronauts. Manned exploration missions, e.g., to Mars, present even greater challenges of evidence from robotic missions.

Over the past two decades NASA has developed extensive statements regarding the human factors issues of manned space flight. These statements are in addition to the extensive medical requirements. The NASA Standard 3000 – the Man-Systems Integration Standard (MSIS) is a compilation of evidence from both the profession of human factors (and other sources) and domain knowledge. Military standards such as Mil-Std. 1472 were particularly influential in MSIS development. The basic MSIS standard has been adapted to program-specific statements for Space Transportations System (Shuttle) and the ISS. These basic and derived standards have, like the programs to which they refer, been subject to hostile attacks in the requests for waivers from engineers, programs, and contractors. As the space program matures working groups, tiger teams, and review boards all contribute evidence on which the next generation of standards will be based.

A general challenge to human factors is exemplified by the NASA review processes. Almost all human experiences are dogged by individual, contextual, and temporal variability. For example, an ideal thermal environment is affected by activity, clothing, individual acclimatization/tolerance, and duration of exposure. The requirements for strenuous exercise are different from those of reading. Consequently, an overly specific statement like 72°F will inevitably be inappropriate much of the time. The challenge to a human factors review panel is to address all the, possibly interacting, criteria in coming up with a decision that the engineer can design. Clearly, adjustability is required, but how much adjustability? Further complications arise because of constraints on design or change. For example, a particular intervention may be too costly or not feasible in the time scale of the overall project. The final complication of the review process is that the judges (usually experienced managers) overlay their own experience/prejudice on the decision. The task of the human factors engineer is to apply his/her own principles to the conditions surrounding the review process. It is the responsibility of human factors engineers to be "user friendly" in their own practice. The response: "come back in a year when I have done a comprehensive study" is only occasionally warranted. Similarly, an answer that says 72°F, because the textbook says so is equally naïve.

The human factors community at NASA makes extensive use of digital modeling in the design, evaluation, and real-time mission support phases. The primary contractor – Boeing – made extensive use of anthropomorphic modeling during the early design phases of the ISS. Currently, the Interior Volume Control Working Group uses models of the ISS interior, together with anthropomorphic models to evaluate additions and changes such as sleep quarters, the galley, exercise equipment, and protruding racks that interfere spatially and temporally with routine and emergency activities. The application of digital human modeling in the evaluation of the conditions of work in an assembly task was shown to be more precise, faster, and far less expensive than the alternative of a full-blown trial in the Neutral Buoyancy Laboratory. Modeling was particularly useful in the iterative design and analysis cycle of the crew quarters rack which provides facilities for sleep, computer workstation storage, and privacy.

Lighting models, using the Lawrence Berkeley Laboratories "Radiance" software, are critical to operations, given the changes from extreme brightness to complete darkness every 45 minutes. Differing viewing points for crewmembers and cameras, shadows, and glare compound the difficulties. Just-in-time modeling and prediction of lighting conditions are invaluable to many Extra Vehicular Activity operations. Exterior robotic operations in the rapidly changing day/night cycles make use of both human vision and camera vision for both training and real-time activities. These models are particularly useful to aid decision-making when contingencies change the timeline and hence the lighting conditions for particular activities.

Although there are very few crewmembers, the multiple demands on their time and the many resource constraints, such as equipment, power, materials, and lighting, make activity scheduling a very difficult challenge. The difficulty is compounded by sparse and imprecise evidence regarding the duration of human activities in the microgravity environment and the ever-present challenge of human variability. The problem is being addressed by enhanced data collection approaches and a range of complex and simple scheduling models. The crew work day on the ISS is broken up into three main categories – work (including scheduled science investigations, assembly, maintenance, planning, and communications), sustaining activities (sleep, exercise, eating, and personal time), and responding to contingencies (such as caution and warning signals). The considerable spatial restrictions of the ISS complicate work activities through stowage constraints and spatio-temporal interference. The personal preferences of individual crew members in the highly congested conditions sometimes result in excessive time being spent in finding tools and materials. The crew workday is categorized in detail, however, the variability of times of activities within categories is not well understood. Consequently, steps are being taken to collect better data and develop simulations of activities on a daily, weekly, and mission basis. These models show not only the occasions when the schedule is overbooked but also how different priorities of activities can be used to accommodate this overbooking – such as sacrificing sleep or personal time.

GLOBAL INTEGRATION

Human Factors and Systems Engineering are essential to an effective and efficient design. However all designs of processes and contributing systems are complicated by change and human, situational and temporal variability. A major thrust over the past decade has been attention to common processes of both the design activity itself and the resulting product and manufacturing processes. Unfortunately times and best practices change, so the processes must be flexible to assess and accommodate these changes. These challenges are particularly evident in international operations, where economies of common processes often conflict with different national practices that have been established over many years. Of particular value in the human factors area is the establishment of a common communication currency that enables comparisons to be made between widely differing alternatives and conditions and which resolves the ubiquitous importance weighting problems.

Global integration efforts are often the source of conflict between the efficiencies of common processes and the perception of what are best practices. "Not invented here" is often an underlying motive. The challenges of competition in the automotive industry, coupled with the explosion of computing and telecommunication facilities have combined to fuel the fires of globalization. Manufacturing organizations seek out high-quality, but lower cost labor markets. Also, it is not efficient to have an engineering design center in every country – why develop essentially the same product separately in multiple markets? But engineers worldwide are conservative and resistant to imposed change. In these cases, there is no substitute (other than dictatorship) for extensive face-to-face interactions among the design teams, including the human factors engineers.

The ISS faces similar challenges. The program has very important political underpinnings, the costs are extremely high and national identities need to be clear. And there are other constraints – only three crew members at a time, relatively few modules and only occasional opportunities to visit. The management of such a program is not limited to the handful of astronauts and cosmonauts; there are very large support staff in operations, engineering, medicine, and science management. The ISS is at once a miracle of systems and safety engineering and at the same time a management nightmare.

DESIGN AND PROCESS CHANGES

The mechanisms of dealing with change are described in different organizations as a request for waivers, change requests, or engineering change orders. In many cases, these requests are appropriate, albeit due sometimes to poor planning or unclear requirements and specifications statements earlier in the process. Often, however, they are seen as frivolous – made only to accommodate the failure of a supplier to be able to deliver on earlier agreements. Where waivers are processed on an individual basis, they may not comprehend the implications on other aspects of the process or the trickle-down effects to other subsystems. Cost is a common reason for the change and the systems engineer and the human factors engineer must work with the managerial accounting community to establish a common basis for the rational processing of waiver requests.

Human factors cannot be practiced without engineering – the people who design the systems – and management – the people who make the policies. Sometimes policies are imposed from elsewhere – through technical standards, government regulations, or labor agreements. Some human factors practitioners feel that it is their duty to convey policy, especially where engineering and management do not have the appropriate information to decide on policy. On occasion, human factors specialists may substitute dogma for policy. The notorious 5th percentile is a prime example. It is a useful concept, often with good rationale, but it is widely misunderstood and often inappropriately applied.

A major problem is that someone who represents the 5th percentile on one measure is unlikely to hold that relative position on another; furthermore, when accommodation is based on multiple dimensions then it may be difficult to define who or what is a 5th percentile. Monte Carlo simulation methods may be applied to somewhat relieve this problem. Another difficulty is that the implications of a design decision may be more or less important. Thus in the case of a highly sensitive design decision, it may be appropriate to accommodate the 1st percentile. On other occasions, design for the average may be an adequate approach, given that the dimension in question is not related to an important outcome. An example of a highly sensitive design decision would be the walking speed of old people crossing a busy road.

OUTCOME AND DESIGN SCALES – COMMON CURRENCIES

Human Factors engineers are made aware of the processes and theoretical underpinnings of scaling methods from their earliest training in the statistical methods applied in the broad context of human variability. Scaling systems abound – percentages and Yes/No are separated by Lickert-type scales of varying degrees of resolution. A prerequisite of any scale is the establishment of anchors for both endpoints and intermediate thresholds. The fuzzy classification reflects reality but is often a practical inconvenience. Appropriate resolution is always needed. The zero to ten scale is probably the most universally familiar one and has stood the test of time. It usually has adequate resolution and can easily be linked to a response categorization.

Given this ten-point scale it is relatively easy to visualize a nonlinear mapping function that covers the full range of outcomes from ideal to unacceptable. Examples include lighting, noise, temperature, spatial, and force scales of design specification ranges. It is also possible to comprehend single and complex variables, although, for engineering design purposes the evaluation will ultimately have to identify individual variables for change. Where the relationships are not monotonic, as in the choice of an optimal temperature, then it is convenient to use two scales – one for hot, the other for cold. There may also be multiple outcomes – some of which might be conflicting. For example, a spatial scale related to controls, such as vehicle pedals, may have movement time and inadvertent actuation conflicts. Such conflicts point toward the importance of consensus processes in the establishment of mapping statements and cut-offs. The reality of human variability is such that a single mapping function will never be precisely "right". Again the inclusion of human population accommodation policy in the consensus decision is essential to assure "buy-in" of all concerned as the design process develops.

DESIGN CONSENSUS

The development of human factors design standards is best pursued through a consensus process, using the common currency described earlier. The credo that standards should be data-driven is over-simplistic. Policy, scientific logic, technical feasibility, and experience must all contribute to the establishment of a standard. It is also essential that representatives of the customers – internal or external – who will have to apply or be affected by the standard should be involved. In this way up front agreement in both the principles and the values related to the standard will help to assure "buy-in" and less demand for waivers. Human factors standards are also iterative in that they should be verified, validated, and evaluated as the project or program evolves. A clear example lies in the establishment of speed limits for different road conditions.

Design trade-offs should be made with a full global view of all the relevant information, preferably with a common communication currency. The choice of cut-off points on individual variables can be used to assign explicit weightings. A broad range would imply wide tolerance (less importance) and a narrow range of greater importance. The common currency outcome prediction scale facilitates the process of amalgamation. At the simplest level a count of the number of variables in each of the outcome ranges produces an index or profile that reflects the general nature of the problem. The count can also be used as a decision aid – for example, decision policies could be "no reds" or not more than "5 yellows".

Addition (as opposed to counting) is rarely justified, although this is the preferred method for some checklists. However, the case for multiplication in the amalgamation process may be justified where interactions are likely. Such situations are best handled by two-dimensional matrices using the same common currency described here. The special case of interactions between basic variables and time may also be approached in this way. This big picture decision aid can be used to indicate before and after change situations, comparisons between alternatives or progress of a project through the design process.

The common currency scale can be viewed as an estimate of the "probability of failure on a single transaction". For example, the probability of "failure to accommodate" of a 12-inch wide seat would be of the order of 0.7. Similarly, the probability of failure of a 2-mm high font, given an elderly reader population could be 0.99.

At the other extreme, the probability of failure of a 24-inch diameter escape hatch might be 0.05.

COORDINATED DECISIONS

The decision process must also involve the benefit of the transaction, the number of transactions per unit time, the number of people affected, and the various costs of failure. To complete the assessment an evaluation must be made of the costs, benefits, and probabilities of alternative designs and outcomes. Where possible, objective evidence (including data) should contribute to the probability and costs/benefits and exposure estimations. Where individual, personal decisions are made then subjective probability and cost estimates may be sufficient – this is, of course, the basis of most naturalistic decision-making.

The hypothetical example of our choices of transport to work – car, bus, or tank – can be used to illustrate the quantum decision process. Assuming a decision horizon – a day, year, or project lifetime – all measures can be reduced to base ten arithmetic. In the case of choice of transportation mode based on cost and safety then the analysis shows that we should ride the bus. However, if we are the president of a country in political turmoil then we may wish to revise our probability, exposure, cost, and outcome estimates and at least buy an armored car. This quantum arithmetic approach is appropriate for a cursory analysis and exploration of the sensitivity of the different elements of the decision to changes. Greater resolution may be obtained by including multipliers and decimal components, while still adhering to the basic decision logic. However in this case it may be appropriate to use some computational aid.

SYSTEMS INFERENCES

It is important to reiterate that it is the role of management or governments to communicate policy, given human factors evidence of outcome likelihood and effect. The role of consensus in the establishment of design standards was also addressed. In this respect, it is important to address the realities of false consensus and the sometimes inappropriate or overly weighted influence of experts. The opinion of experts should always be weighted heavily, but only in the area of their expertise. A better approach is to use independent and interdependent "voting" processes, with sufficient allowance for discussion as the standard or decision scenario is developed.

The design of boats, trains, cars, and space vehicles as most other human factors opportunities necessarily involves teams of one kind or another. Generally, the teams consist of an exhaustive and exclusive "set" of people with technical and domain knowledge. Individual team members may have discrepant objectives. Team dynamics create challenges to both the effectiveness (accuracy) and efficiency (speed) of standard, risk, and design decisions. Greater effectiveness and efficiency can be achieved through the application of common currency and clear visual aids to comparison, trade-off, and decision-making.

Even the decision processes common in contemporary design projects face the challenges of over/under reliance on expertise. On occasion, the efforts to substitute process for expertise may also be counterproductive. Overly enthusiastic attention to "process" can be cumbersome, where simple experience may be sufficient. One hundred years ago craftsmen built outstanding automobiles, slowly. Nowadays, processes result in the high volume production of automobiles. Hospitals used to be run by medical experts but now the processes of HMOs have diverted the purpose to profit rather than a caring motive. Soccer is essentially a game of experts, football has experts that are bound by processes. Music was once the realm of experts, now it is relegated to simplistic marketing processes. We went to the moon on the backs of experts; processes will get us to Mars.

Human factors is alive and well early in many design processes. The appropriate place for human factors is as a branch of engineering, not as an after the event "ergocop" in safety or consumer protection functions. Human factors is rightly a component of systems engineering, it can contribute important knowledge and tools both to the designs themselves and to the design processes. One important contribution is the establishment of a common currency for communication of human factors and implications of design.

CONCLUSIONS

The ergonomist has useful contributions to offer at all stages of a product life cycle by representing the "voice of the customers" – user, maintainer, manufacturing employee – in the design process. The ergonomist is effectively and efficiently supported by a tool kit that not only includes in depth analytic tools but also simple communication tools such as E4S4, 6Us and 2Ms and QFD.

REFERENCES

Akao, Y. (1990). *Quality Function Deployment: Integrating Customer Requirements into Product Design*. Productivity Press, New York, NY.

Donabedian, A. (1988). The Quality of Care, *JAMA* 260(12): 1743–1748.

Peacock, B. (2004). "The Semantics of Human System Design," SAE International Conference on Environmental Systems, Denver, CO.

Peacock, B. (2014). Ergonomics of Design. In: Badiru, A. B. (Ed.), *Handbook of Industrial and Systems Engineering*. CRC Press, Boca Raton, FL.

Peacock, B. (2019). *Human Systems Integration*, Self-published manuscript, Fernandina Beach, FL.

Peacock, B. (2020). *How Ergonomics Works,* Self-published manuscript, Fernandina Beach, FL.

Peacock, B. (2021). *Ergonomics Tools and Applications*, Self-published manuscript, Fernandina Beach, FL.

Peacock, B., and W. Karwowski (1993). *Automotive Ergonomics*. Taylor and Francis.

Peacock, B., and G. Orr (2001). "A Checklist for Industrial Ergonomics Checklists," IVth IIE Annual Applied Ergonomics Conference, Orlando, FL.

Peacock, B., and M. Resnick (2010). The 6Us, *Applied Ergonomics*, Summer 2010.

ReVelle, J. B., J. W. Moran, and C. A. Cox (1998). *The QFD Handbook*. Wiley, Hoboken, NJ.

4 Semantics and Syntaxes in System Design

INTRODUCTION

Before designed systems can be integrated in the manner espoused by the DEJI Systems Model, designers and users must understand the semantics and syntaxes that govern the systems environment. The technology of system design has grown over the past half-century from roots in human factors engineering manufacturing process engineering, product design, and, more recently, software systems design (Peacock, 2019, 2020, 2021). Contemporary complex systems are characterized by many agents: human, hardware, and software subsystems collaborating in an integrated organization to carry out a mission in the context of environmental uncertainty and time constraints. The technology has developed its own esoteric jargon and in some instances, the tools of system design have become so cumbersome that they add unnecessary complexity, cost, and delays to the process. The purpose of this chapter is to explore the semantics of system design and some of the more useful tools in the context of space exploration, with occasional detours into automobile manufacturing for explanatory purposes. The chapter concludes with some simple rules for human system design.

The fundamental semantic challenge is to separate the concepts of entities and activities, agents and functions, systems and processes, and nouns and verbs. Entities, agents, and systems are described by nouns. Activities, functions, and processes are described by verbs. A system (noun) can only be designed by specifying its characteristics – adjectives – in a quantitative way. Give the engineer a number! Systems by themselves are inert. It is only when they interact with other systems that a purposeful process occurs. Processes are described by verbs and quantified (qualified) by adverbs (Peacock, 1995).

Systems, subsystems, components, and elements may be comprised of hardware, software, humanware, or organizationware. Extravehicular activity requires a suit, software for control and communications, an organizational structure for supervision, and last but not least, the person in the suit and his or her support entourage. Extravehicular activity per se has no specific purpose – activity is a generic term for a collection of purposeful functions, such as inspection, assembly, maintenance, translation, manipulation, protection, etc., that can be measured. Driving to work requires a car (nowadays with a lot of software), a driver, a road, and an organizational structure, including the highway patrol. Driving to work is also dependent on many supporting processes, such as buying, taxing, and insuring the car; filling it with gas; and maintaining it. Failure of any subprocess can have outcomes that vary from catastrophic to a minor inconvenience. See Evans et al. (2000) for pertinent statistical inferences and Herzberg (2008) for motivational aspects.

DOI: 10.1201/9781003328445-4

PROCESS OUTCOMES

These two activities (verbs) – EVA and driving – have common general purposes. First, there is quality or effectiveness – achieving a stated objective. Next, there is efficiency or productivity – consuming the minimum amount of resources, such as fuel, money, or time, to achieve the objective. The third general purpose of all processes is safety – there should be no (or minimal) harm done to any of the collaborating systems, except where the conversion of systems is the purpose of the process, as in propulsion. Harm in this context includes both acute damage to a system and cumulative damage, such as wear of moving parts, radiation sickness, or undue fatigue. Parenthetically, all processes result in a change to some or all of the contributing subsystems, such as "normal wear and tear" or in the case of the human subsystem, temporary fatigue or learning. A unique purpose of complex processes that include human subsystems is that of satisfaction. People or their organizational supervisors must find some intrinsic satisfaction in the activity and not be overly affected by the dissatisfiers such as discomfort or insufficient rewards.

Processes take place in environmental contexts, which add uncertainty to the outcomes. Space exploration involves extremely hostile environments and the lack of complete or timely information regarding these contexts can be catastrophic. The weather or the other drivers may complicate the process of driving to work. These uncertain extrinsic contexts may interact with any of the intrinsic subsystems and where these subsystems are vulnerable, process failure or subsystem damage may occur. A solar flare may tax the effectiveness of the radiation protection subsystems and a wet road will attack worn tires (Haushalter, 1971). In the case of the human subsystem, a cold environment will interfere with finger dexterity, and a surfeit of information, as occurs with cell phone use in busy traffic, will overtax the human attentional processes (Peacock, 2003).

HUMAN ERROR

Two apparently contradictory opinions about design outcomes are voiced by Casey (1993) and Petroski (1992). Petroski argues that "To Engineer is Human" and that engineers will usually learn from their mistakes; he sees that engineers continually break new ground and are faced with constraints and trade-offs, which lead to risky decisions. He cites a series of catastrophic civil engineering failures and their "after the event" explanations. Some critics at the 1992 Human Factors and Ergonomics Society Annual Meeting argue that Petroski failed to understand that human error, both in design and execution, is predictable and therefore preventable. Casey presents descriptions of a series of notable catastrophes that were caused by execution or foreseeable design errors for which there are clear human factors explanations. Examples include interface design shortcomings in an X-ray machine and a trust in automation (radar) issued by the commander of the U.S. Fleet in 1926. The system design factors that gave rise to operator overload error at Three Mile Island are legendary. More recently, the NW 255 plane crash in Detroit can be attributed to the crew failing to perform a correct procedure, together with a failure of a warning system. Beaty (1969) presents a perceptive assessment of "The Human Factors in

Aircraft Accidents" and Hancock and Desmond (2001) identify the vulnerabilities of the human operator to "Stress, Workload and Fatigue". Finally, Robinson (1993) in his fictional account of the settling of "Red Mars" eloquently identifies the psycho-social and political vulnerabilities of complex systems involving people.

THE GRAMMAR OF DESIGN SYNTAX

These concepts can be drawn together using the familiar grammatical construct of a sentence:

- I want to operate my car.
- Some people want to drive their cars to work in less than half an hour while listening to their voicemail.
- Qualified drivers want to drive their racecars quickly and safely around the wet, winding racetrack.
- Well-trained astronauts want to capture a large satellite with a robotic arm.

These sentences articulate process requirements with varying degrees of specificity. Designers of the processes – driving, capturing – need more information to satisfy their customers. The first task is to identify the customers and their perhaps differing requirements. The end user – I, the driver, the astronaut – may not be the only customer. Other customers include trainers, maintenance engineers, managers, legislators, and the general public and they may have differing, perhaps conflicting requirements. Sometimes the customer may be an individual that has a tailor-made [space] suit. On other occasions, the end user may be one of a large population of users for minimally adjustable hardware. (If the glove doesn't fit, try the next size up and if that's too big, tough!)

A requirement must be articulated as an adverb that may be evaluated, assessed, or validated by objective or subjective methods. The adverbs quickly and safely can be assessed subjectively and for this assessment to be useful, it should reflect the consensus of all the customers. One way of improving the reliability of requirement assessment is to provide verbal or numerical anchors to the assessment statements. For example, the requirement "quickly" could be quantified by a speed or a time to cover a fixed distance, under controlled environmental conditions. The "safely" requirement is more problematical and may be quantified in a safe/not safe scale, by articulating a continuum of possible outcomes – such as mission failure, crewmember fatality, and subsystem damage – or by adding probabilistic statements that can only be assessed reliably in the light of experience with the operational system or through simulation. Parenthetically, the best available estimate of the probability of failure of the Space Shuttle mission process is 2/113; however, this only predicts an expected value, and where confidence limits are placed around this point estimate, using the Binomial, Poisson, or Normal distributions, the 95% confidence level is of the order of an unacceptable 1/30. Such historical estimates are always suspect as learning occurs in most human-managed processes and possible failure modes are eliminated in the light of experience, thus changing the system design and process reliability. In the car driving context, the biggest process reliability change would be to eliminate drunk driving which contributes to almost half of the 40,000 fatalities a year in the United States, but on balance, driving is a very reliable process and most drivers who are "under the influence" don't have accidents and most drivers who drive faster than the speed limit don't get caught.

RISK ASSESSMENT OF SEMANTICS AND SYNTAXES

The technology of risk assessment has progressed over the past few decades and there exist various standard processes for linking outcomes and likelihoods. These methods typically use nonlinear probability scales and ordinal outcome or severity scales, which are sometimes converted to a common currency, such as dollars. A fundamental shortcoming of these risk assessment approaches is that they do not usually address the trade-offs that must be made with positive outcomes – benefits. Where a common currency approach is adopted, it is possible to develop key ratios that relate to costs and benefits (Peacock, 1998). A common metric in space flight engineering is equivalent system mass (ESM); this also fails to comprehend trade-offs between costs and benefits and is therefore an insufficient decision tool. More sophisticated analytical processes are essential if we are to comprehend how space flight trade-off decisions are made. Another challenge is related to the costs of the development of countermeasures as well as the countermeasures themselves. For example, the development of a planetary surface suit that is both protective and offers good mobility and where there may be acceptable trade-offs between protection and mobility may have very high development costs and conflicting operational advantages and disadvantages. Very few car buyers purchase Hummers in order to increase their personal safety but the evidence is clear that vehicle mass is a major contributor to accident outcomes. The issue of the trade-off between "production and protection" is also very apparent in high-volume manufacturing and materials handling processes. The shareholders and management want "productivity", whereas the union fights for "protection" and this trade-off has been escalated to the highest circles in the country with the debate about ergonomics standards (Peacock, 1993).

THE DESIGN PROCESS

The engineer cannot do his job effectively without requirements that can be validated. In other words, the requirements statements must contain verbs and their associated adverbs that define process performance and the conditions or tests under which this performance is to be evaluated. For example, a sports car may be expected to go from 0 to 60 in 6 seconds. A suited astronaut may be expected to travel 100 meters over a planetary surface in 5 minutes, with a heavy load of equipment.

Given these validatable operational or process verbs and quantitative adverbs, together with contextual information, the engineer and operations designers are in a position to start addressing the systems that may be needed to satisfy these requirements. Designers create things – nouns – and can only design them with quantitative information – adjectives. For example, 0 to 60 in 6 seconds may be achieved with a big heavy car with a big powerful engine or with a small, aerodynamic car with a small, efficient power train. The planetary surface astronaut's task may be achieved with a rucksack or a golf cart. Once the engineer has the process requirements, he/she can then explore the systems (nouns) and their characteristics (adjectives) to develop concepts that may satisfy the requirements.

Unfortunately, the design process does not always work in this tidy way. The customer may not articulate clear requirements but may seek to specify design options and impose requirements after the fact. For example, the customer may ask for a small, aerodynamic sports car and may be disappointed when his luggage doesn't fit in the trunk. An exploration program manager may specify a lunar rather than an orbital launch platform. Conversely, the engineer sometimes seeks requirements that fit his predetermined design specifications, much like the health care specialist with a limited set of interventions may seek diagnostic information to justify those actions. Such conservatism is sometimes justified as the system design characteristics may be well evaluated. The challenge occurs when the system is expected to meet new requirements. "If your only tool is a hammer, very soon everything begins to look like a nail". These possibly unfair references often occur because of a shortage of research and development funding but in the long run, the new challenges of long-duration space travel will require new technologies.

QUALITY FUNCTION DEPLOYMENT

An adaptation of Quality Function Deployment (McHugh, 1986) can provide the discipline of separating systems, processes, nouns, verbs, adjectives, and adverbs. Quality Function Deployment employs a series of matrices that transfer information from market research through product design, manufacturing, and production processes to sales and the rest of a product life cycle, including maintenance and recycling. The vertical axes of the matrices contain information about customer requirements and their quantitative adverbs, often obtained by benchmarking tests. For example, a space suit user may expect good shoulder mobility and the adverb may require this to match unsuited shoulder girdle function (an impossible task with current hard upper torso technology). A vehicle maintenance function may involve visual, hand, and tool access and the quantitative adverb may expect spark plug change in 5 minutes – a process performance standard derived from comparison with other similar vehicles.

The horizontal axes contain descriptions of the systems (nouns) and their quantitative adjectives. These are system design specifications. For example, maintenance access may require a cone with a minimum diameter of 20 centimeters. Radiation protection may require a material thickness of x millimeters. Eventually, the engineer must design the system with these quantitative values. Give the engineer a number. Unfortunately, no single number will ever be "correct", at least where human subsystems are concerned. A tradition in engineering has been the inclusion of tolerances in specifications – a range of values around a point that is acceptable. Commonly, the engineer may assume that if he or she stays within the upper bound of the tolerance range, then the implications in terms of performance will be acceptable. Unfortunately, tolerances have a way of "stacking up" and although all subsystem designs may be within tolerance, the total system may fail. For example, a space suit may have sets of different-sized modules that accommodate a range of expected crew member segment sizes, but because of human body size and shape complexity, including imperfect correlation of segment sizes and changes due to microgravity, the performance of an EVA activity may be compromised.

An alternative to traditional "tolerances" is the use of loss functions. This involves the identification of a target value that will be ideal and not interact adversely with collaborating subsystems and a nonlinear function that "penalizes" deviations. As the system design develops, these penalties are amalgamated and a total system score is calculated (Peacock, 2001). The decision process for system acceptance is based on a policy statement regarding the total system score and identification of those subsystem deviations where the greatest impact may be made regarding process performance. For example, a vehicle interior may specify loss functions for headroom, shoulder room, knee room, and eye height as well as many other parameters (Roe, 1993). In the final assessment of the perception of interior spaciousness or performance in a standard entry – Egress Test – the design compromise will optimize the amalgamation of these multiple loss functions.

Formal testing of the relationship between the individual (or sets) of system adjectives (independent variables) and process performance outcomes or adverbs (dependent variables) is the very basis of human factors engineering and its regression or analysis of variance tools. A shortcoming of this reductionist approach is that experimental management of many interacting and concomitant variables is often prohibitive, because of system complexity. An interesting alternative approach is described by the paradoxical statement that "if a nonconforming system passes a [process] performance test, then the system can be considered to conform". Or to use a familiar truism: "the proof of the pudding is in the eating".

HUMAN FACTORS IN DESIGN

This paradox envelops the relationship between human factors engineering and their designer and user customers. When human factors enter the design process late with usability tests of the total operational process, it is often too late or too costly to rectify fundamental system design shortcomings. For example, ergonomics intervention in automobile manufacturing may influence workplace design, tool selection, and task content but cannot change the main design problem of inaccessibility of a particular component. The same is true of the maintenance of space hardware; if a suit is to be maintained on a remote planetary surface, there will be very different challenges from those encountered in a well-equipped workshop on earth. Conversely, when human factors is involved in the life cycle requirements planning early in the design process, it is more likely that a comprehensive set of performance requirements will lead to a corresponding set of system design loss functions and the sequence of evaluations as the design matures.

System design specifications can be verified and process performance requirements can be validated in an appropriate context. These important design evaluation processes are effective only if reliable testing processes accompany requirements and specifications. A generic phase of the design process can be described by analogy with the familiar educational process. The first component is the articulation of performance requirements – will the existing students have obtained knowledge that fits them for their next course or phase of their careers? The proof of the pudding is in the starting salaries of graduates or better still, the final examination should include an evaluation (validation) of performance in analogous situations. Curriculum or course design specifications flow from the outcome requirements. If the outcome requires problem-solving capability, then the course curriculum should specify practice in problem-solving. Verification of the curriculum, like verification of system design specifications, should be straightforward if the specifications have been articulated clearly and reliable tests have been planned and implemented. All too often classes are designed based on historical specifications rather than customer requirements. The limitation of this analogy is that the educational and design processes are extremely complex and involve many subsystems, including teachers (engineers), classroom facilities (design facilities), students (internal customers), and employers (external customers). However, the discipline of process performance requirements (verbs and adverbs) first followed by system design specifications (nouns and adjectives) will assure a more satisfactory outcome.

The root of the design and education challenge lies in human variability and adaptability. Students may succeed despite their professors; vehicle customers may tolerate poor quality if the styling is exceptional; and astronauts may succeed in their tasks despite design shortcomings. Conversely, unprepared students may fail despite good professors and facilities; poor drivers may fail despite well-engineered systems; and astronauts (or their support entourage) may fail if their training, experience, or readiness to perform are insufficient to meet novel or emergency situations. Examples of the former performance successes, despite subsystem failures, are found in the Apollo 13 and Skylab solar array incidents. Evidence of the latter failures was observed in the Progress collision and the Soyuz/Salyut tragedy (Casey, 1993).

Design for human variability may be addressed in several ways. The obvious way is to reduce the [human] variability by meticulous attention to "humanware design" – selection, training, assignment, and performance monitoring. Historically, NASA has had great success in this respect although performance monitoring has always been a bone of contention among crewmembers who are reluctant to publicize their shortcomings. An analogous process in professional sports does not suffer from this shortage of evidence. The sports pages are full of the most detailed performance statistics of these highly talented, selected, trained, and paid athletes. At the other end of the design spectrum, consumer product design, including automobiles and their usage contexts, must accommodate a wide variety of minimally talented, marginally selected, inadequately trained, and rarely monitored users. Only catastrophic failures are documented and the usually forgiving context allows recovery from gross human error and minimal monitoring of inappropriate behaviors.

Design for highly talented human operators is easier and more forgiving than for the broad population of consumers. But this can lead to complacency and overreliance on the human operator to accommodate for design shortcomings. It should be noted that even the best operators suffer from human fallibilities, such as inattention, fatigue, overload, and debilitation (Hancock and Desmond, 2001). Picture a good driver finding his way through a strange city, in the fog, on icy roads, to an important meeting deadline. Translate this into an astronaut, debilitated by a long interplanetary journey (or EVA), wearing a cumbersome suit, finding his or her way to a safe haven with limited consumables. The focus of system design must acknowledge expected use and foreseeable misuse. An automobile must be designed to protect an inebriated driver in the event of a high-speed collision. Space hardware must be validated in similarly challenging contexts.

Automobile design has a considerable advantage of an enormous amount of data. Space exploration is relatively data poor. Consequently, space system design must take advantage of contemporary modeling, simulation, and analog facilities. These facilities, to be predictive of human performance in space, must address human shortcomings as well as their successes. It is one thing to winter over in the Antarctic and suffers from frostbite, or run out of air in NEEMO and has your buddy lend you his spare regulator. It is altogether different with a minimally redundant crew on their way to a distant planet when the doctor gets a toothache, a solar flare erupts, or a piece of software misbehaves. The Advanced Integration Matrix (AIM) program aims to answer the challenges of expected use and foreseeable misuse with a comprehensive suite of digital and analog simulations and an extensive repertoire of what if questions, with particular reference to the many sources of human performance variability.

SIMPLE RULES FOR PROCESS AND SYSTEM DESIGN

- Differentiate between process requirements and system specifications.
- Develop tests for specification verification and requirements validation.
- Develop a comprehensive picture of system design interactions and process performance outcomes.
- Develop digital simulations of mission process performance and carry out sensitivity analyses of hardware, software, humanware, and organization-ware subsystem design ranges.
- Use contemporary tools such as Failure Mode and Effects Analysis, Fault Tree Analysis, Human Reliability Analysis, Quality Function Deployment, and Discrete Event Simulation to evaluate expected use and foreseeable misuse.
- Comprehend human variability on all dimensions, including physical, cognitive, and psycho-social.
- Design in redundancy and forgiveness – make space travel as safe as driving to work.

1. **Industrial engineering, ergonomics, and production lines**
    a. During the first part of the last century, Henry Ford, Frederick Taylor, and Lillian Gilbreth were among the leaders of the process of scientific management. Their focus was the measurement, simplification, and standardization of manual work, particularly in product assembly. They demonstrated unequivocally that these processes can greatly improve product quality and productivity. Toward the end of that century, the Toyota Production System continued this prescriptive approach to job design but added the principles of continuous improvement and partici-patory quality teams. The production line can now be found in most industries around the world such as automotive, electronics, textiles, consumer products, call centers, and food processing. Unfortunately,

these tremendous gains in productivity and product quality come with a human cost and some of this cost is due to the misapplication or narrow application of physical ergonomics by a focus on the twin issues of force and posture/movement and the removal of "nonvalue-added work". These ergonomics applications certainly contributed to continuous improvement in productivity and product quality through greater work intensity and repetition. But these gains were offset by increases of work-related musculoskeletal disorders and many less tangible cognitive and social detriments, such as vigilance decrement and boredom. These issues may appear less important where labor is readily available, cheap, mobile, and dispensable but ironically they become more important as the employee pool becomes more stable.

b.   There are fortunately a number of variations on the theme of the production line. The first engineering approach is through mechanization and robotics for repetitive and forceful work, although these processes often require human operators to complete the task cycle through such activities as materials input, operation initiation, and parts removal. The second administrative approach is through job structuring, rotation, and enlargement, perhaps involving a sequence of work cells within the production line. These human-centered approaches can create a more knowledgeable, flexible, and stable workforce, especially where the job assignments include such tasks as inspection, materials, maintenance, and a share in the supervisory tasks of training, monitoring, health and safety, etc.

c.   A human-centered or socio-technical systems approach to the design of production lines can maintain these gains in productivity and product quality while providing a safe, healthy, and satisfying employment. One key component of this participatory approach is a reversal of Taylor's scientific management view that the worker "should do and not think". There is no doubt that line workers develop considerable task knowledge and that this knowledge can be harnessed for the benefit of all stakeholders. However, where this participative approach is mismanaged, sub-optimization and conflict replace optimization and cooperation.

d.   These concepts of human-centered manufacturing will be supported by case studies in automobile assembly, textiles manufacturing, bookbinding, and medical claims processing.

2. **Ergonomics research and ergonomics design are two sides of the same coin**

a.   This chapter discusses three ergonomics research and three ergonomics design examples to demonstrate the similarities between the two processes. The ergonomics research topics all focus on running and relate to gait biomechanics, the physiology of fatigue and aging, and the effects of terrain on performance. The design topics all focus on transportation – the design of mass transit railways, cars for the elderly driver, and aviation displays.

b.   Research and design can both be described by a control model involving inputs, outputs, feedback and feed forward, and adaptation and learning. In research, the hypothesis is an anticipation or feed-forward of the relationship among selected technological, environmental or operational variables, and a human response. Design involves the control of the technological (or operational) context to withstand uncertainties in the operational or environmental contexts. Reliable designs maintain their performance levels over the life cycle of the product, given expected and predictable contexts. Resilient designs can withstand unpredicted and sometimes extreme contexts, often by "failing safe".

c.   A major challenge in research is the validity of the investigation design and the results. In the gait biomechanics research, the investigation addressed the hell and mid-foot strike variations. However, as these factors are affected by stride length and running speed as well as individual variation and shoe design, the conclusions may not be widely generalizable. Indeed the major challenges of most human factors investigations are sample selection and size.

d.   Similar challenges of validity are found in design. There are the intended users of a product and sometimes but not always predictable misusers. Also there are the intended conditions of use and possibly more challenging unexpected conditions. For example, a family car may not be able to protect the inexperienced, inebriated, or incompetent driver in heavy traffic. Similarly, the family car may not perform well in off-road conditions or even on a road with ice or flooding.

e.   Research into the physiology of fatigue and aging usually relies on selected or pseudo-random samples representing the ranges of interest. The analysis of differences relies on means and variances. However, these are simply reflections of the samples and not necessarily the age variable of interest – it is possible to select a sample of 60-year-old who will run faster than a sample of 20-year-old. One way out of this sampling dilemma is to use age-based records to model the age effect. Such analyses show very clearly the age effect without being contaminated by such things as aptitude, training, or illness. Similar approaches may be used with cognitive variables.

f.   The design of mass transit systems presents both physical and operational challenges for passengers. The physical design of features such as seats, grab rails, and exits may be entirely suitable in uncrowded conditions but in rush hour and emergency conditions, the abilities and behaviors of a few passengers can disrupt the intended objectives of stability and mobility.

g.   Human fatigue research shows deterioration in physiological and cognitive functions as a function of time and activity. The Weibull distribution is used to model fatigue and wear out of both mechanical systems and human systems. The fatigue curve in long-distance running is very similar to that of a car, albeit with different time scales. Both curves can be changed by maintenance or training but only where the demands are limited. Resilient designs rely on operational factors, such as alternating periods of activity and rest/maintenance.

3. **Simple methods for system design, product evaluation, and accident analysis**

   a. An unfortunate trend in ergonomics teaching and sometimes ergonomics practice is the use of complex analytical methods to address either simple or complex problems. The practice of ergonomics does not always have the luxury of time – either in accident investigation (forensics) or product design. It is rarely possible to have the luxury of sophisticated laboratory equipment, a large representative cohort of subjects, and a few months to conduct an experiment, sufficient to obtain a 100% reliable answer. However, ergonomics practitioners must not run the risk of simply offering their own biases.

   b. This dilemma may be addressed through the use of some simple analysis and modeling tools that are sufficiently accurate and quick and easy to use. Many of these tools have roots in industrial engineering applications in manufacturing and service operations. The contemporary processes of Six Sigma and Lean Analysis have also incorporated some of these tools into their practice.

   c. The first challenge of design or accident analysis is to describe the full context. All too often designers and investigators converge too rapidly on a blinkered viewpoint. Both designers and investigators often converge on "human error" and do not explore sufficiently the causes or the missed prevention or mitigation opportunities. The investigator of a car accident may simply point to "the driver not paying due care and attention" to the road ahead rather than address the cognitive capture associated with an over the complex instrument panel. Similarly, the investigator of a building construction accident may simply point the finger at a careless worker rather than investigate how thermal stress and fatigue caused a cognitive lapse in tying off his safety harness or why a secondary safety net was not incorporated in the system design.

   d. The first tool addresses the purpose of design – E3S3+R&R – Effectiveness, Efficiency, Ease of Use, Safety, Security, and Satisfaction, plus Reliability and Resilience. The 5Ws – What, Where, When, Who, and Why – and How provide a broad context. Next, the 5Ms come into play – Man (and Woman), Machine, Materials, Methods, and Measures. Another way of looking at breadth is through the 4Ts – Topic, Technology, Time, and Team or the 3Es – Event, Equipment, and Environment. The 5Is – Information, Interaction, Interference, Interdependence, and Independence address how all the elements of these analyses combine. The 4Cs – Causation, Communication, Calculation, and Control are all processes that address processes involved in the operation of a product or process. The 5Us are then applied – Utility, Use (Misuse), User (Misuser), Usage, Utilization, and Usability. These sets of questions provide breadth to analysis, evaluation, or investigation. The 5Whys provide depth – asking a branching series of why's will usually get to the root cause of a problem. The 5Ps address the life cycle of a product or process – Product Design, Process design, Production design, Procedure design, and People design. Finally, when the physical or informational context becomes too complex, the application of 5S can greatly simplify and streamline an investigation. All of these tools can be articulated as Concept Maps or Activity Cycle diagrams which differentiate between Entities (or Resources) and Activities (or Processes). These simple tools should be used to complement each other.

    e. This workshop will describe these tools with examples from a wide variety of products and processes. Class members will be provided with kits and worksheets to aid them in the use of these simple tools. The final part of the class will involve small team collaborations in the use of these tools for design, evaluation, and accident investigation.

4. **Ergonomics, safety, and resilience engineering**

    a. A simple, but sometimes naïve view of ergonomics is that it is "human-centered design". The problem with this definition is that most products or processes have many people who are stakeholders and these different stakeholders may have different requirements. Therefore, there will always be trade-offs in design.

    b. A somewhat trite but striking example is with the application of human factors to military equipment and operations design. Clearly, there are different "customers". Similarly, there may be conflicts between the efficiency and effectiveness of security systems. As screening at airports becomes more comprehensive, the long lines grow longer.

    c. Ergonomists have researched the speed – accuracy trade-off for many years. This trade-off can be easily demonstrated in both inspection and manipulation operations.

    d. Traditional vehicle manufacturing systems are built with very robust equipment geared to carrying out the same operations for many cycles and years. But when customer demands change, these robust systems are not agile; they cannot be quickly modified for changes in the product or operations. In the construction industry, the provision of effective safety systems may greatly compromise mobility, and thus tempt the user toward circumvention.

5. **The Laws and Rules of Ergonomics in Design (from HFES Journal, "Ergonomics in Design)**

    a. The practice of ergonomics involved human physiological, cognitive and social science, domain knowledge, and a portfolio of tools. The Laws and Rules of Ergonomics in Design is a practical textbook that describes some 26 different aspects of human factors practice with descriptions of some of the theories, tools, and applications.

        i. Tight Targets Take Time, Blind Ones Sometimes Take a Little Longer: Fitts' Law

        ii. The More I Practice, The Better I Get – De Jong's Law

        iii. My Arms Are Getting Tired – Rohmert's Law

        iv. Gas Happens – The Gas Laws

    v. Murphy's Law: If It Can Happen, It Will
    vi. Wrong Number: They Didn't Listen to Miller
    vii. JNDs, SDs, HSDs, and DNDs: The Weber-Fechner Law
    viii. Bias in Human Judgment: Is Your Halo Slipping? – The Halo Effect
    ix. What Kind of Shape Are You In? Anthropometry and Appearances
    x. Stand (Sit) Straight Up: The Functional Anatomy of Posture
    xi. Pay Attention
    xii. Newtonian Moments – Biomechanics
    xiii. Expectancy and Compatibility – Equipment Design
    xiv. Hotter Than Houston: Body Temperature Control
    xv. Rule-Based Ergonomics
    xvi. Remember Hawthorne – The Hawthorn Effect
    xvii. Unwanted Energy – Vibration
    xviii. Warning: Do Not Use While Sleeping – The Role of Facilitators
    xix. Who's Agenda Is It Anyway? – Robert's Rules of Order
    xx. Boomer, Sooner and Donders – Reaction Time
    xxi. You've Got to Attend to Everything – Mental Workload
    xxii. A Look Back: Ergo is More Difficult than Nomos – Recent History of Ergonomics
    xxiii. Just a Moment – Biomechanics
    xxiv. Ethics and Ergonomics: Customer Satisfaction
    xxv. Product and Process Evaluation Using the Six Us
    xxvi. Time for Bed: Shift work

  b. This interactive workshop will discuss a selection of these topics and evaluate the theories, tools, and applications that are used by practicing ergonomists.

6. **HIM and HER – Human information management and human error reduction**

  a. Accidents usually have a cognitive cause and a physical result. Therefore, the prevention of accidents requires attention to the failures of human information management and human error reduction. However, after the sequence of events leading up to an accident has initiated a system failure the reduction or mitigation of the physical effects usually requires physical intervention. For example, a driver may be driving too quickly at night (perhaps due to a memory or decision lapse) and fail to see another vehicle enter from a side road – a sensory, attentional, or perceptual failure. The outcome of these initial cognitive failures may require a significant application of skill at the last moment but where this fails, the driver is dependent on the defenses designed into the vehicle such as crush space, seat belts, air bags, and a friendly interior to reduce or mitigate the unwanted outcomes of the accident. It is hoped that the forgetful driver had also remembered to pay his insurance premium. He may also wish that he had remembered not to drink alcohol before driving or decided not to use a cell phone while driving.

  b. This whole sequence of events highlights many Human Information Management processes, such as sensing, attention, perception, short- and long-term memory, decision-making, and sensory-motor control as well as anticipating, planning, communicating, controlling, and so on. The reduction of the likelihood of this accident could depend on just one element of the Human Information Management System. Consequently, Human Error Reduction opportunities must be addressed throughout the Human Information Management process.

c.   The perpetrator of this accident may not be the only one to blame. Errors may have been made by the designer of the vehicle or road system. Traffic control devices and driver selection and training may also provide opportunities to improve system robustness.

d.   This workshop will explore the human capabilities and limitations throughout the Human Information Management sequences. Next Human Error Reduction strategies will be proposed to improve human performance by equipment and operations design. These HIM and HER analyses will be applied to a variety of contexts including transportation, industrial inspection, maintenance, and education.

REFERENCES

Beaty, D. (1969). *The Human Factor in Aircraft Accidents.* Tower Publications, New York, NY.

Casey, S. (1993). *Set Phasers on Stun: And Other True Tales of Design, Technology, and Human Error.* Aegean Pub Co, Santa Barbara, CA.

Evans, M., N. Hastings, and B. Peacock. (2000). *Statistical Distributions.* Wiley, New York, NY.

Hancock, P. A., and P. A. Desmond. (2001). *Stress, Workload and Fatigue.* Lawrence Erlbaum. Mahwah, NJ.

Haushalter, F. L. (1971). *Rules for Safe Driving and How Your Tires May Save Your Life.* Vantage Press, New York, NY.

Herzberg, F. (2008). *One More Time: How Do You Motivate Employees?* Harvard Business Review Press, Boston, MA.

McHugh, J. E. (1986). *Quality Function Deployment.* American Supplier Institute, Dearborn, MI.

Peacock, B. (1993). "Production with Protection," *Plenary Address at the National Association of Manufacturers Conference on Ergonomics,* Nashville, TN.

Peacock, B. (1995). "The Grammar of Design", Human Factors and Ergonomics Society Annual Meeting.

Peacock, B. (1998). EDAN – Ergonomics Decision Analysis, *Applied Ergonomics Conference.*

Peacock, B. (2001) "Measurement in Manufacturing Systems" in Charlton and?

Peacock, B. (2003). "Pay attention", Ergonomics in Design.

Peacock, B. (2019). *Human Systems Integration,* Self-published manuscript, Fernandina Beach, FL.

Peacock, B. (2020). *How Ergonomics Works,* Self-published manuscript, Fernandina Beach, FL.

Peacock, B. (2021). *Ergonomics Tools and Applications,* Self-published manuscript, Fernandina Beach, FL.

Petroski, H. (1992). *To Engineer Is Human.* Vintage Books, New York, NY.

Robinson, K. S. (1993). *Red Mars.* Bantam Books, New York, NY.

Roe, R. (1993). Occupant Packaging. In: Peacock, B and Karwowski, W. (Eds.), *Automotive Ergonomics.* Taylor and Francis, London.

5 Human Reliability in Engineering Ethics and Safety

INTRODUCTION

Inputs (systems) – must be designed (Peacock (2019, 2020, 2021).

Human inputs: affected by selection, training, assignment, abilities, limitations, motivation, attention, fatigue, etc.

Equipment and materials: can be designed, must be resilient with regard to users and context.

Context/environment: cannot be designed, must be addressed, e.g., by barriers or shields.

Regulations may be applied to Human, Technology, and Environmental subsystems: passed driving test, energy efficient car, and daytime driving.

Process

The interaction of two or more systems with a measurable outcome in terms of process performance and the change in state of one or more contributing systems.

Outcomes
Generally measured in terms of:

Effectiveness: quality – matching customer requirements.

Efficiency: optimal use of resources (money, time, materials, energy, people, etc.).

Ease of use: resilient to varied users, usage and contexts.

See 6Us handout.

Add Elegance (Aesthetic appeal).

Safety: systems resilient to catastrophic failure and wear.

Process/system failure mitigated to reduce severity of unwanted outcome.

Security: process/systems resilient to accidental or malicious interference by third parties.

Satisfaction – (all) human users (customers) should be satisfied by their experience with the process/systems.

There may be compromises.

Add Sustainability (reliability and resilience).

DOI: 10.1201/9781003328445-5

Feedback, adaptation, and learning control

Feedback – mechanism for communicating outcomes (error) to modulate inputs:
Flying in wind under Visual Flight Rules (VFR) conditions.
Catching a ball.
Balancing on one foot.
Heart rate (what about anticipatory heart rate increase?).
Adaptive – automatic adjustment to inputs based on pre-defined context:
Thermostat – heating/cooling changes to pre-selected conditions.
Jockeying in queue behavior.
Diabetes medication.
Learning – behavior modification and performance improvement with experience/practice:
Hitting a golf ball.
Driving.
Taking examinations.

Feedforward (anticipation)

Prediction of the effects of context on the process behavior and modulation of inputs (subsystem changes) accordingly.
Environmental, technology or regulatory context, etc.
Necessary for the design of resilient systems – systems that can withstand the effects of intended and unexpected context and time.
Market research.
Weather planning.

Feedforward information may be erroneous or at best probabilistic.

What will the other driver do?
Will it rain/snow/freeze today?
Will the technology subsystem (e.g., car) deteriorate over time or without maintenance?

Human beings usually make considerable use of "feed forward"/anticipation.

This activity often leads to timely actions that may be in error due to uncertainty in the anticipation process as in choosing a menu item based on a verbal description, preparing answers to questions at an interview, designing an advertisement aimed at a subset of customers, or selecting a technology for fuel-efficient cars.

Market research is a mechanism used to predict customer needs and wants in the future. However, as product (e.g., car) design takes a few years and operates in a competitive context, these customer requirements may be a "moving target".

Decisions regarding process inputs are usually made with reference to the cost of resources, such as money, time, people, equipment, fuel, and materials.

The Socio-Technical System Design philosophy will face decisions by managers who may be more focused on technology than the vagaries/requirements of multiple customers/stakeholders.

Decisions are usually the prerogative of management.

Decisions in the design process are usually made in the progress review meetings where the components, including HFE advice, are presented in the context of the big picture.

These decisions will be biased by the managers'/committee prejudices and the ability of the engineer to sell his or her point of view.

Effective communication is a learned skill – practiced in the classroom in preparation for the workplace.

Concept/idea of someone involved in a design or purchase process.

An economical family car (requirements)

SEMANTIC AND PHYSICAL ENCODING

> *Semantic encoding* – translating the idea into some known language, diagram, model, etc.
>> Using automotive jargon – "a mid-sized, base-level sedan" – relative statements/requirements.
> *Physical encoding* – Translating the conceptual model into an explicit physical form, such as drawing, writing, or speaking.
> Four seat, four door, sedan with cloth, automatic, Quad 4, entry-level IP.
> Note the use of jargon and abbreviations.

There are many variations/interpretations of this initial verbal set of high-level specifications.

Transmission, noise, added information:
> Transmitting the idea to the intended (unintended) audience.
> The transmission may involve a series of translations by individuals with different priorities.
> Note that the idea may not be clearly articulated (lost in translation – omissions, additions, changes).
> Note also that there may be external physical or informational "noise" during the transmission process.
> The full message may not reach the intended recipient for a variety of technical reasons.

Physical and semantic decoding by multiple customers with different priorities:
> Manufacturing, maintenance, sales, drivers, regulators, etc.
> Receiving and understanding the information.
> The receiver may not physically receive/sense the message.
> Note that understanding requires knowledge of the language and a reception framework – the translation may be biased by the receiver.

Consolidation, retention, forgetting, action

The receiver must consolidate (fit into his framework), remember (or forget), and translate the information into action.

The eventual set of high-level specifications could now be:

Five seat – mid-level vehicles usually have five seats, not four.

Sedan heard as "van" (physical decoding).

Automatic referred to gearbox but added windows, door locks, and seats – typical of less economical vehicles (semantic decoding/added information).

Four doors translated into two conventional front doors plus two rear sliding doors – common in vans.

Quad 4 engine, typical of small cars, was converted to V6 based on the common choice of engine for minivans.

Feedback, adaptation, learning, and iteration

The originator of the idea needs feedback to modify the idea/concept (see control model above).

The feedback cycle should reduce communication errors. However, if feedback is not available, the communication may lead to designs that don't satisfy the initial intent (requirements).

Design is vulnerable to communication failures and participant inconsistencies.

A problem similar to this actually occurred with an attempt to design a front-wheel-drive Camaro.

CASE STUDY – THE LIFE CYCLE OF A CAR

PRODUCT DESIGN

The mission/purpose will vary enormously depending on the functional requirements of the intended user/buyer/customer – a sedan, truck, sports car, luxury car.

The seven ages of "carman":

Teens – wheels (Civic), 20s – style (Camaro), 30s – function (minivan), 40s – prestige (Buick), 50s –lavish style (Corvette), 60s – comfort and safety (Cadillac), and 70s – wheels (Civic).

The mission/purpose should consider many other customers.

Manufacturers, maintainers, shareholders, and regulators.

Within each general lifecycle stage there will be subcategories, each aimed at emphasizing particular aspects of the vehicle.

These sub-requirements will be based on generally accepted customer standards.

The design process will include many iterative steps of "design – make – test – decide".

These cycles will be at both the component/subsystem level and at the system level.

Concept selection will be based on many "weighted" criteria.

Use, manufacturing, safety, maintenance, and disposal.

Concept selection is an imprecise process carried out around a conference room table.

"Votes equal opinion time salary" – seniority is equivalent to wisdom?

The front-wheel-drive Camaro:

Conflict between engineering and marketing and management.

Hard or soft seats in a Caprice.

An opportunity for a psychophysical investigation.

The ACCESS Car:

A marketing mistake?

An engineering opportunity.

A human factors driven process.

Intermediate shaft installation.

Transatlantic disagreement.

Battery location (engine compartment or trunk).

Engineering and manufacturing conflict.

The proliferation of warnings.

Conflict among human factors, marketing, and legal staff.

MANUFACTURING AND PRODUCTION DESIGN

DFM/DFA (Design for manufacturing and assembly):

Aimed at productivity, quality, and worker comfort and convenience.

Access, easy targets, force, posture, and fastener repetition reduction are the general aims.

See "Tight Targets Take Time" handout.

Cars with pressure for a low cowl height will create engine compartment packaging challenges, which in turn lead to accessibility problems in assembly and maintenance.

A decision to sequence the seat installation after the doors have been installed can lead to longer cycle times, mutilations during the seat transfer into the vehicle, and difficult access for installing the seatbelts and the seat secure bolts.

Major allocation of function decisions.

Mechanization and automation.

Articulating arms are very useful for heavy components/subassemblies.

Often found tied to a pillar for intermediate-weight components.

The job is possible without the arm.

The job may be faster without the arm.

The repeated load may give rise to injury.

Robotic undercoat and paint spraying is the norm. However, robots cannot easily access certain inside-facing areas such as the bottom of the doors which need to be painted by human operators who have to sustain awkward postures throughout the job cycle, giving rise to quality and injury problems.

Cleaning the paint booths of residue is a largely residual manual task. Pulling grates is a difficult and physically stressful task.

TOOL SELECTION

Threaded fasteners are usually torque controlled. Task completion often induces a stressful torque reaction, giving rise to injury. This is sometimes reduced by a torque bar but alternative technologies such as hydraulic/pneumatic/electric pulse tools can remove the torque reaction problem with no loss of quality (torque control).

Inline, pistol grip, or right angle tools can sometimes be used to allow more convenient arm postures, depending on the amount of torque and the location of the fastener. Inappropriate tool selection can cause discomfort and injury.

Tools may be supported by balancers but these may interfere with task access and so may be discarded by the operator.

Modular design for model differentiation/subassembly content:

Major trend to increase module content, thus reducing the final assembly operations:
  Steering columns include lighting, windshield, cruise control, HVAC, entertainment, and navigation functions.
  The module becomes heavy and awkward.
  The residual intermediate shaft (between the steering column and steering box) installation is a major source of difficulty for the operator.
Windshield wiper motor and brake booster install.
These two components are hard to reach in the center/bottom of the engine compartment, respectively.
Task can be made easier by product design (for the windshield wiper motor) bringing it out board and by assembling the brake booster module on a different station on the production line.
Spare tire in the bottom of the trunk is both difficult for assembly and difficult for the driver who needs to change a wheel, but a convenient place for packaging.
Product design solutions include the mini spare, which is lighter and may be packaged at the side of the trunk for easier access.
Question – Does the car owner need a spare tire or a cell phone to summon help?
Seats/seatbelts.
Difficult install postures – seatbelts can be designed to be integral with the seat, given appropriate structural modifications, which in turn leads to a much easier assembly task.
Layering and Fastener orientation for operator access.
This is a packaging and component design problem. Product design engineers should spend time on the line installing their own components to appreciate the line operator's difficulties.
Vehicle carrier systems.

Many opportunities in manufacturing design to improve operator posture:

Overhead rail – bring low- and under-body work to an accessible height more convenient than working in pits.

Tilting – rotate the vehicle 45, 60, or 90 degrees to improve visual and hand tool access.

Skillets – operator adjustable vertical height.

PRODUCTION OPERATIONS

Production targets affect staffing levels and choice of shift system (1, 2, or 3 shifts).

Shift work should be based on operational, human, and technological system needs:

Production targets
Physiological and social requirements
Access to equipment for maintenance
Task design
Will vary according to line speed and work area footprint
Learning curves for job content and "experienced worker standard" assignment
Balance of nonvalue-added work – carrying, walk back, etc.
Stock/components/fastener/hand tool presentation
Aimed at reducing error, nonvalue-added time, and improving comfort and convenience
Work team design/task allocation
Job enlargement/rotation and team assignment philosophy
Rotation and enlargement strategies
Quality, productivity, and safety monitoring
Methods engineering

DISTRIBUTION AND SALES

Class discussion/exercise of customer requirements and design includes the following:

Substantial human contribution
Order management
Transport
Protection
Long-distance driving/railways/container ships
Brochures, warranties, financing, insurance, taxes, and incentives
Salesperson employment strategy
Incentives and salary

PRACTICE ACTIVITY

Discuss the human factors issues related to the Use or Operations phase of an automobile; make use of one or more of the models described earlier in this class. Address the following issues:

Buyer/driver/passenger
Road ways
Traffic
Taxes
Garaging/parking
Adverse environmental conditions
Night and day, fog
Snow and ice
Heat and cold
Traffic noise
Vibration
Road condition
Service and maintenance – class discussion/exercise
Context of maintenance
Tools
Training of maintainers
Support manuals
Spare parts
Distribution strategies
Disposal
Green car
Design/materials/manufacturing cost constraints
Used car market
Warranties
Spares availability
Concurrent engineering – a delivery opportunity for Socio-Technical System
    Design
All life cycle stages, customers, and stakeholders need to be accommodated
Multiple overlapping steps
Feedback and iteration
Technical memory
Evaluation

SYSTEMS SAFETY TOOLS

System Safety may be addressed by a variety of qualitative or descriptive techniques that may be applied to accident investigation or failure prediction.

Manufacturing industry adopted these techniques, largely from the Japanese automobile industry (The Toyota Production System) to investigate and design systems, products, environments, tasks, and processes. Over the past decade, many of these methods have been incorporated into the practice of SIX SIGMA.

5Ws

Perhaps the simplest of all tools is to ask Who, What, Where, When, How, and Why whenever an accident occurs. This approach will help the investigator to avoid focusing on the obvious "What" by asking more questions about the background and conditions surrounding and preceding the accident. At the design level, a similar simple process will help to broaden the scope of analysis of the use of a product or process.

5S

The 5S process is another method adopted from the Japanese but in reality, it has been used for generations by mothers telling their children "there is a place for everything and everything should be in its place" in reference to tidying up their room. The purpose of 5S is to eliminate unused equipment and materials from the workplace and to create an orderly safe place to work effectively and efficiently.

The method of application of 5S is to have at least two people involved, one is the owner of the room, workplace, or factory and the other is there to question the honesty of the owner when asked "when did you last use that —?" You may not like being asked these questions but it really works and saves the industry millions of dollars by reducing "just in case" inventory.

5 WHYS

The 5 Whys' process can also be used in accident investigation and in design. It is often used in conjunction with other processes, such as the 5Ws, 5S, 5Ms, and 6Us. It is a very simple way of getting to the root cause of an accident. It is possible that inquiry pathways may branch and link up. Concept mapping is a useful adjunct, particularly because of its utility in adding links containing detail to background material.

6Ms

The 6Ms method is quite like Ishikawa (Fishbone) diagrams and Cause and Effect diagrams.

The method of application is to:

Agree on what actually is the problem
Systematically list the different causes of a problem in each branch
List all feasible causes, however, remote
Analyze in detail (5 Whys') all possible causes and sequences for each node
Discuss causal pathways
Simplify the diagram by amalgamating less populated branches
Split up overpopulated branches
Add likelihoods and risks

THE 6Us AND 2Ms

The 6Us is a process for analyzing qualitatively the many factors that affect the success and safety of a product or process. See Table 5.1.

TABLE 5.1
Product or Process Evaluation Sheet

Product Attribute	Product or Process Description	Why? Why? Why?	Product or Process Analysis	Why? Why? Why?
Utility	Why is the product or process useful?		Why should the product be improved?	
Usage	In what way will the product or process be used		In what context will the product be used?	
Utilization	How often and by how many people will the product or process be used?		What are the probability and frequency of failure?	
User	Who is the intended user?		How can the intended user be selected or trained to use the product or process?	
Misuser	Who is the expected misuser?		What kinds of users will be associated with these failure modes?	
Usability	How easy is it to use the product or process?		How could the product or process be changed to make it easier to use?	
Misuse	How easy is it to misuse the product or process?		How could the product or process be changed to prevent and mitigate the effects of misuse?	
User Error	What kinds of failure modes can be predicted?		What are the consequences of failure?	

FAILURE MODES AND EFFECTS ANALYSIS (FMEA)

Sometimes extended to Failure Modes Effects and Criticality Analysis (FMECA).

RISK ASSESSMENT CODES

The second step in FMEA is to develop Risk Assessment Codes that relate the probability of system failure to the severity of the consequences of the failure. A Risk Assessment Code of 1 is clearly a cause for concern

THE SHEL MODEL

Developed by Edwards to indicate the factors associated with system operation and failures. The model emphasizes the Interactions, Interdependencies, Interruptions, Interfaces, and Integration of these factors:

Liveware – the people
Software – the organization
Hardware – now should include software
Environment – physical social and temporal context

SWISS CHEESE MODEL

The Swiss Cheese model was developed by Reason. The logic is that for every unsafe act, there are predisposing causes, such as the equipment, context, and operator training. Subsequently in most work operations, it is the responsibility of the supervisor to implement a safe working environment, including training and monitoring of operators. Finally, the behavior of the organization's management regarding safety cascades down through procedures and communications as well as up through active monitoring. Ultimately, although the clear cause of an accident or system failure may be because an operator made an error, it is usually possible to trace a causal pathway through the various layers of context and support structures.

HUMAN FACTORS ANALYSIS AND CLASSIFICATION SYSTEM

The Human Factors Analysis and Classification System (HFACS) was developed by Shappell and Weigmann (2000), mainly for use in aviation accident investigation, although it is applicable in many other complex system safety situations, including design. HFACS was developed from Reason's Swiss Cheese model of accident causation.

HFACS addresses and dissects Unsafe Acts, Predisposing causes, Unsafe Supervision, and Organizational Safety Climate.

UNSAFE ACTS

Unsafe acts are divided into errors – usually due to a failure of the operator information processing and action execution performance – and violations, which usually may be a one-off shortcut in a procedure or a habitual disregard of proper procedure.

PRECONDITIONS

The preconditions for unsafe acts address most of the factors associated with ergonomics in design. They include equipment and environmental contexts; it is also possible to include social, operational, and temporal contexts. They also include the condition of the operator and operations team in terms of "readiness to work".

UNSAFE SUPERVISION

In most operational contexts, individuals have someone who selects, trains, and assigns them to a task and supervise their attitude, behavior, and performance. In large organizations, these responsibilities of the supervisor are usually explicit. However, in some organizations, considerable autonomy may be ascribed to the individual operator with supervisor involvement only following an incident or accident.

The failure of supervision may include inappropriate training or assignment, including planned deviations from safe procedures. Also, the supervisor may be delinquent in monitoring or failing to correct known problems. Often supervisory violations are due to a conflict between productivity and safety interests.

ORGANIZATIONAL INFLUENCES

It is well recognized that the priorities of most organizations are effectiveness (product or process quality) and efficiency (optimal use of resources, such as people, money, and time materials). Even so many organizations state in their mission statement that "Safety is our first priority". Or "Our employees are our biggest assets".

Despite these priority issues, it is clear that some organizations are better at instilling a positive safety climate than others through strategies such as formal responsibilities and regular communications (meetings, statistics, etc.) regarding safety issues. Such organizations may be particularly hard on middle management/first-line supervision when system safety failures occur.

The complexity of accidents and design for safety is such that it does not lend itself to deterministic, quantitative analysis tools. Statistical tools may be used for collecting, analyzing, and modeling safety data, but the analysis of system failure and design for safety usually relies on ad hoc, domain expert-centered discussion, which can be guided by the use of a set of qualitative analysis tools. These tools include the 5Ws, 6M, 5s, 5 Whys, 6Us, FMEA, SHEL, Swiss Cheese Model, and HFACS.

HUMAN RELIABILITY

Human Reliability is not an exact science. Rather it is an attempt to apply techniques of systems analysis, mathematical logic, probability, human factors, and simulation to the prediction of system (process) performance as affected by human contributions.

The basic unit of human performance is the TRANSACTION – a (similar) activity that is definable and repeatable – turning on a switch, driving to work, and taking a test. A transaction has an intended outcome (success) and possible failures. The expected success rate of a transaction is its RELIABILITY. This can be estimated by repeating the transaction many times.

There may be variability associated with the transaction – the *information* content, the *time* taken, the *conditions*, the *frequency* of repetition, and the *person* carrying out the transaction. All of these "Performance Shaping Factors" affect the likelihood of success or failure.

The failure of a transaction may be because the conditions or timing are beyond the capabilities of the operator. The Three Mile Island nuclear power plant accident was exacerbated by the fact that seventy warnings went off at once – and the operator had too much information to handle in the time available. A bad landing in an airplane may be because of wind shear or a sudden gust of crosswind.

A RESILIENT system is one that succeeds despite the conditions.

The time of occurrence and the conditions (performance-shaping factors) of a transaction may be predictable. The red light follows green, the command "GO" follows the warnings "ready – set", it will be dark on the roads at night, and a pilot will hear a weather forecast if he tunes in to CTAF. These predictable conditions provide the operator with the "mental set" or preparedness to participate in the transaction.

Human error often occurs when the conditions and timing of the transaction are unpredicted or unpredictable.

The success of transactions by human operators is greatly affected by **practice**.

The occurrence of a transaction may be due to the design of the system – such as a conveyor belt or the behaviors of other, external activities, such as other traffic. The operator may also initiate a transaction at a time of his/her own choosing; sometimes the transaction may be initiated when the conditions are not appropriate – we may start a drive to work when the roads are icy. Often the transaction may take place in parallel with other transactions – driving and using a cell phone.

The information needed to carry out the transaction may be limited. The lighting may be poor or it may be noisy; we may not pay attention to or understand the information or the needed information may not be available. All of these informational issues will give rise to risky decision-making and increase the likelihood of transaction failure.

The OUTCOME may affect the individual, the equipment, a third party, or the environment. Other unwanted effects of failed transactions may include loss of business. The costs of a failed transaction may include equipment replacement, operator or third-party medical costs, environmental cleanup, lost production time, and legal costs.

Transaction (a VERB) measurement (an ADVERB) is usually in terms of time and accuracy (or error). There may also be QUALITATIVE outcomes. The consequences of a transaction will be the change in state (ADJECTIVE) of a resource (NOUN – person, machine, time, environment, money, etc.).

Human error probability may be expressed as a decimal (0.01), a fraction (1/100), or an exponent 10^{-2}. The subjective estimate will be based on the observer's observations and biased by his experience or motives. Objective estimates are based on the observation of many transactions. The reliability of the estimate will be affected by the number of observations and the conditions under which the transaction takes place.

PERFORMANCE-SHAPING FACTORS

Performance-shaping factors include human, equipment, environmental operational, and temporal factors. Performance-shaping factors are Adjectives that describe the nature/state of a subsystem.

Human Information processing is the main cause of human failure due to the many complex stages in the process. For example, a simple "failure" such as a slip or trip may have underpinnings in sensing:

you may not see the hazard, attending
you may be looking elsewhere, perceiving
you may see the hazard but not recognize it as such, calculating
you may misjudge the height of the hazard, and so on

Forgetting is a very common cause of failure, particularly in high cognitive workload conditions. Why would anyone forget to fasten their seat belt or switch off the lights before leaving? At the task implementation end of the process, it is common to miscommunicate items such as codes or input an inappropriate control response to a task, such as steering or braking.

It should also be noted that learning is a major contributor to human information processing activity. Comparison of a novice musician with a skilled one shows enormous differences. The classification of human performance into knowledge, rule, and skill-based levels provides a clear demonstration of the variability of human variability in information processing. Highly skilled operators are very flexible in their response to unusual conditions – just picture the many ways a soccer player can bring the ball under control or a pilot can deal with difficult wind conditions or equipment malfunctions.

Human operators may perform reliably under normal conditions but fail because they cannot adapt to changing or extreme conditions. For example, in dim light it is difficult if not impossible to read; similarly, the visual system becomes dysfunctional in the presence of glare. High ambient noise makes the accurate perception of audible commands impossible. It is difficult to use a tool accurately if it is vibrating. Cold heat and humidity have a considerable influence on sustained motor performance. All of these factors will turn an otherwise accurate action into a failure.

It is very common for people to carry out activities in collaboration with others – either directly or indirectly and the relative status, knowledge, or attitude of the other person or persons will have considerable bearing on performance. For example, customer service agents have to deal with irate and uninformed customers, team members must play specific roles in games such as basketball, and one always has a "boss" to please who is hard to please and always in a hurry. Consequently, the social context will greatly influence human performance and may precipitate categorical failure. Consider, e.g., a bus driver who is being berated by an irate passenger or challenged by an aggressive maneuver of another driver.

Mental workload may be described as information divided by time. However, because of the human ability to learn, people need less time to process the same amount of information when they are experienced in the task at hand. Also, workload may be affected by other performance-shaping factors as described elsewhere in these notes. Thus, a more appropriate equation is as follows:

$$\text{Mental Workload} = (\text{Information} \times \text{Context})/(\text{Time} \times \text{Experience})$$

There are many other temporal factors that affect performance. For example, sleep loss, fatigue, and vigilance decrement will reduce human performance capability. Why do medical interns work 24-hour shifts? The recent highly publicized oil spill around the Great Barrier Reef was precipitated by a tired sailor.

Equipment design factors can have a considerable effect on human performance, including both speed and accuracy of transactions. This is the most common focus of Ergonomics.

Workplace factors may also affect performance by design features, such as equipment location, work surface height and orientation, seating, access, activity sequences, reach, fit, tool, and materials storage.

For manipulative tasks, features such as target size will greatly affect performance.

Vision is greatly affected by font/object size and contrast. Communication is affected by sound level and ambient noise. Sensory performance in general is affected by contrast and change.

Perception and understanding are affected by display and control design and panel/screen layout.

Displays may be analog, digital, representational, or pseudo-realistic; they may vary in their physical characteristics, such as size, font, contrast, and color coding.

Controls may be discrete or analog and vary in their type (knobs, levers, buttons, etc.), range of motion, gain, linearity, layout, etc.

Panels may contain controls and displays with functional arrangement and grouping according to importance, sequence and frequency of use, etc. They may also contain labels, and with computer displays readily available explanatory information.

Human performance is also affected by the cognitive (mental model of the system) and emotional interfaces.

A DISCOUNTING MODEL

It is clear that human performance is a very variable concept. Consequently, it is very difficult to accurately describe human reliability. However, there have been many attempts (Swain) to create tables that can be used with similar information regarding mechanical and electrical components. An alternative approach using a discounting model (similar to the NIOSH Lift Equation) applies a variety of performance-shaping factors to human physical capability. A similar approach can be developed for cognitive ability.

RELIABILITY BLOCK DIAGRAMS

Develop a logical description of the sequence of activities/transactions in the process

Draw reliability block diagram for INDEPENDENT systems

The system RELIABILITY – probability of SUCCESS – of a system with two subsystems in SERIES is:

$A \rightarrow B$
$.9 \rightarrow .8$

$R_S = R_A * R_B$, where R_A and R_B are the reliabilities of sub-systems A and B, respectively – Both A and B succeed

$R_S = .9 * .8 = .72$

The unreliability – the probability of system FAILURE is

$F_S = F_A + F_B - F_A * F_B = (1 - R_A) + (1 - R_B) - (1 - R_A)(1 - R_B)$
$F_S = .1 + .2 - (.1 * .2) = .3 - .02 = .28$
$F_S = 1 - R_S = 1 - R_A * R_B$
$F_S = 1 - .72 = .28$
Or $F_S = 1 - .9 * .8 = .28$

For two subsystems in PARALLEL

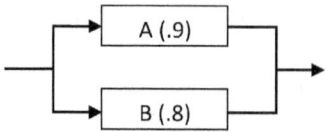

The probability of PARALLEL system FAILURE is

$F_S = F_A * F_B$ – Both A and B Fail
$F_S = .1 * .2 = .02$
$F_S = (1 - R_A) * (1 - R_B)$
$F_S = (1 - .9) * (1 - .8)$
$F_S = .1 * .2 = .02$

The RELIABILITY of a redundant/parallel system (Independent subsystems in parallel)

$R_S = 1 - F_S$ – Either A or B succeeds
$R_S = 1 - .02 = .98$
$R_S = (1 - F_A) + (1 - F_B) - (1 - F_A) * (1 - F_B)$
$R_S = .9 + .8 - .9 * .8$
$R_S = 1.7 - .72 = .98$

HUMAN RELIABILITY MODELING PROCESS

It must be emphasized again that many reliable people believe that, because of the subjectivity and unreliability of elemental human reliability estimates, the attempt to link human reliability to engineering reliability is likely to be unreliable. (Think this one through!)

Here is a process for Human Reliability Modeling
Assign ordinal performance-shaping factors
Assign error probabilities (HEPs) to each basic transaction
Use tables
Published statistics
Expert opinion
Epidemiological data
Empirical data
Best guess (median) with optimistic and pessimistic range values
Weight these probabilities with performance-shaping factors
Use a discounting model
Construct Reliability Block Diagrams
Estimate system reliability
Assign cost/utility functions to system/process failures
See the material on Quantum Risk Analysis (Study Unit 4)

Test the sensitivity of the analysis to changes in HEPs as affected by system
    redesign alternatives or changes in performance-shaping factors
In Quantum Risk Analysis terms use ±1 quantum (exponent) as the confidence
    interval
Evaluate expected, best, and worst case
Make recommendations for system redesign

CASE STUDY – OIL SPILLS

The EXXON Valdes created an oil spill in Alaska in 1989 that cost billions of
dollars and is still an active issue. In 2010, the Shen Nang 1 ran aground on the
Great Barrier Reef. In 1985, an investigation into the human factors issues of LNG
transportation resulted in the following (much abbreviated) analysis. Fault Tree
Analysis is a natural extension of reliability block diagrams and will be dealt with
in a later class.

ROLE OF PROFESSIONAL ASSOCIATIONS

The following segment is adapted from Mark Crawford (2022a). Interdisciplinary
collaboration and coordination provide a good alignment of the views of industrial
engineering and the guidelines of mechanical engineering. This chapter is based
on an article by Mark Crawford on the seven concerns of the American Society
of Mechanical Engineering regarding ethics and safety in engineering. Mechanical
engineers, like engineers in other professional engineering associations, often deal
with short production timelines, impacting safety and forcing them to make ques-
tionable ethical decisions.

Mechanical engineers often design innovative consumer products in fast-paced,
regulated industries. Speed to market is a constant pressure, putting MEs in a difficult position. Often, MEs need to insist on taking the time needed to be certain a
product is safe and well-designed, despite heavy pressure from management or the
customer to shorten the process to meet production or delivery timelines.

Many safety-critical decisions arise during the normal course of engineering
design that must be answered through the application of ethics. To determine the top
ethics/safety concerns that MEs face today, we consulted an expert in the field, P.E.
Kenneth L. d'Entremont. He is an associate professor and lecturer in the Department
of Mechanical Engineering at the University of Utah-Salt Lake City and author
of *Engineering Ethics and Design for Product Safety* (McGraw Hill 2021). Below are
his top seven ethics/product safety issues MEs may face in their engineering projects.

A LACK OF ETHICS EDUCATION

The curriculum of a sound mechanical engineering undergraduate program is packed
with required topics arising from new developments and technologies. These include
thermodynamics, fluid mechanics, solid mechanics, FEA (finite element analysis),
CFD (computational fluid mechanics), and solid modeling. This heavy emphasis on
technology developments can push ethics into the background.

However, engineering societies have long included the need to maintain the public's safety, health, and welfare in their codes of ethics.

"While many might hold out engineering as a profession, many practicing engineers consider it a career and have few, if any, interactions with these engineering societies", d'Entremont said. "There is little incentive for mechanical engineers to become professional engineers. For these people, engineering is a way to make a living, support a family, and obtain modest promotion over time within a business organization".

Few engineers are reminded of their duties as engineers to protect the public. Fortunately, most engineers do this on their own, "but engineers should also be explicitly reminded of this and given opportunities, such as classroom case studies, to sharpen their ethical thinking skills", d'Entremont said. "They may not have this ethical-reflection opportunity after they take on real-world jobs with tight schedules on real products that will affect the public".

A LACK OF SAFETY-ORIENTATED COURSES

There is little actionable mention of safety within the engineering curriculum as with ethics. Engineers are instructed not to hurt people and only make safe products; engineering students generally have no problem with these directives but are given no guidance on achieving this goal.

If any instruction is provided, "it is usually the pablum of simply following codes and standards", d'Entremont said. "Many of us who have developed standards understand their limitations and minimal requirements. Such standards often lack critical information for design engineers making crucial design decisions affecting overall product safety".

Furthermore, the idea of safety is often mixed with the concepts of reliability, compliance, quality, and availability. They are all related but have distinct properties. A reliable product may perform perfectly as designed but not as intended. A low-quality product may never operate due to failure but then be perfectly safe because of it.

UNFORGIVING DELAYS AND SET-BACKS

The current hyper competitive business climate rewards innovative products delivered by record-short schedules. Any necessary redesigns for any reason, including improving product safety, can be met with negativity from management and executive leaders. "An engineer's career might come to an end if an important deadline is missed", d'Entremont said.

For example, a company might decide to ship a product when the revenue is needed for the financial report instead of when that product is thoroughly tested and actually ready for customer use. Depending on the company culture, MEs may be reluctant to challenge the decision because they fear being reprimanded, penalized, or losing their jobs.

MANAGEMENT RESPONSIBILITY AND ETHICAL DECISIONS

Decisions made by the management directly affect the "psychological safety" felt by company personnel when confronted with an ethical issue. A company where its employees, including engineers, operate out of fear is headed for disaster. "People make terrible decisions when they are afraid", d'Entremont said. "Such poor decisions may not come to light until sometime later, but their effects can be disastrous to a company's bottom line and its reputation". Managers, directors, and executives often make important decisions by simply doing nothing. "The status quo of only rewarding those managers and engineers when they meet their performance goals, without regard for public welfare, has indeed decided that ethical decisions are secondary to financial ones", d'Entremont said. "The entire workforce has been trained – via stimulus and rewarded response – to behave only in this manner into the future. Regrettable decisions and actions should be expected to follow".

The Fallacy That Any Standard Is Better Than No Standard

Many of the people involved in standards development know that not all standards are helpful or are developed by competent people with pure motives. Some standards have been developed to support the current state of products without pushing designers and manufacturers to make further improvements. Sometimes, well-intentioned people have extrapolated their own limited knowledge during standards development to apply to products in different industries. In addition, some companies involved in standards development activities have used standardization efforts to stifle competition.

"Many engineers will be involved in standards development during their careers", d'Entremont noted. "They should think critically about the motivations, goals, and methods used in their standards activities".

An Engineer's Allegiance

Engineers involved in day-to-day engineering and standards-development activities may ask themselves to whom they owe their allegiance. By being employees of companies, they know that they must satisfy their employers but they also must – as people – do what they think is right. Ethical conflicts can arise for engineers when they serve on a standards-development committee. They may ask, "do I serve at the will of my employer or for the good of the public?" This is further complicated by the definition of "member" for each standard organization. Sometimes members are the engineers' employers, while at other times, individuals are the members, regardless of the employer. "Although the latter is the better of the two membership models, it would be naïve to neglect the pressure a member's employer will likely put on an engineer to represent commercial interests", d'Entremont said. "Some standards committees read a preamble stating that members are chosen based on their abilities, not their employers – this is a commendable practice".

The Lack of Design Standards and Codes

The necessarily slow pace of standards development means that designers of innovative new products will never have standards and codes to dictate their design decisions. Although discrete sections of existing standards can be cobbled together to help guide designers, there will be no authority behind them and probably best serve as necessary conditions rather than conditions for product performance and safety.

OUTLOOK ON NEW TECHNOLOGIES

"Because of the above realities – which will only likely expand with new technologies and accelerated product-development schedules – engineers will continue to be challenged ethically by their designs, hazards, risks, intended uses, and potential misuses, once consumers start using them", d'Entremont said. "It is necessary to objectively evaluate their designs with respect to consumer safety and speak up when needed. When this happens, hopefully, these ethical engineers will be supported by their companies".

HOW TO MAKE ETHICAL DECISIONS IN ENGINEERING

Crawford (2022b) emphasizes the danger of thinking that the field of engineering does not lend itself to philosophical questioning. The truth is, engineers and philosophers alike need to ponder the ethical implications of new technologies like artificial intelligence and 3D printing. A variety of challenges face the design engineer at a corporation that designs, manufactures, and distributes innovative consumer products worldwide. Many safety-critical decisions arise in the normal course of engineering design that may have to be answered through the application of ethics. For an up-to-date look at the relationships among engineering ethics, product design, and product-safety engineering, ASME.org interviewed P.E. Kenneth L. d'Entremont, associate professor and lecturer in the Department of Mechanical Engineering at the University of Utah in Salt Lake City, and author of the new book *Engineering Ethics and Design for Product Safety* (McGraw Hill, 2021).

Q1: What are the ethical responsibilities of engineers today? Has this changed over the last few years?
Kenneth d'Entremont: Today's engineers really have the same responsibilities that prior generations of engineers have had. These are distilled down by the ASME into its "Code of Ethics for Engineers". The first Fundamental Canon is "Engineers shall hold paramount the safety, health, and welfare of the public in the performance of their professional duties".

The current pace of engineering is such that more and more work is asked of engineers over a shorter time period than ever before. Competitive pressures and the practice of Lean have further increased the workload of engineers in the corporation. This trend shows no signs of lessening.

A result of all this is that engineers are now faced with designing and developing more products with new, untried features and higher performance levels than ever before and having to do so within a time schedule unforgiving to design setbacks, such as necessary redesigns. Thus, engineers will face more ethical decisions than ever before.

Q2: What is the connection between engineering ethics, product design, and product-safety engineering?
K.E: In product design – especially with innovative, new products – there will likely be little to guide the design engineer regarding an acceptable level of safety. If a company is designing all-terrain vehicles (ATVs), for instance, the existence of

decades of prior ATVs helps the engineer assess the new product's safety. So long as the new ATV is as safe, or safer, than prior ATVs, then the design is probably sufficient, although that engineer should not neglect further design changes that could improve user safety.

Most engineers understand the need to provide safety to product users and wish to meet those needs. Innovative products are frequently 10 years ahead of any industry standard or regulation. Designers, therefore, are "working without a net". They must depend upon their engineering judgment and ethics, and those of coworkers, to direct their product design.

Ultimately, the ethical engineer will satisfy her/himself that the product is sufficiently safe. A product cannot be absolutely safe, but products can be safe when used by responsible people in their intended manners. Although misuse should be considered by design engineers, the fact that some new products have no history of use makes the consideration of misuse quite difficult.

Q3: How does product-safety engineering differ from other areas of engineering?
K.E: Although several aspects of engineering-design projects have objective criteria – speed, weight, size, output – there will be hard-to-quantify aspects of the final design that may not be explicit. Product safety is often one of them. The project-management team may presume that product safety will be sufficient without providing pass-fail criteria to design engineers. Even when consensus can be reached within the engineering team, there may be significant disagreement with consumer advocates and regulators. This should be expected since the final assessment of safety depends upon value systems, not upon science. Because value systems are different, separate groups of people will arrive at differing levels of safety sufficiency.

Q4: What are the biggest challenges facing MEs today regarding engineering ethics and product safety?
K.E: A primary challenge for mechanical design engineers is the pace at which innovative products are being introduced. On top of that, the competition is fierce so that development cycles are incredibly short, so more has to be accomplished in less time. The modern consumer has likewise been evolving and is smarter than ever, as well as much more demanding.

Engineers and engineering teams are faced with incredible technological challenges with schedules that might limit development and testing time prior to the necessary ship date. All must go according to plan. Any setback can seriously impact meeting an aggressive business plan. Practicing engineers may be faced with deciding between a business goal and an ethical obligation if a product is not ready for shipment. In cases such as those discussed above, the engineering team may decide to ship a product anyway or assume the role of engineering leadership and postpone shipment, despite the objection of the executives who are responsible for profit and loss instead of product safety.

Another issue that could cloud the practice of product-safety engineering is the concept of product liability. Even attorneys struggle with product-liability practice. The engineer has no chance whatsoever. The guidance I would give to engineers about product liability is to ignore it and focus on designing and manufacturing the safest of products. Nothing can prevent a company from being sued for its products. Whether or not a company is sued is not important; what is important is whether or not anyone gets injured. Engineers should focus attention on this.

Q5: How do you know when you have designed enough safety into a product? Can you have "overkill?"

K.E: Yes, it is possible to design-in too much safety. If this happens, the proper function of the product is negatively affected. Either the product no longer performs its intended function, or the user/operator of the product must now use the product in a less-safe manner – perhaps pressing down harder on the product during use or holding the product incorrectly to achieve the same effect. There must be a balance – or equilibrium – between product function and product safety. Of course, it can be difficult to determine that balance point, especially with innovative, new products which have never before existed. There will be no prior database of what has been acceptable by users. Competitor products do not exist for benchmarking. Lacking explicit design guidance in the forms of prior art, standards, regulations, and competitive benchmarks, design engineers are left to depend upon their professional ethics to decide what constitutes a safe product.

REFERENCES

Crawford, M. (2022a). https://www.asme.org/topics-resources/content/7-concerns-regarding-ethics-and-safety-in-engineering

Crawford, M. (2022b). https://www.asme.org/topics-resources/content/how-to-make-ethical-decisions-in-engineering

Peacock, B. (2019). *Human Systems Integration*, Self-published manuscript, Fernandina Beach, FL.

Peacock, B. (2020). *How Ergonomics Works*, Self-published manuscript, Fernandina Beach, FL.

Peacock, B. (2021). *Ergonomics Tools and Applications*, Self-published manuscript, Fernandina Beach, FL.

Shappell, S.A. and D.A. Weigmann. (2000). The Human Factors Analysis and Classification System-HFACSD. Aviation Medicine Reports No. OT/FAA/AM-00/7. Office of Aviation Medicine, Washington, DC.

6 Tools for System and Process Analysis

1. Introduction

It is impossible or impractical to separate the social aspects of design from the rigorously technical aspects of engineering (Peacock, 2019, 2020, 2021). The field of industrial engineering permits us to enmesh human social considerations with engineering principles. This chapter focuses on the Socio-Technical Systems Design.

The purposes of ergonomics, like the purpose of all system designs, embody the following elements.

a. **E3S3 outcomes**

   1. **Effectiveness** – the product or service meets customer quality expectations.
   2. **Efficiency** – productivity – optimal use of resources (people, money, materials, equipment, energy, etc.).
   3. **Ease of Use** – human interaction with the product or service should be convenient, comfortable, and error-free.
   4. **Safety** – the system (product, service) should not fail and cause harm to the user, associated hardware, the environment, or the organization.
   5. **Security** – the system should be resilient to malicious or accidental interference by third parties.
   6. **Satisfaction** – all users of the system should be satisfied with their experience and be motivated to continue to use the system.

b. **Scope of ergonomics**

   i. Body (Physical), Mind (Informational), and Soul (Social)
   ii. All people bring all three interacting components to all processes (see Figure 6.1)

c. **STS, Macroergonomics, and HFACS foundations**

   i. The concepts of STS were developed by the Tavistock Institute whose social scientists introduced and integrated the social component in a formal way following earlier specialized approaches to physical and information ergonomics and various attempts to address the social dimensions of work.
   ii. Macroergonomics, a term conceived by Hendricks and Kleiner, articulated a top-down STS approach to the design of complex systems, coupled with bottom-up participation. The process of "participation" by rank-and-file employees may be through paternalistic management or negotiations, sometimes adversarial pressure from trade unions.
   iii. HFACS (Human Factors Analysis and Classification System) was conceptualized by Shappell and Weigman as an accident analysis

DOI: 10.1201/9781003328445-6

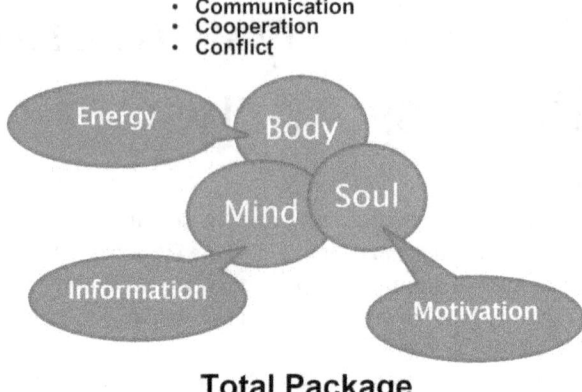

- Communication
- Cooperation
- Conflict

Total Package

FIGURE 6.1 Total package representation of mind, body, and soul.

process, initially developed for the aviation domain but which can also be applied to any other domain and design. HFACS is a comprehensive approach that addresses:

1. the unsafe act (human failure);
2. preconditions for the unsafe act (including hardware and context);
3. supervision;
4. and the organizational safety climate.

2. **Terms and Definitions**
   a. **Microergonomics** – The scientific study of human characteristics, capabilities, and limitations applied to the design of products, equipment, services, and environments.
   b. **Macroergonomics** – A top-down Socio-Technical Systems approach to the analysis and design of complex system integration.
   c. **Ergonomics** – The amalgamation of macro- and microergonomics. It may be argued that ergonomics is by definition comprehensive and should not require any qualifiers, such as physical, information, or macro, as these only serve to fragment the approach. However, where some level of operational focus is convenient, care should be taken to address the possible interactions of other factors in task behavior and performance.

3. **Micro- and Macroergonomics**
   a. Microergonomics deals with limited scope issues including:
      i. anthropometry and workplace design;
      ii. biomechanics and manual materials handling;
      iii. work physiology and physical fatigue;
      iv. sensory processes and information display;
      v. attention processes and information display;
      vi. cognitive processes and information display and processing;
      vii. motor skills analysis and task design;

      viii. design of controls;
       ix. communication, cognition, and control theory related to process design;
        x. environmental analysis and human performance;
       xi. social context of behavior and performance.
  b. Macroergonomics is a top-down/bottom-up Socio-Technical Systems approach to the design of work systems and the application of the overall work system design to the design of human – job, human-machine, and human-software interfaces.

4. **Shortcomings of Traditional Design of Complex systems**
  a. **Technology-centered design**
      i. Frequently the strategy of developing technological solutions to problems results in an incomplete product or service that may be ineffective, inefficient, difficult to use, unsafe, insecure, and unsatisfying to the human user. Often such designs require considerable human intervention to be effective and may fail if the human requirements are not met.
         1. "Automated/E-Ticket" airline check-in requires considerable "help" for problems and passengers who do not understand the process.
         2. Many "automated machines", such as presses, require human operators to feed in raw materials and remove finished products and scraps. These tasks tend to be repetitive and sometimes dangerous, and may also lead to quality and machine down time issues.
         3. Early grocery stores had the storekeeper bring the goods, pack them, and take the customer's money. (Sometimes they even delivered the goods.) As the industry grew with more customers, the only role of the shopkeeper was to stock the shelves and take the money – the customers collected the goods from the shelves. The microergonomics task of the cashier is onerous and fatiguing and may lead to errors and customer dissatisfaction.

            Contemporary supermarkets have automated checkouts, but store personnel are always on hand for problem-solving and to monitor customer honesty.

  b. **"Left over" approach to function and task allocation**
      i. Where mechanization or automation is incomplete, humans are assigned to the residual tasks.
         1. Airline baggage handling has many residual human links in the process, leaving the human vulnerable to injury, especially the counter clerks, who are predominantly female and not physically capable of dealing with the heavy bags.
         2. The HK Mass Transit Railway is "fully automated" – the train starts and stops itself at each station and adjusts for headway variance. But there is an operator (in Singapore, the bus driver is called the Captain) to deal with residual issues related to

passenger behavior and track/vehicle/system discrepancies and to let the passengers know that there is a driver in charge, not a computer.

3. UAVs (Unmanned Aerial Vehicles) or UASs (Unmanned Aerial Systems) do not have an onboard pilot. They are controlled remotely from the ground with more or less elaborate teams of operators. This is an advantage in conducting wars remotely; also civilian surveillance (traffic, pipeline, forestry, agriculture, border security, etc.) may have many advantages of eliminating the need for human pilots. However, the ground-based "pilot/operator" is not committed and may be distracted or susceptible to many human failure modes, such as fatigue, perceptual errors (with the onboard video systems), and less commitment to safety.

c. **Failure to consider socio-technical complexity**
   i. Simplistic system designs may not be resilient in their interactions with human and environmental conditions.
      1. Large lecture classes and "objective" grading, aimed at productivity, are not conducive to effective learning – the students at best must follow up outside class or at worst not attend and just "read the book".
         a. Participative and interactive strategies, with inquiry components, supported by offline study have been shown to be effective for the transfer of knowledge and the development of useful outside world skills.
      2. Complex systems require humans to detect, assess, and counter human variability.
         a. Aviation security requires sophisticated automation plus human sensing to detect intruders with malicious intent.
            i. Anticipatory systems that combine information from many human and automatic sensors will always require human-aided automation.
         b. Many computer-based information system interfaces are surrounded by paper notes to supplement user short-term memory. The "windows" and "applications" concepts although powerful tools still require human integration for the system to be successful and error-free.

d. **Criteria for STS design**
   i. Consider the system as a whole, including the context of operation and the intended users and foreseeable misusers.
      1. Use FMEA, the 5 Whys, and HFACS to track the root cause of an apparently simple accident, such as a slip or trip on a sidewalk. (These techniques will be discussed in depth later in the course.)
      2. Describe a private university from the STS viewpoint using a concept map.
         a. Describe the requirements of the various constituencies.

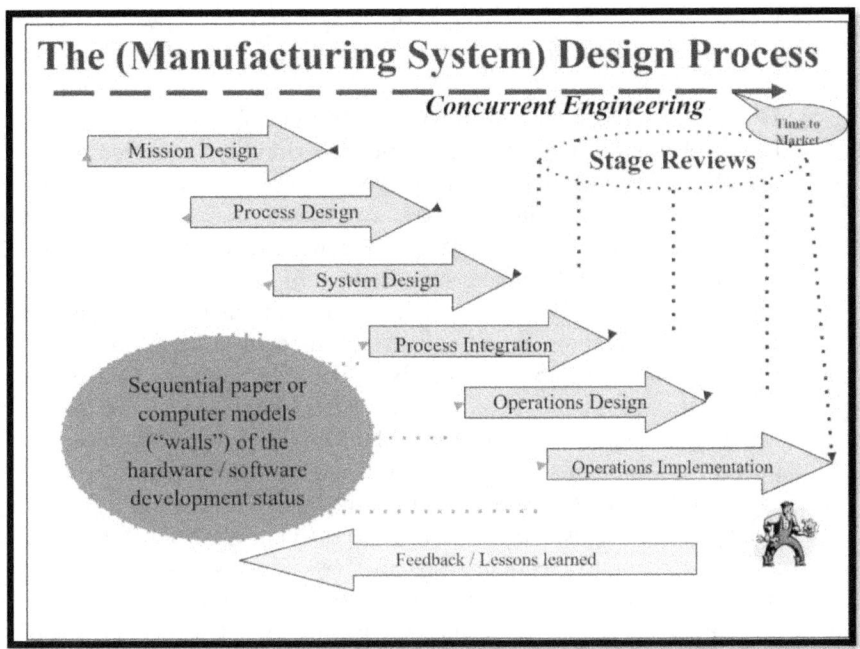

FIGURE 6.2 The manufacturing design process.

    ii. Consider human contributions at all times
        1. There is no such thing as automation
            a. There are always designers, managers, users, maintainers, customers, etc.
            b. Read Kurt Vonnegut's "Player Piano" – a satirical account of the automation utopia
    iii. Use a concurrent design approach
        1. Design the system (see Figure 6.2), the development/manufacture of the system, system implementation, and system operations monitoring as a concurrent exercise, with iterative modifications.

5. **History of STS**
    a. **Taylor** – scientific management, prescriptive task design, "line workers should not be expected to think", short-cycle work.
    b. **Ford** – the production line, work simplification, productivity focus, help for the line worker by reducing nonvalue-added activities, such as fetching and carrying.
    c. **Hawthorn** – effect of social factors (attention) on job performance that confounded the main experimental variable of lighting levels.
    d. **Maslow** – the hierarchy of human needs (physiology, safety, social, esteem, and self-actualization). Maslow argued that this was a monotonic progression with the lower levels needing to be satisfied before the

higher ones could be achieved. Many people do not achieve the higher levels in their jobs, sometimes because attention to the lower levels is incomplete.

e. **Herzberg** – intrinsic motivation and hygiene factors (dissatisfiers). Herzberg argued strongly that people are primarily motivated by the intrinsic content/challenge of their work. Other factors, such as environment, supervision, and even pay, only serve to dissatisfy the worker.

f. **Tavistock Institute** – formal approach to the study of integrated Socio-Technical Systems following observations of the failed implementation of mechanization and automation. The initial investigation was related to the introduction of automation in coalmining, which served to divide and separate team members who had hitherto considerable group cohesion.

g. **Macroergonomics** – a term coined by Hendrick to describe a broad, integrative, human-centered approach to complex system design. Many of the case studies referred to in the book focus on three issues – management buy-in to the need to address employee needs other than wages, worker participation at various levels of decision-making in the organization, and broad attention to many microergonomics issues.

   i. It may be argued that the microergonomics interventions may sometimes be cosmetic and subject to the "Hawthorn" effect, whereas other microergonomics factors may, in fact, address the primary needs of workers.

   ii. The participation and microergonomics contributions in the automobile industry during the 1990s had full management support, albeit with pressure from the government and the unions. However, the bottom-up micro solutions did not always address the more fundamental problems of product/component design and the inherent shortcomings of production line work.

      1. See the handout on "Fatigue and the job cycle".

h. **Woodward** – described the structure of organizations from the STS viewpoint (unit, batch, mass, and process production).

i. **Deming** – a statistician credited with large improvements in quality in Japanese industry argued the case of "the honest worker" who just wanted to do a good job, and also distinguished between common causes of variation and special causes, which required attention.

j. **Volvo** – teams of assemblers followed a unit along the assembly line. The vehicle rotated for better access to the underbody. Widely supported by Swedish-organized labor with its participative approach to job design.

k. **Saturn** – a joint GM – (Independent) Union-managed facility. Work teams with job rotation, equitable shift systems, and final assembly on height adjustable palettes for easy access. Many innovative processes, tools, and vehicle designs with assembly in mind are coordinated at the outset. Very high levels of worker participation at all levels of management. Ergonomics teams are very active in problem-solving.

The system broke down in the late 1990s when the central union (UAW) was voted in by the employees and the Saturn division became like all other GM divisions.

l. **General Motors, Lansing Craft Center** – in general assembly, the vehicle does not move on a production line, rather it is surrounded by a team, baskets of components, tools, and fasteners where it is completely assembled and tested before moving to final inspection. The whole GA process takes about an hour per unit with – substantial employee input to product, manufacturing process, and task content design.

m. **Toyota Production System** – a formal top-down, prescriptive (e.g., 5S – Sort, Set in order, Sweep, Standardize, and Sustain) job design process that makes use of operator teams (e.g., quality circles) and knowledge related to process quality, productivity, and safety. System performance is monitored by visual controls, often taking the form of control charts. Introduction of the andon cord which allowed anyone to stop the line when a problem arose – all affected workers, including skilled trades and engineering – would congregate to resolve the problem.

n. **South West Airlines** – employees first, no first class, first come first served seating. Many strategic cost-saving policies – only short-haul domestic flights, fleet limited to B737, the introduction of winglets to reduce drag, wake turbulence and improve handling, no pre-assigned seating, no first class, lower salaries but better benefits, etc. Most successful N American airline.

o. **General Motors Manufacturing Ergonomics Program** – development of a center of ergonomics expertise in corporate manufacturing engineering. Parallel development of training center in UAW Center for Health and Safety.

Development of checklists/analysis tools. Training and assignment of salaried and hourly ergonomics specialists in every plant. Training and assignment of ergonomics specialists in car programs. Development and deployment of reactive and proactive ergonomics programs. Deployment of programs throughout North America and Europe. Task force approach to difficult problems, such as installation of the intermediate shaft, hoses, batteries, wheels/spare wheel, wind shield wiper motor, seats, exhaust system, and wiring connectors. Some are addressed by engineering interventions and some by administrative controls.

p. **Politicization of Ergonomics** – accelerated with a tripartite agreement between the U.S. Department of Labor, The UAW, and the Big Three Automobile manufacturing companies following previous agreements in the meatpacking and retirement home industries. This is followed by the pursuit of an ergonomics standard by the U.S. Department of Labor and managed by joint management and worker/union representatives to reduce the incidence of work-related musculoskeletal disorders.

This standard was generally supported by the ergonomics community as a formal way of introducing ergonomics methods both reactively and proactively. This was a bottom-up "political" process in contrast to the top-down (management manages and employees participate) process of the Toyota Production System. The ergonomics standard was introduced by the Democrats and immediately canceled by the Republicans. Throughout the developments, management set up opposition through the National Association of Manufacturers and the U.S. Department of Commerce (see hand out – an alternative ergonomics standard).

6. **Costs and Benefits of Macroergonomics**
   a. **Benefits**
      i. The benefits of macroergonomics are through a combination of management commitment and employee participation leading to many microergonomics improvements which in turn lead to:
         1. Improved effectiveness (quality)
         2. Efficiency, improved productivity, optimal use of resources
         3. Ease of use, reduced human errors, and rework
         4. Safety, reduced costs by accident prevention, reduced injury, and damage to equipment and the environment
         5. Security reduced susceptibility to accidental and malicious system interference
         6. Satisfaction, improved quality of work life and motivation, and improved customer satisfaction
         7. Reduced skills and training
         8. Reduced waste
         9. Reduced maintenance – first-line maintenance by operators (TPM)
      ii. Profit (and cost) sharing is the ultimate level of participative management of organizations especially when rewards are tied to tangible contributions.
   b. **Costs of macroergonomics**
      i. Added form of bureaucracy – nothing is done without full participation – more meetings
      ii. Time needed for analysis and design
      iii. New equipment
      iv. Reorganization barriers
         1. Interference with production during the reorganization
         2. "Not invented here", "We have always done it this way"
         3. Job responsibility changes up and down the management chain
      v. Increased training for rotation and enlargement (vertical and horizontal) skills
      vi. Difficulty in the objective assessment of macroergonomics interventions

7. **Macroergonomics Implementation**
   a. **Macroergonomics aspirations** are rarely the province of the macroergonomist, rather top management must become attuned to the principles of macroergonomics, by whatever name.
      i. The Industrial Relations/Human Resources departments grew to assist management with personnel issues, such as hiring and

firing, wages and benefits, health and safety, and negotiated work conditions. The HR department rarely got involved in production operations.

ii. Macroergonomics sees a more collaborative rather than prescriptive world but relies on the established processes/departments for multiple, specific microergonomics interventions.

iii. Top-down/bottom-up philosophy

    1. Top management must support and employees/customers should participate at all levels of decision-making.

    2. Alternatively, macroergonomics can be implemented top-down with employee cooperation but without "vertical enlargement".

iv. GM macroergonomics program was a participative effort, supported by top management and union leadership with the center of expertise within the manufacturing/industrial engineering, and within vehicle programs.

    1. Involvement of GM Europe hampered by considerable resistance to top-down (Detroit-centered) process; resolved by collaboration.

b. **Macroergonomics implementation** varies according to organization structure(s).

  i. Product centered

    1. Manufacturing (e.g., automobile, electronics, textiles, and plastics)

    2. Retail – labor differentiated between purchasing, processing, transportation, and the retail front end with shelf stocking and checkout operations

    3. Construction – many skills brought together (serially) to create a single product, organized by a general contractor who arranges just-in-time materials delivery and sequential structure and services operations

    4. Efficiencies in product-centered organizations generally lead to mass production strategies and work simplification

    5. Macroergonomics implementation in vertically differentiated organizations must come from the top, with employee participation

      a. Alternative model of joint responsibility leads to conflicts of process and domain/technical ergonomics knowledge.

  ii. Function centered

    1. Hospital

      a. Differentiation of knowledge, skills, and activities – specialties – both medical and service departments (therapy, X-ray, labs, etc.).

      b. Growth of independent and competitive (for budget, space, and equipment) departments.

      c. Specialist medical knowledge and skills require very different resources – costly equipment, operating rooms, etc.

        d. Specialist service departments (X-ray, biochemical testing, rehabilitation, etc.) also compete for budget and growth.

        e. Growth of hospitalist/general medical practitioner to deal with all other aspects of the patient's situation.

2. In aviation, there must be cooperation among the captains, the rest of the flight crew, air traffic control (center and en route), dispatch, maintenance and fueling, passenger management, and their respective organizations.

        a. Crew Resource Management (CRM) was developed to assure effective collaborative activities among all the human, technical, and administrative "resources".

3. Macroergonomics implemented in horizontally differentiated/function-centered organizations requires the buy-in/participation of the experts/specialists, who defend their own territory aggressively. Difficulties of implementation arise due to professional as well as administrative hierarchies, and operational focus around knowledge and experience.

        a. CRM concepts are applied in both aviation and medicine and face resistance from a long history of expert-centered management traditions that vary with national/ethnic traditions.

iii. Hybrid

    1. University

        a. Specialized departments/degrees/subjects

           i. Can become product centered with vertical separation

        b. Competition for growth, space, equipment, and budget

        c. Student may be narrowly trained

        d. Professors become specialists in order to publish

           i. They become professional experts

              1. Collaborate (and sometimes compete) with professional peers

              2. Lead their junior colleagues and graduate students

              3. Function-centered/independent laboratory structure

          ii. Must teach more generally – basic material

              1. This requires team work

              2. Experts are subservient to administration for such things as classroom/time slot allocation, teaching load, and examination format

        e. Pressure on the curriculum as knowledge in a subject area grows

           i. Further subdivisions, new function-centered organizations

          ii. Conflict over priorities arises

        f. General education requirements became targets for removal/minimization

           i. Humanities, business, arts, language, etc., for engineers

          ii. Technology awareness for arts and business students

         iii. Computer and communications literacy

        g. Technique specialization

           i. Research methods

               1. Laboratory data capture and statistical analysis skills develop in a function-centered format around the expert

           ii. Teaching methods – influence of education process experts on teaching methods

               1. Leads to balance teaching (top-down) and learning (participative and inquiry)

               2. Balance varies between subject and level

          iii. Computer skills overlaid – all participants require common computer skills plus specialist computer package knowledge, such as Statistics, MATLAB, and Simulation.

      2. Large vertically differentiated organizations (e.g., automobile manufacturing) also rely on horizontally differentiated technology centers that may overlap:

        a. Safety, ergonomics, and industrial hygiene

        b. Robotics, paint, and welding

        c. Styling, engineering, marketing, and program management

c. **Complexity**

    i. Vertical and horizontal differentiation occur with growth

      1. Departments, levels, and titles

        a. Manager, director, vice president, president, etc.

        b. Leading to competition for promotion

      2. Horizontal differentiation may be accompanied by overlapping sub-specialties

        a. For example, ergonomics, safety, and industrial hygiene (see above)

        b. Leading to the development on small independent hierarchies and opportunities for horizontal competition for resources

    ii. Integration/coordination/communication challenges

      1. Development of parallel hierarchical committee structures

        a. Overlap and separation of responsibility among permanent committees and limited duration task forces create opportunities for conflict

        b. Competition of committee hierarchy with line management for authority

   iii. Varying degrees of formalization – well-defined structures, processes, and outcomes

      1. Rigid structures breed integrated teams (interdisciplinary, cross-functional, and product development teams)

      2. High degrees of formalization with vertical differentiation usually lead to limited life task forces rather than standing committees

    iv. Centralization
1. Perceived better control by higher levels
2. Counter argument of "autonomy with responsibility"
3. Major challenges for international companies
   a. General Motors' best practice/common process policies presented challenges between international centers and technology centers
   b. Considerable pressure for regional autonomy

    v. Hierarchy
1. Vertical separation
   a. (Compare General Motors (19 levels) and the Catholic Church (4 levels))
   b. Span of control
      i. Flat organizations
         1. Project leads – individuals have different roles in different projects
         2. Become unwieldy as the organization grows
            a. Needs for specialized departments
            b. Human "need" for "promotion"
            c. Administrative functions better dealt with by a small span of control
      ii. Deep hierarchies (vertical differentiation) causes and shortcomings:
         1. Usual result of company growth
         2. Usually occurs in product-focused organizations
         3. Promotion of managers an unspoken primary purpose
         4. Vertical and horizontal communication difficulties
         5. Parallel committee hierarchies
            a. Committees fight with line management for control

d. **Distribution**
    i. Departmental separation
1. Autonomy – departments seek independence
2. Separation of functions – line and support organizations
   a. Line organization, personnel, quality, safety, accounting, etc.
3. Overlaps – departments try to grow in space, responsibilities, budgets, and influence

    ii. Geographical separation
1. Different products/models at different plants
2. Component suppliers – local and international
3. Specialized central technology and support centers and distributed production facilities
   a. Central – Design, engineering, marketing, personnel, safety, etc.
   b. Distributed – components and assembly

          c. Travel budget increases!!
          d. Communications costs and time
          e. Pressure to set up local technology/service centers in horizontally separated units
   iii. Subcontracts – many advantages
      1. Lower labor costs
      2. Geographically separated
          a. Often overseas or nonunionized
      3. Lower overheads
      4. Specialized component knowledge
      5. Conflict on price with OEM (Original Equipment Manufacturer)
          a. Supplier cost engineering squeeze
          b. Disadvantages
             i. Quality control may suffer
            ii. Component cost will creep
           iii. Specifications changes cause large cost increase
      6. May be many layers – OEM, first-, second-, and third-tier suppliers
          a. Communication and transportation challenges
  e. **Communication and contemporary technology**
    i. Essential part of distributed organizations
    ii. Meetings
      1. Face-to-face, teleconference
      2. Chats
   iii. Asynchronous communications
      1. E-mail
      2. Blogs and social networks
   iv. Intermediate technology seen as barrier to efficient operations/communications when compared with face-to-face meetings.
      1. Proliferation of e-mails, voice mails, etc., is a major time consumer for managers who are unnecessarily "copied" on messages.
      2. Face-to-face meetings also have difficulties due to geographical separation, scheduling problems, and uncontrolled divergence.
 f. **Efficiency**
    i. Autonomy with responsibility is often seen as a major motivator leading to the establishment of small independent groups.
    ii. Local autonomous units are internally efficient but may not see the big picture resulting in sub-optimization.
 g. **Macroergonomics contributions**
    i. Systematic way of articulating the structure, process, and outcomes of large systems
    ii. Macro ergonomist as an advisor to management
      1. Must use domain knowledge to complement macroergonomics knowledge and tools
      2. Should identify the "low-hanging fruit" to motivate management to continue support
          a. Quality, productivity, satisfaction, and safety

   3. Must be succinct – managers in industry and business do not
      have time to read the details of the communications from all
      their individual, committee, and departmental reports
      a. General Motors and NASA HF instituted a process of
         one-page report (supported where necessary with back up
         material)
8. **Handouts**
   a. Purpose and Scope of Ergonomics in Design
   b. Concept mapping
   c. Job redesign in the bindery
9. **Case Study – Bookbinding**
   a. The problem
      i. The book bindery at HKU put hard copies of paperback books and
         annual collections of journals
      ii. "work in progress" could be many weeks
      iii. Low "status" of the blue-collar bindery staff in a white-collar
         (library) environment
      iv. Equipment bottlenecks
      v. Inefficient project management/scheduling
         1. Informal priorities due to status differences between academic
            department heads and bindery manager
      vi. Poor environmental context – heat, noise, and glare
      vii. Narrow job responsibilities based on seniority
   b. The solutions
      i. Reorganized storage and categorization of binding requests
      ii. Just-in-time delivery of raw materials/journals/books
      iii. Batch work flow
      iv. Introduced team structure
         1. Team carried a batch of similar material (based on group
            technology principles) through the whole process
      v. Vertical job enlargement – all staff were trained to carry out all
         stages of the bookbinding process
      vi. Slow job rotation (later changed from 1 day to 1 week at the request
         of the bindery staff)
      vii. New equipment to resolve bottleneck problems
      viii. Addressed environmental issues (carpets, blinds). These "microer-
         gonomics" interventions were much appreciated
      ix. Placed "white collar" buffer between bindery and library staff to
         resolve informal priorities issue
      x. Across-the-board pay raise (pay for performance) – very much
         appreciated
   c. The results
      i. Large increase in productivity
      ii. Greatly reduced "work in progress" delays
      iii. Generally improved morale

  d. The conclusions
    i. Major success of Macroergonomics intervention
    ii. The "Hawthorne effect" questions remain
10. **Self-Test Questions**
  a. List and describe the six general purposes of ergonomics, with examples (1)
  b. Define Macroergonomics
  c. Give six examples of microergonomics
  d. List three common failures of not considering STS issues
  e. List ten key people/companies that contributed to the development of Socio-Technical Systems
  f. List ten costs and benefits of applying Socio-Technical Systems approach
  g. Give three examples each of Product and Function centered organization designs
  h. Describe three forms of complexity in large organizations
  i. How can large organizations be "distributed?"

List three key points in communicating macroergonomics advice to management and implementing macroergonomics issues.

SOCIO-TECHNICAL SYSTEM DESIGN

STUDY UNIT 2

1. **Introduction**
  Socio-Technical System Design has many process similarities to technical system design, as in Figure 6.3.
2. **Process and System Design Fundamentals**
  a. Should "participants", "customers", and "stakeholders" be used interchangeably?
  b. The Grammar of Design – design as a communication process – is accomplished more effectively and efficiently, with less error and rework if participants (including customers) adhere to a common language. The following operational definitions are presented to support this process.
  c. Systems and processes
    i. A system is described by a noun and measured (qualified, quantified) by an adjective.
    ii. A process involves the interaction of two or more systems. It is described by a verb and measured (qualified, quantified) by an adverb. A process will usually result in a change to one or more of the participating systems.
  d. Systems analysis addresses Structure, Process, and Outcomes
    i. Structure – the tangible components of a system
    ii. Process – the interactions among system components
    iii. Outcomes – the change in the state of one or more system components as a result of the process

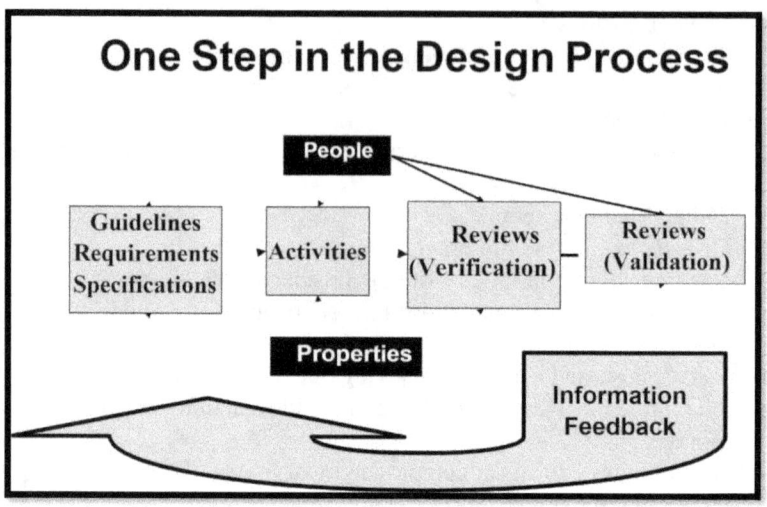

FIGURE 6.3 One-step in design process.

   e. Requirements and specifications
      i. A process has performance requirements as set by the various cus-
          tomers; performance requirements are measured by reference to
          some standard or by comparison with other processes.
          1. Performance may be measured by a change in subsystem/
              component state
          2. Requirements will generally be classified into
              a. E3S3 – Effectiveness, Efficiency, Ease of Use, Safety,
                  Security, Satisfaction
              b. Often a customer may desire high levels of all of these
                  outcomes but may have to settle for a compromise
      ii. A system has specifications, usually quantitative, that is necessary
          for design
          1. Adjectives! – big engine, dry road, trained driver, and restric-
              tive speed limits
   f. Validation and verification
      i. Processes are validated by being implemented in a realistic context
          (environment, users etc.) or simulation
          1. Driving involves human, vehicle, environmental, and regula-
              tory subsystems
          2. Driving quickly involves all these subsystems
          3. Driving "quickly" is relative to other occurrences of the driv-
              ing process which has different subsystem values
          4. Driving safely involves all of these systems and may not occur
              if one or more of the subsystems is "out of tolerance"

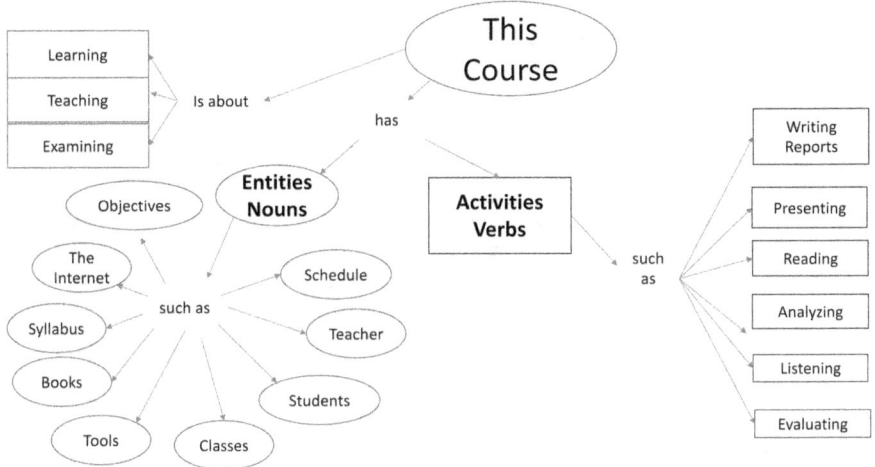

FIGURE 6.4   Concept maps.

Driving quickly and safely depends on high levels of all the subsystems.
  ii. Systems are verified objectively by measurement of key features
       (adjectives) and comparison with the system design specifications
       (with tolerances)
       1. Big engine – V8
       2. Trained driver – attended and passed a safe driving course
       3. Restrictive speed limits – 50 kph
3. **Concept Maps**
  a. Concept maps (Figure 6.4) are a diagrammatic way of describing a
      complex situation, system, or process showing key sequences, interac-
      tions, and links to supporting information.
      i. http://cmap.ihmc.us/conceptmap.html
      ii. http://cmap.ihmc.us/download/
  b. Adaptations of concept mapping will be used throughout the course to
      assist in organizational structure and process analysis.
  c. An operational discipline in concept mapping is to separate activities (pro-
      cesses) from entities (systems) and apply "grammar of design" concepts.
4. **Activity Cycle Diagrams**
  a. A graphical way of describing the flow of entities/resources around a
      complex network of activities and queues based on embedded logic.
      Can be used for complex system description or as a basis for discrete
      event simulation.
  b. Each entity will move around a different activity cycle consisting of
      activities and queues (times when the entity (system) is not being used).
  c. Activity cycle diagram concepts will be used throughout the course to
      describe complex systems and processes.

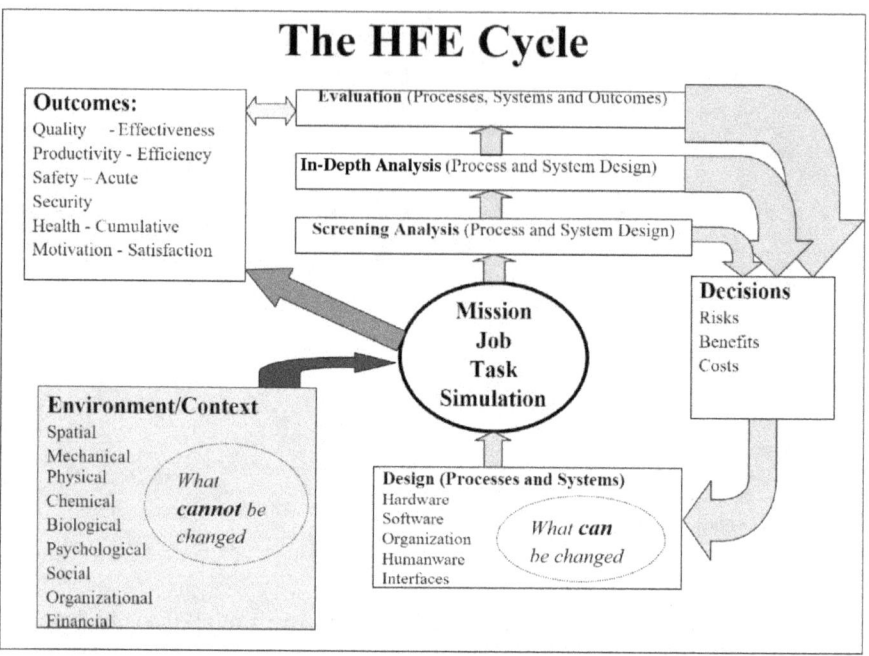

FIGURE 6.5   The HFE cycle.

d. Check the activity cycles and branching logic of the following activity cycle diagram that shows the resources (entities), activities, and pathways associated with a bar or cafe.

5. **Design Process**
   a. **The Human Factors Cycle**
      i. The human factors cycle (Figure 6.5) is superimposed upon the general process model (inputs – process – outputs) by adding both process and outcome analysis and the decision process used to modulate the inputs based on process and outcome analysis. Furthermore, the inputs are separated into those factors that can be changed and those environmental/context factors that usually cannot be changed but must be considered in system and process design.
   b. **Design as a Control Problem** (see Figure 6.6).
      i. Inputs (systems) – must be designed
         1. Human inputs – affected by selection, training, assignment, abilities, limitations, motivation, attention, fatigue, etc.
         2. Equipment and materials – can be designed, must be resilient with regard to users and context
         3. Context/environment – this cannot be designed but must be addressed, e.g., by barriers or shields

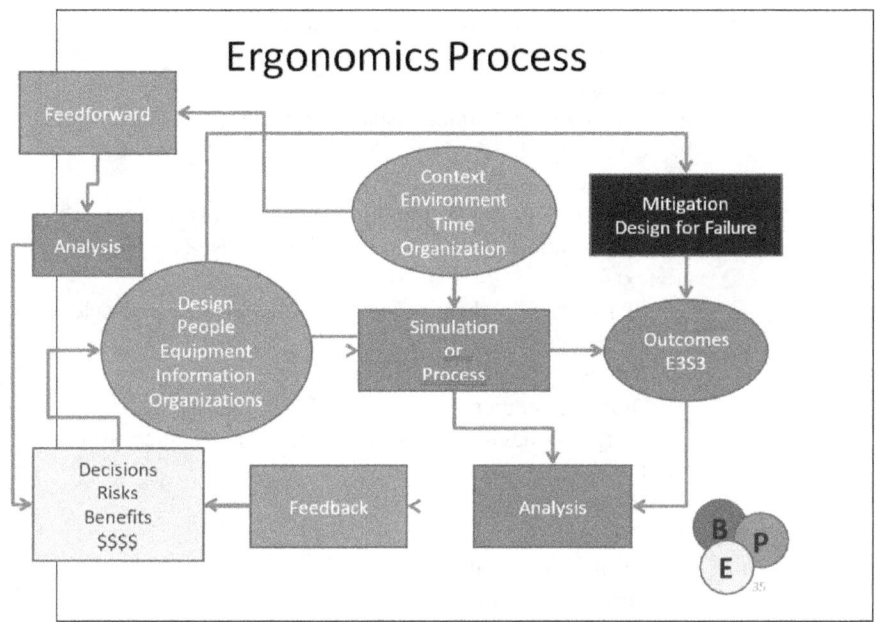

FIGURE 6.6 Ergonomics process.

    4. Regulations may be applied to human, technology, and environmental subsystems

        a. Passed driving test, energy efficient car, and daytime driving.

ii. Process

    1. The interaction of two or more systems with a measurable outcome in terms of process performance and the change in the state of one or more contributing systems

iii. Outcomes

    1. Generally measured in terms of:

        a. Effectiveness – quality – matching customer requirements

        b. Efficiency – optimal use of resources (money, time, materials, energy, people, etc.)

        c. Ease of use – resilient to varied users, usage, and contexts

           i. See 6Us handout

        d. Safety – systems resilient to catastrophic failure and wear

           i. Process/system failure mitigated to reduce the severity of the unwanted outcome

        e. Security – process/systems resilient to accidental or malicious interference by third parties

        f. Satisfaction – (all) human users (customers) should be satisfied with their experience with the process/systems

           i. There may be compromises

iv. Feedback, adaptation, and learning control
   1. Feedback – mechanism for communicating outcomes (error) to modulate inputs
      a. Flying in wind under Visual Flight Rules (VFR) conditions
      b. Catching a ball
      c. Balancing on one foot
      d. Heart rate (what about anticipatory heart rate increase?)
   2. Adaptive – automatic adjustment to inputs based on pre-defined context
      a. Thermostat – heating/cooling changes to pre-selected conditions
      b. Jockeying in queue behavior
      c. Diabetes medication
   3. Learning – behavior modification and performance improvement with experience/practice.
      a. Hitting a golf ball
      b. Driving
      c. Taking examinations
v. Feedforward (anticipation)
   1. Prediction of the effects of context on the process behavior and modulation of inputs (subsystem changes) accordingly
      a. Environmental, technology, or regulatory context, etc.
   2. Necessary for the design of resilient systems – systems that can withstand the effects of intended and unexpected context and time
      a. Market research
      b. Weather planning
   3. Feedforward information may be erroneous or at best probabilistic
      a. What will the other driver do?
      b. Will it rain/snow/freeze today?
      c. Will the technology subsystem (e.g., car) deteriorate over time or without maintenance?
   4. Human beings usually make considerable use of "feedforward"/ anticipation
      a. This activity often leads to timely actions that may be in error due to uncertainty in the anticipation process as in choosing a menu item based on a verbal description, preparing answers to questions at an interview, designing an advertisement aimed at a subset of customers, or selecting a technology for fuel-efficient cars.
      b. Market research is a mechanism used to predict customer needs and wants in the future. However, as product (e.g., car) design takes a few years and operates in a competitive context, these customer requirements may be a "moving target".

     vi. Decisions regarding process inputs are usually made with reference to the cost of resources, such as money, time, people, equipment, fuel, and materials.

        1. The Socio-Technical System Design philosophy will face decisions by managers who may be more focused on technology than the vagaries/requirements of multiple customers/stakeholders.

        2. Decisions are usually the prerogative of management.

        3. Decisions in the design process are usually made in the progress review meetings where the components, including HFE advice, are presented in the context of the big picture.

        4. These decisions will be biased by the managers'/committee's prejudices and the ability of the engineer to sell his or her point of view.

           a. Effective communication is a learned skill – practiced in the classroom in preparation for the workplace.

  c. **Design as a Communication Problem** (Figure 6.7), **Using Car Design as a Case Study**

     i. Concept/idea of someone involved in a design or purchase process

        1. An economical family car (requirements)

     ii. Semantic and Physical encoding

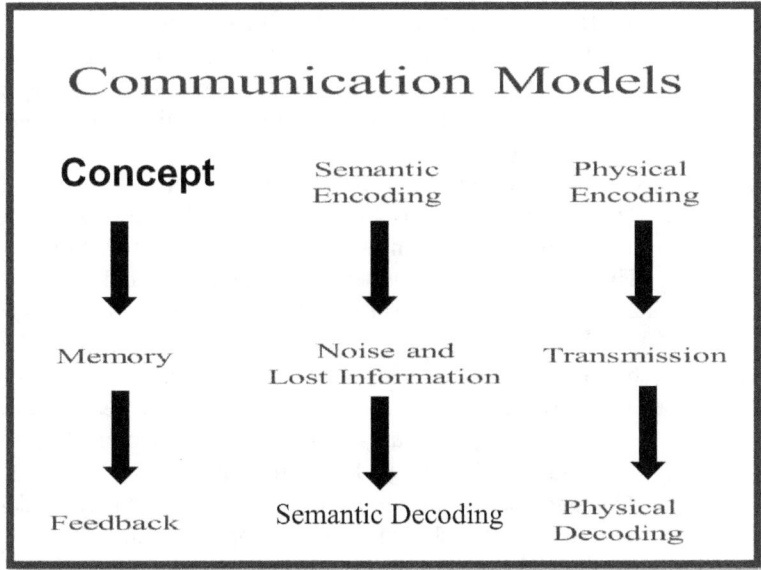

FIGURE 6.7 Communication models.

       1. *Semantic encoding* – translating the idea into some known language, diagram, model, etc.
         a. Using automotive jargon – "a mid-sized, base-level sedan" – relative statements/requirements
       2. *Physical encoding* – translating the conceptual model into an explicit physical form, such as drawing, writing, or speaking
         a. Four seat, four door, sedan with cloth, automatic, Quad 4, entry-level IP
            i. Note the use of jargon and abbreviations
            ii. There are many variations/interpretations of this initial verbal set of high-level specifications

iii. Transmission, Noise, and Added information
       1. Transmitting the idea to the intended (unintended) audience
         a. The transmission may involve a series of translations by individuals with different priorities
         b. Note that the idea may not be clearly articulated (lost in translation – omissions, additions, changes)
         c. Note also that there may be external physical or informational "noise" during the transmission process
         d. The full message may not reach the intended recipient for a variety of technical reasons

iv. Physical and Semantic decoding by multiple customers with different priorities
       1. Manufacturing, maintenance, sales, drivers, regulators, etc.
         a. Receiving and understanding the information
         b. The receiver may not physically receive/sense the message
         c. Note that understanding requires knowledge of the language and a reception framework – the translation may be biased by the receiver

v. Consolidation, Retention, Forgetting, and Action
       1. The receiver must consolidate (fit into his framework), remember (or forget), and translate the information into action
       2. The eventual set of high-level specifications could now be:
         a. Five seat – mid-level vehicles usually have five seats, not four
         b. Sedan heard as "van" (physical decoding)
         c. Automatic referred to a gearbox, but added windows, door locks, and seats – typical of less economical vehicles (semantic decoding/added information)
         d. Four doors translated into two conventional front doors plus two rear sliding doors – common in vans
         e. Quad 4 engine, typical of small cars, was converted to V6 based on the common choice of engine for minivans

vi. Feedback, Adaptation, Learning, and Iteration
       1. The originator of the idea needs feedback in order to modify the idea/concept (see control model above)

    2. The feedback cycle should reduce communication errors. However, if feedback is not available, the communication may lead to designs that don't satisfy the initial intent (requirements)

  vii. Design is vulnerable to communication failures and participant inconsistencies

    1. A problem similar to this actually occurred with an attempt to design a front-wheel drive Camaro

6. **System Life Cycle in Product Design**

The major focus is on car manufacturing from the socio-technical and microergonomics perspective:

- Mission/Purpose
- Concept(s) design
- Concept evaluation and testing
- Concept selection
  - Design for use (Usability Testing)
  - Design for manufacturing and assembly
  - Design for service and maintenance
  - Design for disposal
  - **Design for SAFETY**
- Manufacturing and Production Design
- Production
- Distribution and Sales
- Use
- Service and Maintenance
- Disposal

  a. Consider the life cycle of a car

    i. *Product design*

      1. The mission/purpose will vary enormously depending on the functional requirements of the intended user/buyer/customer – a sedan, truck, sports car, or luxury car

        a. The seven ages of "carman"

          i. Teens – wheels (Civic), 20s – style (Camaro), 30s – function (minivan), 40s – prestige (Buick), 50s – lavish style (Corvette), 60s – comfort and safety (Cadillac), 70s – wheels (Civic)

        b. The mission/purpose should consider many other customers

          i. Manufacturers, maintainers, shareholders, and regulators

      2. Within each general lifecycle stage, there will be subcategories, each aimed at emphasizing particular aspects of the vehicle

        a. These sub-requirements will be based on generally accepted customer standards

      3. The design process will include many iterative steps of "design – make – test – decide"

        a. These cycles will be at both the component/subsystem level and at the system level

4. Concept selection will be based on many "weighted" criteria
   a. Use, manufacturing, safety, maintenance, and disposal
   b. Concept selection is an imprecise process carried out around a conference room table
      i. "Votes equals opinion time salary" – seniority is equivalent to wisdom?
         1. The front-wheel-drive Camaro
            a. Conflict between engineering and marketing and management
         2. Hard or soft seats in a Caprice
            a. An opportunity for a psychophysical investigation
         3. The ACCESS Car
            a. A marketing mistake?
            b. An engineering opportunity
            c. A Human factors driven process
         4. Intermediate shaft installation
            a. Transatlantic disagreement
         5. Battery location (engine compartment or trunk)
            a. Engineering and manufacturing conflict
         6. The proliferation of warnings
            a. Conflict between human factors, marketing, and legal staff
      ii. *Manufacturing and Production Design*
         1. DFM/DFA (Design for manufacturing and assembly)
            a. Aimed at productivity, quality, and worker comfort and convenience
               i. Access, easy targets, force, posture, and fastener repetition reduction are the general aims
               ii. See "Tight Targets Take Time" handout
            b. Cars with pressure for a low cowl height will create engine compartment packaging challenges, which, in turn, lead to accessibility problems in assembly and maintenance
            c. A decision to sequence the seat installation after the doors have been installed can lead to longer cycle times, mutilations during the seat transfer into the vehicle, and difficult access for installing the seatbelts and the seat secure bolts.
         2. Major allocation of function decisions
            a. Mechanization and automation
               i. Articulating arms are very useful for heavy components/subassemblies
                  1. Often found tied to a pillar for intermediate-weight components
                     a. The job is possible without the arm
                     b. The job may be faster without the arm
                     c. The repeated load may give rise to the injury

  ii. Robotic undercoat and paint spraying is the norm. However, robots cannot easily access certain inside-facing areas such as the bottom of the doors which need to be painted by human operators who have to sustain awkward postures throughout the job cycle, giving rise to quality and injury problems.

   1. Cleaning the paint booths of residue is a largely residual manual task. Pulling grates is a difficult and physically stressful task.

 b. Tool selection

  i. Threaded fasteners are usually torque controlled. Task completion often induces a stressful torque reaction, giving rise to the injury. This is sometimes reduced by a torque bar but alternative technologies such as hydraulic/pneumatic/electric pulse tools can remove the torque reaction problem with no loss of quality (torque control).

  ii. Inline, pistol grip, or right angle tools can sometimes be used to allow more convenient arm postures, depending on the amount of torque and the location of the fastener. Inappropriate tool selection can cause discomfort and injury.

  iii. Tools may be supported by balancers but these may interfere with task access and so may be discarded by the operator.

 c. Modular design for model differentiation/subassembly content

  i. A major trend to increase module content, thus reducing the final assembly operations

   1. Steering columns include lighting, windshield, cruise control

    HVAC, entertainment, and navigation functions

    a. The module becomes heavy and awkward

    b. The residual intermediate shaft (between the steering column and steering box) installation is a major source of difficulty for the operator

   2. Windshield wiper motor and brake booster install

    a. These two components are hard to reach in the center/bottom of the engine compartment, respectively.

       b. A task can be made easier by product design (for the windshield wiper motor) bringing it out the board and assembling the brake booster module on a different station on the production line.

       c. A spare tire in the bottom of the trunk is both difficult for assembly and the driver who needs to change a wheel but a convenient place for packaging.

          i. Product design solutions include the mini spare, which is lighter and may be packaged at the side of the trunk for easier access.

          ii. Question – does the car owner need a spare tire or a cell phone to summon help?

  3. Seats/seatbelts

       a. Difficult install postures – seatbelts can be designed to be integral with the seat, given appropriate structural modifications, which in turn leads to a much easier assembly task.

  4. Layering and Fastener orientation for operator access

       a. This is a packaging and component design problem. Product design engineers should spend time on the line installing their own components to appreciate the line operator's difficulties.

 d. Vehicle carrier systems

    i. Many opportunities in manufacturing design to improve operator posture:

      1. Overhead rail – bring low and underbody work to an accessible height more convenient than working in pits

      2. Tilting – rotate the vehicle 45, 60, or 90 degrees to improve visual and hand tool access

      3. Skillets – operator adjustable vertical height

   iii. *Production operations*

    1. Production targets affect staffing levels and choice of shift system (1, 2, or 3 shifts)

    2. Shift work should be based on operational, human, and technological system needs
       a. Production targets
       b. Physiological and social requirements
       c. Access to equipment for maintenance
    3. Task design
       a. Will vary according to line speed and work area footprint
       b. Learning curves for job content and "experienced worker standard" assignment
       c. Balance of nonvalue-added work – carrying, walk back, etc.
    4. Stock/components/fastener/hand tool presentation
       a. Aimed at reducing error, nonvalue-added time, and improving comfort and convenience
    5. Work team design/task allocation
       a. Job enlargement/rotation and team assignment philosophy
    6. Rotation and enlargement strategies
    7. Quality, productivity, and safety monitoring
    8. Methods engineering
  iv. *Distribution and Sales* – class discussion/exercise of customer requirements and design
    1. Substantial human contribution
    2. Order management
    3. Transport
       a. Protection
       b. Long distance driving/railways/container ships
    4. Brochures, warranties, financing, insurance, taxes, incentives
    5. Salesperson employment strategy
       a. Incentives, salary?
  v. *Use* – class discussion/exercise
    1. Buyer/driver/passenger
    2. Roadways
    3. Traffic
    4. Taxes
    5. Garaging/parking
    6. Adverse environmental conditions
       a. Night and day, fog
       b. Snow and ice
       c. Heat and cold
       d. Traffic noise
       e. Vibration
       f. Road condition

      vi. *Service and maintenance* – class discussion/exercise
         1. Context of maintenance
           a. Tools
         2. Training of maintainers
           a. Support manuals
           b. Spare parts
              i. Distribution strategies
     vii. *Disposal*
         1. Green car
           a. Design/materials/manufacturing cost constraints
         2. Used car market
           a. Warranties
           b. Spares availability
  b. Concurrent engineering – a delivery opportunity for Socio-Technical System Design
      i. All life cycle stages, customers, and stakeholders need to be accommodated
      ii. Multiple overlapping steps
     iii. Feedback and iteration
     iv. Technical memory
      v. Evaluation
      vi. Aided by adhering to the discipline of the "Grammar of Design", including control and communication models
     vii. Aided by the use of concept mapping and activity cycle diagrams

7. **Handouts**
  a. The Grammar of Design
  b. The Purpose of Design
  c. Concept maps
  d. Activity Cycle Diagrams
  e. Paper Airplane Design Exercise

8. **Self-Test Questions**
  a. Describe the major components in the "Grammar of Design" concept (Section 2)
  b. What is a concept map? Draw one (Section 3)
  c. What is an activity cycle diagram? Draw one (Section 4)
  d. Describe the main elements of the Human Factors Cycle; give an example (Section 5a)
  e. Describe the design with a control model; give an example (Section 5b)
  f. Describe design with a communication model; give an example (Section 5c)
  g. Describe a product life cycle; give an example and elaborate on one of the stages (Section 6)
  h. Develop a classroom game, similar to the paper airplane exercise to demonstrate various steps in a product life cycle from the STS viewpoint (Section 8)

1. **The Tavistock Studies**
   a. Traced back to the studies of Trist and Bamforth (1951) of the introduction of new technology into deep seam Welsh coalmines.
      i. The original, largely manual, methods involved teams of coal miners doing a broad spectrum of jobs
      ii. The introduction of mechanical cutters resulted in job specialization and much less worker interaction
      iii. The result was lower productivity and inflexible processes that were vulnerable to subsystem failures
      iv. A compromise method reverted to team structure while still using contemporary technology; this resulted in improved productivity and worker satisfaction
2. **The Convergence Hypothesis**
   a. This hypothesis suggests that technology will dictate work organization
   b. The Tavistock Institute studies refuted this hypothesis and demonstrated clearly that new technology could and should be adaptable to different work structures.
   c. Studies of SE Asia textiles (spinning and weaving) also concluded that the technology was amenable to different/traditional work cultures in different SE Asian countries, although Japanese-owned companies were more prescriptive in their work structuring.
3. **Joint Causation**
   a. Systems consist of technological, environmental, personnel, and organizational subsystems that interact to satisfy the "voice of the customers"/ customer requirements
      i. The external environment (physical, social, economic, political) is normally not changeable
         1. Introduction of advanced large-scale farming methods in developing countries
         2. Advanced workplace safety practices may not be effective in countries with an undeveloped regulatory framework
         3. Introduction of energy-saving electric/hybrid vehicles before recharging infrastructure is developed
         4. Introduction of small, energy-efficient vehicles into a country accustomed to "gas guzzlers"
      ii. The other three subsystems (technological, personnel, and organizational) must be designed to be resilient to changes in the external environment through contingency allowances to mitigate the adverse effects of variation in the external environment
4. **Edwards (SHEL) Model of System Failure – Similar to/Forerunner to the Joint Causation Model**
   a. Elwyn Edwards described these subsystems graphically with his SHEL model
   b. The "Software" in Edwards model referred to "organization" and not to the "software" with which we are now familiar

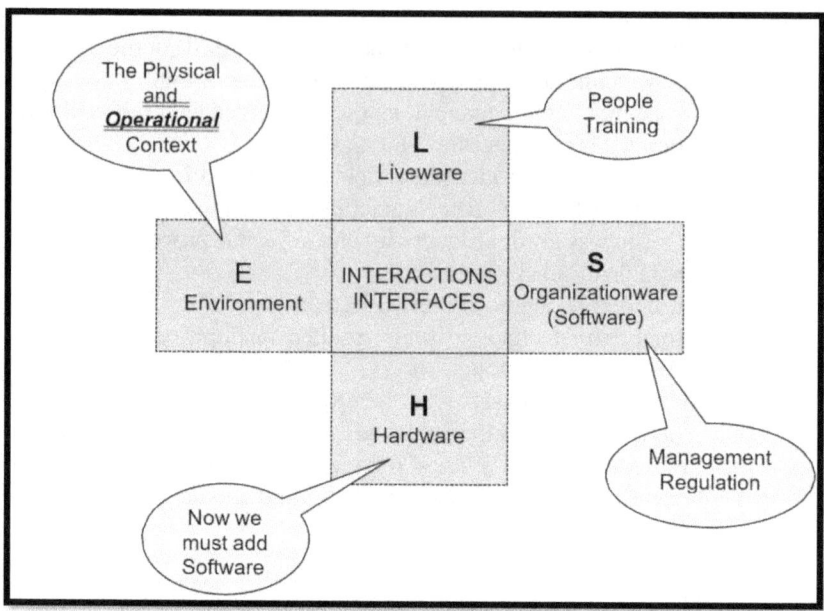

FIGURE 6.8 Systems interactions and interfaces.

    c. Hardware (plus software), Liveware, Operations (including management), Environment (including temporal and social factors). See Figure 6.8.

    d. 4Is - Interfaces, Interdependencies, Interactions, Interferences

        i. Complexity and failures occur with the 4Is which are in turn affected by the design of the particular subsystem

5. **Recognition of the Roles of the Technological, Human, and Operational Subsystems**

Since these early studies, various system analysis tools have been introduced that are broad in scope and address the Purpose and Scope of (Macro) Ergonomics.

The purposes of ergonomics are the same as the purposes of all processes and systems that involve human users.

Often, there will have to be compromises.

E3: Ease of Use, Effectiveness, and Efficiency
S3: Safety, Security, and Satisfaction

- **Effectiveness** – the product or service meets customer quality expectations
- **Efficiency** – productivity – optimal use of resources (people, money, materials, equipment, energy, etc.)
- **Ease of use** – human interaction with the product or service should be convenient, comfortable, and error-free
- **Safety** – the system (product, service) should not fail and cause harm to the user, associated hardware, the environment, or the organization

- **Security** – the system should be resilient to malicious or accidental interference by third parties
- **Satisfaction** – all users of the system should be satisfied with their experience and be motivated to continue to use the system with their experience and be motivated to continue to use the system
  a. Macroergonomics has three main components – management commitment, employee participation, and attention to microergonomics
     i. Management commitment – if top management does not actively buy into the challenges and opportunities that macroergonomics brings, then there will be no improvement in the outcomes.
     ii. Employee participation – if employees do not participate in system design and decision making, then valuable information will be lost and gains will be minimal.
        1. Note that this contrasts strongly with the scientific management philosophy promoted by Taylor.
        2. Note also that employees may not have detailed knowledge of the technologies that are introduced but they will have considerable knowledge of their interactions with these technologies.
     iii. All opportunities for microergonomics analysis and intervention
        1. Human physical and cognitive capabilities and limitations
        2. Human social needs
        3. Equipment and process interfaces
        4. Environment and work context
        5. Temporal demands on performance
        6. Job, task, and organizational structures, processes, and outcomes.
     iv. Case studies
        1. the joint UAW/GM manufacturing ergonomics process
        2. The OSHA meatpacking guidelines and proposed ergonomics standard
6. **Simple Descriptive Tools for System, Process, and Task Analysis**
   a. E3S3 – common system failure modes
      i. Identify the type of failure and the tradeoffs among the different outcomes
   b. Analyze the 4Is – Interfaces, Interactions, Interdependencies, and Interferences with regard to the various subsystems – Human, Hardware (and software), Organizationware, and Environment
   c. Address the 5Ws and a How – Who (By whom and to whom), What, When, Where, and How
      i. When a system fails, ask all these questions in order to be sure that all aspects of the failure are covered
   d. Ask "Why" at least five times for each of the above categorizations to identify a system failure root cause
   e. Apply these tools both reactively and proactively in the design

Example of the 5 Whys

Why did the car crash?

Because the road was icy
Because the brakes failed
Because the driver was asleep
Because the road was bumpy
Because the public requires 24/7 protection

7. **Failure Modes and Effects Analysis**
   a. This is a semi-quantitative process that is used for accident/incident analysis and during the system design process
      i. It can be used qualitatively but more advanced applications make use of quantitative data on historical failures to estimate failure probability and outcome severity.
      ii. This analysis tool can be used to address the various subsystems in Socio-Technical System Design – Personnel, Hardware, Organization, Environment, and their Interfaces, Interactions, Interferences, and Interdependencies.
8. **Reason's Swiss Cheese Model and HFAC**
   Two other similar models of system design and failure that can be applied to Socio-Technical System Design and system failure:
   a. Reason's Swiss Cheese Model
      i. This model suggests that system failures occur due to a successive failure of unsafe acts, preconditions, supervision, and the organization; it is possible to prevent the accident by addressing any of these components.
      ii. A top-down/bottom-up approach with management commitment and employee participation will be the most effective preventive strategy.
      iii. The model may be adapted to address failure modes of any aspect of a Socio-Technical System.
   b. HFACS (Human Factors Analysis and Classification System)
      i. This approach to system and accident analysis was developed over the past ten years, based on Reason's Swiss Cheese model mainly for application in the aviation industry, but it can be applied, with a small modification to the detailed questions, to any Socio-Technical System.
      ii. The model addresses
         1. The unsafe act itself – was it an accidental/occasional lapse or a habitual violation of the procedures
         2. The preconditions for the unsafe act – human, technological, organizational, and environmental
         3. Supervision – did the supervisory chain address habitual unsafe practices or initiate unsafe work

   4. Organizational influences – does the organizational leader-
      ship stress a safe climate with safety processes and training
      throughout the organization
   iii. See Charlotte, North Carolina crash analysis handout
9. **Related References**
   a. 4s, 5s, and 6s
   b. FMEA
   c. HFACS
10. **Case Study – Data Processing (Electricity Company)**
   a. System and environmental scan
      i. 1972 introduction of Help Desk for UK Midlands Electricity
         Distribution Company
      ii. "Fast Random Access Inquiry Devices"
      iii. Dealt with service start-up/discontinuation, billing, service, and
         repairs
      iv. No GUIs – just list and codes
      v. Hand-held phones
      vi. Many errors and customer complaints
         1. About the problem
         2. About the help desk
      vii. Long wait times and very variable service times
   b. Analysis
      i. Survey and interviews
      ii. Errors are mainly due to software bugs, poorly designed interface,
         absence of error recovery processes, and operator unfamiliarity
         with the system and interface
         1. In particular, supervisors lagged the call center operators in
            understanding the vagaries of the new system
      iii. Frequent users were effective and efficient and had less errors than
         occasional users but there were many occasional users
   c. Solutions
      i. Selection and training
      ii. Data capture on calls to provide statistical evidence of failure
         modes
      iii. Isolate and repair the software bugs
      iv. Improved interface medium term
      v. Provide headsets
      vi. Rotate operators and supervisors around other tasks to get broader
         system knowledge

EXERCISE
   a. Apartment complex design and management
   b. Apply Socio-Technical System analysis of an apartment complex
      i. Use concept mapping
      ii. Describe/sketch physical layouts (structures)

   iii. List/describe hardware (structures) – apartment structure, gas, water, sewer, electricity, cable, doors, elevators, stairways, surrounds, transport services, parking, playgrounds, swimming pool, etc.

   iv. Describe users – owners, tenants, managers, maintenance, and security

   v. Describe processes – purchase/rental, access/security, maintenance, and emergencies

   vi. Identify potential problems and positive experiences (outcomes, E3S3)

   vii. Develop data collection processes – surveys, incident reports, etc.

  c. Recommend changes (re-design) to structures and processes

   i. Address people (training, information support), hardware (costly), and processes (less costly) opportunities

   ii. Consider environment/context (unchangeable issues)

   iii. Institute continuous improvement process

     1. Suggestion plan

     2. Periodic inspections

     3. Regular management/tenant meetings

     4. Notice board/electronic complex communications

11. **Self-Test Questions**

The questions below are helpful to readers for self-test purposes.

a. Describe the original STS coal mining studies (1)

b. Describe the purposes of system design; give examples (5)

c. Describe the components of the SHEL model; give examples (4)

d. Describe the 5 Whys analytic process; give an example (6)

e. Describe Failure Modes and Effects Analysis; give an example (7)

f. Describe Reason's Swiss Cheese model of system failure; give an example (8)

g. Describe the major components of HFACS; give an example (9)

h. Draw a concept map to describe the factors to be considered in the design of an apartment complex (13)

1. **Macroergonomics Methods/Participatory Ergonomics**

a. Quality of work-life programs

   i. Companywide programs involving broad departmental representation to address work content and organization, quality, safety, facilities (cafeteria, fitness, parking, etc.) also self-help

b. Self-directed work teams

   i. Semi-autonomous production teams with free range to assign work among themselves

   ii. Contribute also to product design, process, quality, productivity, and safety issues

c. Cross-functional teams

   i. Usually made up of representatives from specialist departments focused on issues (usability, cost, design) related to a particular product

      ii. May also address safety

     iii. May be set up as ad hoc, limited-time teams, to address particular problems, such as poor supplier quality, in the process damage

  d. Product development teams

      i. Developed during the design process with representatives from marketing, design, engineering, manufacturing, production, and human factors/ergonomics to address the needs of all constituencies during product development

        1. May include employee representatives from the manufacturing plant

      ii. Also called Design for Manufacturing/Design for Assembly teams

  e. Quality circles – flexible, often problem-oriented, teams

      i. Introduced in the Japanese automobile industry to address the real and perceived quality problems with Japanese products in the early 1980s

      ii. Usually cross-functional teams, including line workers

     iii. Had an enormous impact on vehicle quality

     iv. These were management initiated team activities with employee involvement

  f. Participation

      i. Describes the involvement of production workers in "extra production" activity

        1. Participation in manufacturing/production operations/process design

        2. Participation in product design

        3. Participation in training

        4. Participation in quality and productivity discussions and interventions

        5. Participation in safety programs

        6. May be top-down or bottom-up

          a. Management designed and managed teams

          b. Teams based on union contracts, jointly managed by union and management

  g. Mechanisms of participation

      i. Standing committees – related to routine issues or focused work groups to address problems

      ii. Ad hoc and problem-focused teams

     iii. Visual controls

        1. Display of outcome data related to quality, productivity, and safety

        2. Andon chord – the ability and responsibility of any worker to stop the line if any form of problem arises, such as with component quality, safety, or inappropriate behavior

2. **STS Investigation Methods**

  a. Consider data accuracy (no bias) and reliability

  b. Field studies

    i. Realism – studying actual people at work has considerable face validity

   ii. Lack of control of context and independent variables

  iii. Observer effect – the process of observation may affect how the operator does his job

      1. Case study on end-of-press line staffing

         a. Led to $2B strike!

c. Field experiments

    i. Greater control than observational studies – applies formal experimental design methods – control of independent and concomitant variables

   ii. Very difficult to implement in practice

  iii. Realism – retains the face validity of field observation studies, however, there may be bias due to subject/operator favoring one or other of the experimental manipulations

      1. An example would be parallel departments/groups with different processes/equipment but the same objectives

      2. Another example would be the experimental introduction of job rotation or manual materials handling aids (hoists, arms, Cobots, etc.)

   iv. Greater observer effect – may influence behavior and attitudes

      1. Hawthorne effect?

      2. Example - Electronic data processing "pilot team"

         a. Medicare data processing/claims handling

         b. Dissatisfaction with computer system reliability

         c. Keystroke monitoring for productivity

         d. General productivity, quality, and morale problems

         e. Work tasks are subdivided and simplified

         f. Work layout – like a classroom with the supervisor at the front

         g. Cross-functional team implementation

            i. Cross-training period was needed

           ii. Office rearrangement was necessary.

         h. Considerable support from the highly selected pilot team members

         i. Some job specializations remained by choice of the team

         j. Attention to microergonomics issues – carpeting and computer system upgrades.

         k. Much better within team communication to deal with problem cases

         l. Great improvements in productivity, quality, and job satisfaction

         m. Pressure to expand the concept before the trial period was over

   v. Example – early introduction of in-vehicle navigation systems

      1. Technology push, marketing, and management support

      2. HF distraction concerns – conflict with management

        3. Contemporary technology with better interface and functionality contains distraction warnings

        4. Similar to contemporary cell phone/texting issues while driving

          a. Microergonomics issues of driver spare mental capacity

          b. Hands-free red herring

d. Survey methods

    i. Questionnaires

        1. Limit length to maintain the attention of the subject

        2. Focus on particular issues, don't be too broad

        3. Sometimes implement more general periodic QWL surveys, e.g., every six months (system satisfaction scan)

        4. Unambiguous response selection

        5. Allow space for explanatory comments

        6. Population sampling – many dangers of bias if surveys are not designed properly

        7. Response bias – e.g., only people favorable to the subject may respond

        8. Nonpunitive/anonymous incident reporting

          a. Widely used in the U.S. aviation industry to address safety violations

          b. Sometimes mistrust in the use of the information results in under reporting

e. Interviews

    i. Structured questioning – the investigator must have a template or pre-arranged set of questions

    ii. The respondent must be assured of confidentiality/anonymity and be allowed to not answer particular questions or discontinue the interview at any time

    iii. Interviewer bias may creep in depending on the interviewer's style

        1. Example of the relative success rates of different interviewers in a Hong Kong biomedical survey

f. Focus groups

    i. Mediator training is needed to assure

        1. Equal opportunity for all the participants

        2. That the conversation does not wander off subject

        3. That individuals do not use the platform to further their own agendas

    ii. Not more than ten participants per group, otherwise the discussions become unwieldy and difficult to manage

    iii. Examples

        1. Design of remote entry systems for cars

          a. Brainstorming suggested very many applications – car "brains" in a card

          b. Limited feature remote entry system introduced

   2. Use of advanced synthetic terrain display technology in aviation
      a. Considerable technology push
      b. Naïve realism possibilities
      c. Need for "off" mode unless needed (strong focus group message)
      d. Small incremental improvement over simpler systems supported by formal laboratory investigations.
   3. Head-up displays in cars
      a. General engineering push for high-content displays
      b. Focus group support for high-content displays
      c. Laboratory experiments demonstrated considerable cognitive capture/distraction effects
      d. Low content implemented in upscale vehicles.
      e. Market did not respond positively.
g. Laboratory simulations
   i. Allow role-playing by group participants to address particular operational or social issues
      1. They can be fun and productive/informative
      2. They can also be counterproductive if not well controlled
      3. Example – sexual harassment sensitivity training for managers and employees
   ii. Simulations may also involve physical mock-ups of product, equipment, or workplace of interest
      1. Example Styrofoam mock-ups of manufacturing workplaces/equipment
      2. Bamboo mock-ups of Hong Kong Mass Transit Railway passenger compartments and ticket turnstiles to address passenger movement issues
   iii. Tabletop simulations using scaled pieces can be used interactively to address workplace layout issues
   iv. Interactive simulation software, especially with animation, may be used to explore alternative resource allocation strategies, procedures, and layouts
      1. Example – discrete event simulation studies of emergency evacuation from transport category aircraft (see Handout)
         a. Focused attention on blockages, layout, passenger behaviors, cabin crew training, and behaviors
         b. Integrated with staged physical simulations/demonstrations

3. **Work System Structures**
   a. Woodward (1965) studied 100 large companies to detect reasons for relative success
      i. Variable managerial levels (2–12)
      ii. Variable span of control (2–12 at top and up to 90 at the first supervisory level

      iii.   Effect of technology on organizational structures
          1.   Unit (craft work), batch, mass (automobile manufacturers), or process (e.g., gas and oil) production
      iv. Complexity brought vertical differentiation
          1.   Increased number of administrative staff
             a.   Successful companies had moderate vertical differentiation
                i.   Unit – 3, mass – 4, process – 6 levels
          2.   Ratio of administrative to production staff increased with complexity
      v.   Indicators of success
          1.   Unit companies, low complexity, low staff-to-production ratio, and small span of control for first-line supervisors
          2.   Mass production – narrowly defined jobs, high formalization (standardized jobs), and centralization (decision-makers at the highest levels– little bottom-up communication)
          3.   Process production – high vertical differentiation (many levels) supervisors had wide spans of control, relatively low formalization, and centralization – (decision-making resident in lower levels)
      vi. Complexity is often solved by formalization and high vertical differentiation/hierarchies
          1.   Spatial/Geographical separation increased the need for the development of parallel committee structures which led to
             a.   Competition between line management and committees
             b.   Communications difficulties, especially bottom-up communication.
  b.   Technology-centered production (automation) led to high vertical differentiation – hierarchies and associated
      i.   Productivity and quality gains
      ii.   Increased focus on maintenance
      iii.   Workforce retraining/redundancies
  c.   Machine Bureaucracies (p60)
      i.   Usually very prescriptive work
          1.   Inflexible
      ii.   Taylorism
          1.   Work simplification for quality – very short job cycles – easily learned
          2.   Management's responsibility became one of training, monitoring, and problem-solving
          3.   Line workers trained to – "experienced worker standard" – for rate setting
          4.   Line balance – equal workload to keep the production line running smoothly
          5.   Supervision became impersonal – management must be "scientific and objective"

    iii. Industrial Engineering
1. Development of standard times
    a. Using systems, such as MTM and Work Factor
2. Modified workplace layout to reduce nonvalue-added movements of people and materials
    a. Tools were hung on balancers close to the job
    b. Materials/components baskets were brought as close to the line as possible
    c. Conveyors moved the product through the assembly process
3. Job specialization became the norm with job choice being based on seniority in unionized organizations
4. The effect of IE methods was increased and more predictable production rates and more accurate product costing
    a. Later, greater refinements were made with lean and agile manufacturing, and work cell development and processes such as SIX SIGMA based on statistical process control
5. In the mid-1980s, ergonomics was introduced to reduce the adverse effects of repetitive work
    a. Ergonomics contributions included workplace arrangement to reduce awkward postures and movements, job aids to reduce forces, and job rotation to reduce repetition
6. Centralization became the norm in mass production industries with top-down decision-making ("votes equals opinion times salary")
7. Formalization – increased levels of standardized work often leading to very short-cycle times – just a few seconds in component manufacturing
    a. Example – contemporary meat packing
d. Knowledge-centered organizations or professional bureaucracies (p62) have greater horizontal differentiation and are divided up into technology centers
1. Usually highly trained individuals – college degrees in engineering or technology
2. Examples of professional bureaucracies include Hospitals, Universities, and Research organizations, especially within the government
3. Less/minimal top-down control, management deals with policy and resource allocation, and professionals are the technical decision-makers
4. Sometimes professionals become difficult to manage due to their confidence in their own value to the organization
e. Adhocracies (p63)
    i. Matrix organizations
1. Seen as a method to improve flexibility with new programs
2. Organization divided into technology and program centers
3. Technology centers supply necessary skills to programs

        a. For example, robotics, fastening, product engineering, manufacturing, industrial engineering, materials, human factors, paint, and welding

    4. Programs have majority of funding

    5. Technology centers retain some funding for R&D

    6. Employees have the problem of two bosses – their home technology center and the program to which they are assigned

    7. Managers have the problem that employees may not balance their loyalty equally

  ii. Product development teams were introduced during the design process with representatives from marketing, design, engineering, manufacturing, and production to address the needs of all constituencies during product development

  iii. Also called Design for Manufacturing/Design for Assembly teams

f. Degree of skill/professionalism

  i. Work can be described according to the following general categories

    1. Production line – short-cycle work and minimal training

        a. Automobile assembly, textiles

    2. Craft/skilled trades work, longer training, mainly rule-based

        a. Maintenance, plumber, electrician, carpenter, etc.

    3. Creative work, skill-based, but with experimentation

        a. Arts and crafts, acting, and music

    4. Knowledge work – deductive reasoning and problem-solving

        a. Medicine, engineering, and law

    5. Investigative work – inductive reasoning and research

        a. University research

4. **Demographic factors** have a considerable effect on the type of work that an individual performs

a. Age

  i. Child labor in developing countries with minimal education systems, routine work

  ii. Work experience produces efficiency but less versatility

  iii. Work pace diminishes with age

  iv. Increasing problem in developed countries – graying of the workforce

b. Sex

  i. Traditional roles/jobs – e.g., textiles, agriculture, and homemaking

    1. Varies with country

  ii. Move into management/glass ceiling – females generally lag their male counterparts in both mechanical and professional bureaucracies

  iii. Diversity and equal employment opportunity legislation is being pursued aggressively in the United States and Western Europe to level the hiring and salaries of females and minorities

    1. Affirmative action programs

    2. Quota programs

   c. Ethnic origin
      i. Immigrant workers
         1. A reality in most industrialized countries
         2. Cultural and language differences sometimes alienate the community
         3. Training/language challenges lead to immigrant workers being offered only menial and lower paid jobs
         4. Development of ethnic "ghettos"
         5. Turnover – workers return to their home countries
         6. Line "unbalance" to bring new workers up to speed is applied as a pragmatic process in some short-cycle assembly jobs
         7. Developing countries with the ready available trainable workforce from rural regions
            a. China, South East Asia, and Mexico
         8. Lower paid jobs
            a. Assembly, service operations, cleaning, and agriculture

5. **Environmental Factors**
   a. Socioeconomic
      i. Companies locate new plants close to the available/experienced workforce
         1. Automobile industry in the United States
         2. Greenfield plants
      ii. New plants may be launched where there is an abundance of trained or trainable labor
         1. Japanese textiles, electronics, and plastic plants expanded throughout SE Asia
         2. Japanese automobile manufacturing plants spread to the United States
         3. The U.S. and European components manufacturing spread to Mexico, Eastern Europe
         4. The U.S. and European IT facilities moved to India
         5. The U.S. and European manufacturing activities moved to China and Korea
         6. Now China, India, and Korea dominate these industries and build plants in Western Europe and the United States to be closer to the market
     iii. Problems arise for communities following the undulations of the economy
         1. Whole towns may suffer if they rely on a single company that closes
            a. Flint, Pontiac, Nummi
   b. Educational
      i. Industry relies on the educational system to provide sufficient numbers of professional and technical job candidates
      ii. Establishment of local R&D centers around universities and centers of expertise

        1. Silicon Valley in California
        2. Research Triangle in North Carolina
        3. I75 corridor in Michigan

  c. Political
     i. Unions
        1. Originally focused on health and safety and working conditions
        2. Now very much involved in salary negotiations
        3. Also, unions argue strongly for participation in all levels of the organization
     ii. Seniority is the main mechanism for job choice and wages
        1. "Incompetence is no reason for dismissal" – Peter Sellers movie about a Japanese car plant in the United States
        2. May interfere with the team structure, job rotation, and job enlargement
    iii. National and International work standards and regulations
        1. ILO – International Labor Office promulgates employment and safety standards
           a. ILO is a strong proponent of participation in the workplace
        2. ANSI – American National Standards Institution promulgates national product and manufacturing standards
        3. Lesser standards in developing countries

  d. Cultural
     i. Nationality/ethnic differences affect organizational processes
     ii. Management styles
        1. Autocratic – top-down, rule-based
        2. Collaborative/participative style either top-down or bottom-up

  e. Legal/Policy
     i. National laws and standards vary considerably from country to country
        1. Safety laws are generally less stringent in developing countries
           a. Mortality and morbidity statistics reflect these differences
        2. Diversity – equal access and employment opportunity is a major issue in the industrially developed countries
           a. Addresses discrimination on the basis of sex, age, ethnicity, sexual orientation, disability, etc.

6. **Case Study**
  a. ACCESS Car (See handout)
     i. Transportation for the elderly
     ii. Physical access
        1. Seats, seatbelts, step over, storage, controls
    iii. Informational access
        1. Instrument cluster, navigation, lights, entertainment, communication
    iv. Social/operational access
        1. Neighborhood car, Emergency communication system, lease, and rentals

     2. Agent-broker system
        a. "Let me tell you about my grandchildren"

6 Us (and 2Ms)

7. **Self-Test Questions**

a. Describe, with examples, five forms of participation design (1)
b. Describe four kinds of STS investigation methods (2)
c. Describe Woodward's categorization of work systems; give examples (3)
d. Describe a Machine Bureaucracy; give an example (3c)
e. Describe a Professional Bureaucracy; give an example (3d)
f. Describe a Matrix organization; give an example (3e)
g. Describe some demographic factors that should be addressed in Socio-Technical System Design (4)
h. Describe some "environmental" factors that should be considered in Socio-Technical System design (5)
i. What factors should be considered in designing a transportation system for the elderly? (8)
j. Describe the 6Us method of assessing system usability (9)
k. What factors should be considered in designing an Internet-based Social Network for ergonomists? (9)

1. **Review of Systems Theory and Characteristics**
   a. Systems
   b. Processes
   c. Requirements
   d. Specifications
   e. Verification
   f. Validation
   g. Life cycle
   h. Concurrent engineering
   i. Purposes/outcomes
2. **Work Types**
   a. Process control
      i. Monitoring tasks require system knowledge, vigilance, and sustained attention to detect discrepancies/variances in process behavior
      ii. Must have knowledge, rules, and skills to respond to emergencies
         1. Rules address immediate tasks, such as "remove the power source"
         2. Skills require experience and practice
            a. May be obtained through simulator training
         3. Typical tasks are in petrochemical processing and aviation
   b. Craftwork
      i. Work such as skilled trades (plumber, electrician, carpenter, farmer, forester, model maker, and artist) where each job is somewhat unique and requires adaptability of the person based on fundamental training/skill and broad contextual experience.

    c. Job shops
        i. Low throughput or one-off production facilities staffed by a team of specialists
       ii. Varied levels of job skill breadth and system flexibility
      iii. Typical jobs include construction and large equipment manufacturing
      iv. Also found in repair facilities
    d. Short-cycle/production line work
        i. Typical production line work – automobiles, computers, toys, textiles, food processing – with cycles ranging from a few seconds to a few minutes
          1. Task choice by seniority
          2. Job rotation and enlargement opportunities to reduce physical stress, create a more flexible workforce, and increase understanding of the larger product picture
       ii. Also found in routine tasks like Air Traffic Control, Retail Check out and "Help Desks", or call centers
          1. Job cycle may last a few seconds to a few minutes
          2. May require skill and rule-based decision-making

3. **Job Design Methods**
    a. Job rotation
        i. Purpose is to relieve physical and cognitive stress and to increase the flexibility of the workforce by broadening the skill sets.
       ii. Applied in high-frequency assembly (disassembly) work and customer service jobs.
      iii. Operator rotates around five or six tasks every hour or so
      iv. Sometimes rotation may be quicker – every few minutes or slower – every few days.
       v. See the handout on "The Case for Job Rotation".
    b. Horizontal job enlargement
        i. Increasing the duration of a production job cycle by increasing the number of elements
          1. Cycle time may be 5 minutes to an hour
          2. Each operator takes a production/service unit through a series of workstations/operations
          3. Increases knowledge/flexibility of the workforce
          4. Reduces the repetition component of physical and cognitive stress
    c. Vertical job enlargement
        i. Allows operators to participate in tasks other than the direct production/service tasks such as:
          1. Design of tasks, including workplace layout, equipment, tools, methods, and product/component design.
          2. Assignments among the workgroup, including rotations and individual assignments.
          3. Quality and productivity – "there is always a better way"; line operators have a unique insight into the product and process.

      4. Safety – ad hoc and statistical evidence may be applied to the reduction or severity of acute and cumulative morbidity.

  d. Work cells

    i. A collection of machines, operations, and operators around a small area in contrast to the linear production line

      1. The purpose is to improve productivity by reducing the distance and time of product and component handling, and by reducing the in-process storage of products.

      2. Also, quality improvements may be achieved by giving individuals or groups of workers responsibility for more machines/operations.

      3. Work cells may be staffed by individuals or groups.

      4. Similar methods and time analyses and standards are applied to those used in production line work.

4. Socio-Technical Systems (STS) Analysis

  a. Vision, mission, principles, and policies

    i. A Socio-Technical Systems analysis should begin with a scan of the high-level vision, mission, principles, and policies of the organization.

    ii. This scan can identify gaps in the system design such as the organization's concern for employee remuneration, health, safety and well-being, or operator-induced continuous improvement (Kanzei Engineering).

    iii. The scan will also identify the external environment including market, labor, unions, plant locations, suppliers, raw materials, and applicable regulations.

      1. Concept mapping is a good tool for this analysis.

    iv. Policies such as participatory practices should be identified in this high-level scan.

  b. Environmental scan

    i. Descriptions of the physical, temporal, geographical, social, economic, market, competition, regulatory, and demographic context in which the organization exists.

    ii. Note that this context is usually unchangeable so the system design must adapt to both benefits from the environmental context and withstand the adverse effects of the context.

      1. Examples include developing organizations close to resources (people and raw materials) and markets.

      2. Another example might be the avoidance of potentially catastrophic environmental influences – severe weather, earthquakes, and political unrest.

      3. Japanese and European car makers develop plants in the United States – close to the markets to reduce transportation costs and to give the impression that the vehicle is "made in the USA" despite the fact that many components come from overseas and the revenue goes overseas.

    c. Organization scan
      i. This is a more detailed scan of the organization's subsystems
      ii. Descriptions of the products/services, manufacturing processes, and quality management
      iii. Description of the organization structure – divisions, departments, interdependencies, technical, and administrative support subsystems
      iv. Descriptions of the personnel subsystem – hierarchies, supervision, span of control, work assignments, and practices
      v. Description of the statistical subsystem measuring process behavior and outcomes – materials, products, equipment and tooling, quality, safety, and costs
      vi. Description of the environmental subsystem – suppliers and customers, locations, physical environment, transportation, community, demographics, regulations, etc.
    d. System and process analysis
      i. Technology (Hardware and software) – a detailed description of the technology and its functions
      ii. Liveware – a detailed description of the particular work assignments, job design practices, assignments, selection, and training
      iii. Organizationware – a detailed description of the flow of materials, products, people, and information through and around the organization
        1. Use activity cycle diagrams
        2. Also, address shift work and job rotation issues
      iv. Interactions, Interdependencies, Interfaces, and Interruptions – a description of the relationships among technological, personnel, environmental, and organization subsystems
    e. Product/service scan
      i. Product quality
        1. Analysis of the incoming materials quality and outgoing product quality and the quality audits from different parts of the organization
      ii. Productivity
        1. Resource (equipment, people, power, and money) evaluation overall and in different departments
    f. Safety and security scan
      i. Injury/illness statistics overall and by the department
      ii. Survey of hazardous operations using FMEA
      iii. Evaluation of the risk of materials, product, tools, and information theft
    g. Work satisfaction scan
      i. Using interviews, focus groups, and employee surveys to detect perceived and actual shortcomings of the technological, personnel, organizational, and environmental subsystems
      Company structure, process, and outcome analyses

h. Specify organization structural design
    i. Describe department hierarchies, span of control, geographical separation, technology, personnel, and environments/contexts
    ii. Use concept mapping
i. Define process flows and outcomes
    i. Describe interdependencies, interfaces, interactions, and interferences
    ii. Use activity cycle diagrams to show resource allocation and process branching logic
    iii. Establish realistic targets and goals regarding outcomes
        1. Use the evaluation matrix (see below)
j. Describe micro process inputs/resources, process logic, and outcomes
    i. Use activity cycle diagrams
k. Collect and analyze variance data (see Figure 6.9).
    i. Use evaluation matrices for snapshot
    ii. Develop control charts for temporal trends in outcome variance data
    iii. Distinguish special and common causes of variances from targets/goals
l. Profile analysis using the common currency
    i. Develop metrics for all operations (inputs/resources, outcomes)
        1. Use E3S3 – effectiveness (quality), efficiency (productivity/resource utilization), ease of use, safety, security, and satisfaction)
        2. Sample metrics – reject rate, throughput, operator ratings, injury rate, security reports, and worker satisfaction ratings

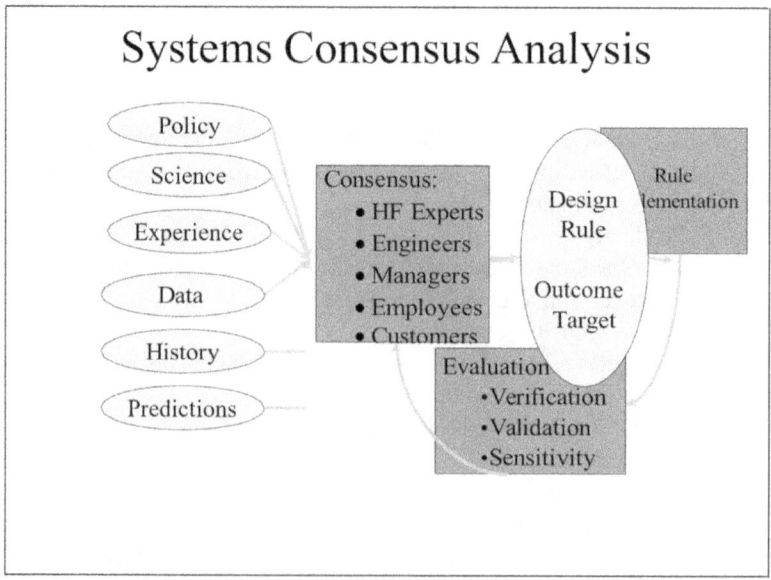

FIGURE 6.9   Systems consensus analysis.

    ii. Convert metrics to (pre-weighted) common currency scale using the consensus process

    iii. Enables all stakeholders (management, employees, customers, shareholders) to view the "forest as well as the trees"

    iv. Score by COUNTING – number of reds, yellow, and greens

        1. Possible decision rule – one red or three yellows is a show stopper

    v. Note that more specific process stages and outcome metrics may be used

    vi. Experience with similar analytical processes (Quality Function Deployment, ISO 9000) demonstrates the danger of too detailed analyses in that the analytic process can become cumbersome, time-consuming, and of diminishing utility.

        1. It is suggested that an upper limit for any evaluation matrix should be not greater than 10×10.

  m. Index of organization performance

    i. Can be calculated using common currency and element weighting

    ii. Can be the simple averages as shown above, given that the "weighting" was applied to each variable by consensus in the original process/outcome assessments

        1. Note that different constituencies will attempt to weigh the different evaluation criteria differently, that is why the consensus process should be used in metric development

    iii. Benchmarking with similar companies using comparable metrics is informative

  n. The index of performance can be further assessed by a risk/consequence process in which each process failure mode score (high in the example) reflects the probability of failure and the consequence of this failure mode in terms of damage to people, equipment/technology, the environment, and the organization is assessed using a similar consensus process or quantitative evidence where available. A typical risk/consequence matrix is shown in Figure 6.10:

    i. Define mission, purpose, and outcome targets

    ii. Develop alternative concept designs

    iii. Evaluate context/environment of use

    iv. Describe organizational context

    v. Define functions and processes

    vi. Evaluate user population characteristics

    vii. Evaluate failure modes/probabilities/consequences (FMEA)

    viii. Allocate functions between people and equipment

        1. Human roles

           a. Refer to Fitts lists

           b. Should be designed to be stimulating and make good use of human capabilities

           c. Not "left over roles"

           d. Machine supervision

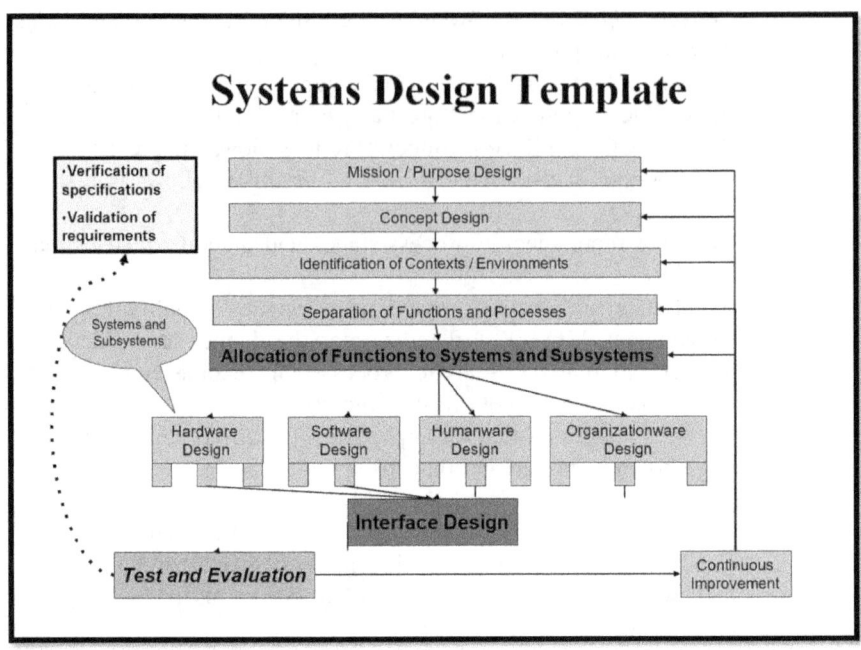

FIGURE 6.10   Systems design template.

  2. Hardware/software roles
   a. Refer to Fitts lists
   b. Routine
   c. High force, high repetition
   d. Hostile environments for human activity
   e. High precision
 ix. Develop subsystems
  1. Hardware
   a. Functions, Costs, reliability, service, and maintenance
  2. Human
   a. Functions, selection, training, and assignment
   b. Develop job/task scope, enlargements, rotations, shift systems, etc.
 x. Develop metrics, collect and analyze data, and develop visual controls using materials, information, production, and products addressing:
  1. E3S3
   a. Quality (Effectiveness)
   b. Productivity (Efficiency)
   c. Safety and Security
   d. Satisfaction, usability/ease of use
o. Address system life cycle
 i. Product, process, and production operations design
  1. Develop outcome targets (E3S3)

2. Identify and rectify failure modes early in the design process
   a. Catch the big/product design fish early

Sequential Evaluations

Product Design
  Process Design
    Production Design
      Operations

  ii. Production operations
    1. Record and analyze outcome data
  iii. Use, maintenance, and disposal
    1. Identify and measure outcomes (E3S3) for each activity
    2. Analyze outcome data
    3. Identify failure modes using FMEA, 5Ws, 5 Whys, etc.
    4. Rectify product, manufacturing, and production operations design and operations root causes
  iv. Develop interventions regarding production, manufacturing, product/component, operator, organizational, and environmental factors
    1. Assess cost-benefit of interventions
  v. Implement redesigns

5. **Case Study 1 – Digital Library**
   a. Identify discrepancies/variances
      i. Identify relationships and root causes of variances
      ii. Develop a variance control matrix
      iii. Identify human, technology, organization, and environmental contributions to variances
   b. Identify inputs and sources – authors, reports, books, papers, lecture notes, etc.
   c. Carry out internal controls for content management
   d. Identify legal issues regarding ownership and copyright
   e. Address technology alternatives
      i. Hardware, software
      ii. Identify special skills needed for this technology
   f. Develop delivery medium
      i. Internet
      ii. Web page design
      iii. Access control
   g. Identify customers

6. **Case Study 2 – Automobile Manufacturing**
   a. Construct a concept map and annotated process flow diagram
   b. Product design for manufacturing, production, maintenance, use, and disposal
      i. Use the 6Us analysis method

c. Process scan
  i. Body
     1. Robotic welding – more consistent than manual
     2. Difficult welds left to the operator
     3. Repetitive sanding to smooth welds
     4. Machine monitoring and maintenance – e.g., tip change
  ii. Paint
     1. Mainly robotic – more consistent than manual
        a. Difficult areas left to human operator
        b. Residual human tasks
           i. Inspection
           ii. Repair
           iii. Cleaning
           iv. Machine maintenance
  iii. Chassis and engine
     1. Mainly threaded fasteners
     2. Automatic torque control
     3. Often difficult access/orientations
        a. DFM/DFA opportunities
  iv. General assembly
     1. Often push fasteners
        a. Exterior and interior trim, wiring bundles
  v. Machine paced work
     1. Andon chords – stop the line for problems
     2. Visual controls
d. Shift work, work-related musculoskeletal disorders
e. Environmental scan
  i. Economic
     1. High-priced units
     2. Considerable external competition
  ii. Political environment
     1. Highly regulated industry
     2. Unionized – UAW
     3. Free international market
  iii. Organizational environment
     1. Very large company
     2. Vertically differentiated
     3. Multiple brand names
     4. Many component manufacturing and vehicle assembly plants worldwide
     5. Three major design and engineering facilities
     6. Matrix organization for engineering
     7. Dedicated design and marketing functions by brands
  iv. Social environment
     1. Plants are often the major employer in the city
        a. Pontiac, Flint, Hamtramck

     b. Substantial hardship following plant closure
     c. Jobs bank – continued employment at a reduced rate if laid off due to production cutbacks

f. People scan
    i. Salaried and hourly
    ii. Hourly workers represented by UAW
      1. Relatively good wage rates
      2. Production workers, maintenance, and materials are the main occupations
      3. Job choice by seniority (from date of hire)
      4. Aging workforce
    iii. Short-cycle work – 40 seconds to 5 minutes
    iv. Some plants have team structures with job rotation
    v. Shift work – flexibility depending on demand

g. Work design scan (General Assembly)
    i. Awkward postures – under and inside vehicle compartments
    ii. Threaded fasteners with torque reaction stress
    iii. Push fasteners with posture, force, and repetition stress, supported on balancers where convenient
    iv. Heavy components/subassemblies may or may not use articulating arms or, more recently, cobots
    v. Tuggers and forklift trucks are used for materials delivery

h. Work-related musculoskeletal disorders
    i. Upper limb tendinitis, carpal tunnel syndrome, back, and shoulder injuries
    ii. Reached epidemic proportions as a push for productivity increases line rates
    iii. Posture and force issues addressed by various ergonomics interventions
    iv. Repetition issues are addressed by work content negotiations and job rotation
    v. Major medical management strategy
    vi. Development of ergonomics teams in all plants
    vii. Development of a major proactive (DFM/DFA) program

7. **Self-Test Questions**
a. Describe with examples
    i. Craftwork (3b)
    ii. Job shops (3c)
    iii. Production lines (3d)
b. Describe with examples
    i. Job rotation (4a)
    ii. Horizontal job enlargement (4b)
    iii. Vertical job enlargement (4c)
    iv. Work cells (4d)
c. Describe five stages of STS analysis (5)

    d. Describe the common currency method of company process analysis (6)
- i. Describe five steps in company process analysis (6, 8). What are the two major components of Risk analysis? (6g)
- ii. Describe the classical system design model (7)

    e. Describe an example of process analysis from the automobile manufacturing industry (11)

    f. Describe an analysis of process analysis for the employment of teenagers with developmental disorders (12)

8. **Macroergonomics Success Stories**
   a. Metrics
      - i. Measurement of the success of macroergonomics intervention may address any or all process outcomes
         1. E3S3
            a. Effectiveness, quality, meeting customers' expectations
            b. Efficiency, productivity, optimal use of resources
            c. Ease of use, intuitive use of the product or service (with or without facilitators) with error avoidance and mitigation
            d. Safety – prevention or mitigation of damage to system or subsystems (technology, human, environmental, and organizational) due to failure of one or more subsystems or interactions between two or more subsystems
            e. Security – prevention or mitigation of system/process failure due to malicious or accidental adverse activities of human subsystem components
            f. Satisfaction – optimal satisfaction of all human stakeholders/customers
      - ii. Note that there may be tradeoffs among these outcomes
      - iii. Metrics best communicated through an outcome summary matrix
   b. E3S3 measurement
      - i. Use quality circles, focus groups, customer feedback surveys, and statistical analysis of quantitative outcomes to address each possible outcome
         1. use random/stratified sampling where possible
      - ii. Be aware of bias from different constituencies
         1. Note that different customers/stakeholders/participants may have different priorities
      - iii. Note that negative feedback (e.g., customer complaints) is more easily generated than positive feedback (e.g., customer loyalty)
         1. Consider E-bay seller feedback ratings
      - iv. Use specific measures of these general outcomes where available
         1. Effectiveness, quality, meeting customers' expectations
            a. Popularity in the marketplace – sales
            b. Customer complaints, warranty, and independent product ratings
            c. Life cycle measures

2. Efficiency, productivity, optimal use of resources
   a. Cost relative to the competition
   b. Direct resource measures – labor, energy, and materials (scrap)
3. Ease of use, intuitive use of the product or service (with or without facilitators) with error avoidance and mitigation
   a. Life cycle measures
      i. Manufacturing, use, maintenance
      ii. Use of need for facilitators
      iii. Forgiving – error trapping, redundancy
4. Safety – prevention or mitigation of damage to system or subsystems (technology, human, environmental, and organizational) due to failure of one or more subsystems or interactions between two or more subsystems
   a. Accident/incident measures – frequency and consequences
      i. Use common currency methods – Quantum Risk Analysis
5. Security – prevention or mitigation of system/process failure due to malicious or accidental adverse activities of human subsystem components
   a. Attractiveness as target
   b. Implications of security failure
   c. Incidence and severity of security lapses
6. Satisfaction – optimal satisfaction of all human stakeholders/customers
   a. The major management/collaborative challenge
   b. Multiple constituencies
   c. Tradeoffs
      i. Use common currency Risk Benefit Evaluation
      ii. Use collaborative/consensus approach to evaluation

c. Textbook
   i. Red wing shoe company
      1. Outcome problem – Work-related musculoskeletal disorders
      2. Administrative controls – Rotation, work cells, etc.
      3. Engineering controls – Addressed posture and force issues by design
      4. Evaluation – Major savings
   ii. Telecommunications – AT&T
      1. Problems – Workers' compensation costs
      2. Microergonomics interventions (low-hanging fruit)
      3. Engineering/Administrative controls
         a. Removed conveyor-paced work
         b. Job enlargement – work cells
      4. Reduced lost days' injuries and workers' compensation

    iii. Foodservice system
- 1. Productivity shortcomings
- 2. Extensive microergonomics improvements
  - a. Participatory approach
  - b. Engineering interventions – Workplace layout and interfaces
  - c. Method changes
- 3. Large productivity increases – increased sales

    iv. Petroleum distribution
- 1. Safety problems
- 2. Participation strategy
  - a. Worker-initiated microergonomics changes
  - b. Safety training
- 3. Safety culture changes
- 4. Reduced accidents/injuries
- 5. Cost savings

d. Quality of Work Life in Sweden
- i. Sweden is seen as a leader in participatory industrial processes
- ii. Government Quality of Work Life organization
  - 1. Many industrial interventions
- iii. Political cancellation of QWL institution by the new government

e. Ergonomics standard in the United States
- i. Introduced following specific activities by OSHA/Unions in meat-packing, automotive, and nursing home industries
- ii. Heavily supported by organized labor
- iii. Rationale was the high prevalence ("epidemic") of Work-related musculoskeletal disorders particularly as related to the back and upper limbs
- iv. Experts recruited by OSHA to develop standard
- v. NIOSH, University and Professional Society technical support
- vi. Proliferation of tools and rules
- vii. Rapid increase in number of ergonomists employed by industry
  - 1. Varied levels of training/certification
  - 2. Industry-sponsored conferences
  - 3. OSHA-sponsored town hall meetings
- viii. Bandwagon interventions by opportunists
  - 1. Backbelts, computer workstations, chairs, and tools
  - 2. Varied credibility/effectiveness
- ix. Opposition established in the U.S. Chamber of Commerce, National Association of Manufacturers, and Center for Office Technology (Computer Manufacturers)
- x. Standard Introduced by Democrats their end, canceled by Republicans (George W Bush) immediately after taking office
- xi. Seen as a shift in the locus of control to a bottom-up process using WRMDSs as the reason
  - 1. The truth lies somewhere in the middle – productivity/efficiency pressure leads to top-down short-cycle work. Interventions to

reduce posture and force factors, but repetition factors increase. Unions resist job rotation based on traditional seniority-based job choice.

2. Some psychosocial and political overlays confound the issue
3. An ideal Socio-Technical System participative approach could address the issue but historical attempts have had relatively short-lived success (Volvo, Saturn, etc.)
4. Management reluctant to share management responsibilities with employees

f. Saturn
   i. Partially successful for 15 years
   ii. Joint Union/Saturn management
   iii. Team structure with rotation
      1. Approach is sometimes defeated by medical restrictions
   iv. Poor shift system – one-week rotation
   v. No integration of design at plant location – still tied to GM engineering
   vi. Innovative design – entry-level vehicle, plastic panels, skillets in a general assembly
   vii. Product failed in the marketplace eventually
   viii. National union infiltration
   ix. Saturn becomes another division of GM
   x. Saturn line discontinued

9. **Formal Studies of Group Activities**
   a. Joint optimization – personnel and technology (plus environment and organization)
      i. Department performance improved by joint optimization (60/40)
   b. Formalization in planning
      i. Compared to down-structured processes with less formal approaches
      ii. No statistical differences
      iii. Compare with "Naturalistic decision making"
   c. Facilitation
      i. Improved participation by group members
      ii. Did not improve performance
      iii. Facilitated brainstorming groups produced more ideas
   d. Decentralized quality control
      i. Push QC down to the line, using Quality Circles and Statistical QC charts (visual controls)
      ii. Indicated that more elaborate 3D charts may not work – "naïve realism"
   e. Engineering design groups
      i. Product development teams
      ii. Concurrent engineering
      iii. Horizontal differentiation
      iv. Experimental investigations are not conclusive

        v. In practice, CE works, provided there is not too much formalization and opportunities for some independent subgroup activities.
           1. "Throwing the design over the wall"
              a. Battery location, Intermediate shaft, windshield wiper motor, and brake booster
      vi. Larger groups costlier than smaller groups
     vii. No significant value of group mediation technology
   f. Virtual group behavior and performance now show considerable success in distance learning
        i. Mediation needed
       ii. Grading reliability challenges

10. Future Directions of STS
   a. International companies
        i. Central responsibility for work system structures
       ii. Local responsibility for micro-process design
   b. Global labor forces
        i. Wage rates much lower in developing countries
       ii. Imported labor – Central America, Southern / Eastern Europe, India
      iii. Paternalism – community support
           1. Automobile manufacturing in the U.S. southern states
           2. Demise of Flint and Pontiac
   c. Transportation costs
        i. Components – many are made on high-rate production lines in third-world countries
       ii. Finished products – proliferation of container vessels
           1. Energy costs borne by customer
   d. Customer locations
        i. Changing from the United States and Europe to producer nations – China and India
   e. Environmental issues
        i. Energy sources
           1. Fossil fuel, sun, wind, and water
       ii. Global warming
           1. Transportation
           2. Manufacturing
      iii. Waste and packaging
      iv. Water
           1. Dams, reservoirs, and desalination
   f. Information technology
        i. Major effect on industry
           1. Computer control of processes
               a. Software reliability and safety issues
           2. Data capture and analysis
               a. Corporate now has all the data
                  i. For example, data capture in gas and oil well drilling

3. Internet
   a. Information access – enormously available and rich source on almost any subject
   b. Teleconferencing – Now commonplace for both synchronous and asynchronous communications with Skype, Facebook, etc.
   c. Telemarketing – from television to mobile devices to computer pop-ups.
      i. Marketing directed to individuals based on their web surfing behaviors
   d. Control of the Internet – censorship
   e. Distraction and productivity
   ii. International spread/subcontracting of software development, call centers, etc.
g. Litigation
   i. Product design failures
      1. Considerable rise in "usability" concepts before and after the event
   ii. Medical error
      1. Major design of product and procedure challenges
   iii. Consumer products
      1. Foreseeable misuse
h. International companies
   i. International standards Litigation
      1. Lead-based paint, toys, pet food
      2. Pesticides
      3. Bhopal
      4. Exxon Valdes
   ii. National standards and regulations
   iii. WHO and ILO influence on product and manufacturing safety
      1. Lead-based paint

CONCLUSIONS

STS design is plausible but threatens traditional organizational design models. It is sometimes seen as politically motivated by the disenfranchised. There is a danger of reversion to informal and formal hierarchies as the organization matures.

REFERENCES

Peacock, B. (2019). *Human Systems Integration*, Self-published manuscript, Fernandina Beach, FL.

Peacock, B. (2021). *Ergonomics Tools and Applications*, Self-published manuscript, Fernandina Beach, FL.

Peacock, B. (2020). *How Ergonomics Works,* Self-published manuscript, Fernandina Beach, FL.

7 Industrial Engineering in Aerospace Systems Design

INTRODUCTION

Industrial engineering (IE) interest in aerospace has been growing rapidly in recent years. The emergence of renewed focus on space-related research, exploration, and development has created new opportunities for IE and its sub-fields, particularly the field of human factors (Peacock, 2019, 2020, 2021). This chapter focuses specifically on human factors in aerospace, as a core part of the theme of this book, entitled Industrial Engineering in Systems Design. This development is conveyed through Badiru (2022), which is echoed below.

As a professional organization, industrial engineers often wonder what is needed to continue to make the profession of IE relevant in a fast-changing world. Former astronaut Buzz Aldrin said it aptly, "We explore, or we expire". Space exploration and industrialization offer one way of preempting human expiration. IEs can place a role in this regard. The rapid emergence of private space flights will expedite the creation and dissemination of new engineering and technology products that may very well help. Looking back on history, IE has contributed to the development, advancement, and sustainment of industries. Practical case examples of the diverse applicability of IE can still be seen in the auto and aerospace industries. The official definition of IE says it befittingly.

"Industrial engineering is a profession that is concerned with the design, installation, and improvement of integrated systems of people, materials, information, equipment, and energy by drawing upon specialized knowledge and skills in the mathematical, physical, and social sciences, together with the principles and methods of engineering analysis and design to specify, predict, and evaluate the results to be obtained from such systems".

Systems, human factors, operations management, energy management, new product development, novel management, optimization, simulation, ergonomics, coordinated design, and so on are all within the purview of IE. Space (as in outer space) is an emerging platform that is opening up to leverage the benefits of IE. Nations around the world aspire to become active players in Space. Those already working and playing up there are not relenting. They are all looking for new methodologies of efficiency, effectiveness, and productivity, which are the bastion of IE. We should not just sit and watch the new frontier opportunities fly or drift away. We should seize the opportunity now and position IE principles as what is needed both on Earth and in Space. The government of the United States saw the Space vision and created a whole new military arm dedicated to Space operations. The U.S. Space Force, carved out of the traditional U.S. Air Force, is now making waves up and down the

DOI: 10.1201/9781003328445-7

spectrum of the galaxy. Eventually, Space Industrialization will emanate from all these efforts. Even many Earth-based efforts are targeting Space sustainment. IE should step up and step forward to lay a legitimate claim to solidifying space industrialization through IE methodologies. Why, Where, When, and How we do it should enter our immediate scope of deliberations and strategies. One key strategy that IE can leverage is the development of new space-themed curricula. Space engineering, as an academic track or specialization option within existing IE curricula, is an immediately-achievable strategy. IE offers a workable mix of qualitative (human-based) and quantitative (technology-based) career paths.

Examples of traditional IE courses that can be modified to cater to space industrialization vision include facility design, operations management, system optimization, product quality assurance, and functional integration. It is my strong belief that many IE courses are directly extendable to a space focus. Military-based industrial engineers are already laying the operational cables and linkages to space. For example, leveraging IE's systems-based methodologies, the DEJI Systems Model® (for system design, evaluation, justification, and integration) has been used at the Air Force Institute of Technology for developing new space-themed certificate programs. Selected existing courses were cannibalized and retooled to address a focus on space. Although this was done at the graduate level at AFIT, a similar academic retooling can be done at the undergraduate level.

As the space industry develops and matures further, full-fledged degree programs will be considered. IE programs around the world should start strategizing along similar academic tracks. The more prevalent such options are available, the more we can proclaim and advance an IE foothold in space operations, whether Earth-based or space targeted. This approach will also open additional career options to IE students, to the extent that those graduates will further help to spread the practice and recognition of IE in space-oriented organizations. Functional attributes that space professionals will need are already embodied in the practice of IE. Examples of skills and/or preparations students might seek for space engineering include systems thinking, human-machine interaction, project management, energy consciousness, math modeling and optimization, quality assurance, ability to work independently, worldview, curiosity, inquisitiveness, and exceptional interpersonal skills.

Space is linked to everything we observe and experience here on Earth. Space-themed IE education can help us figure out where those linkages exist and how we can leverage them for the benefit of humanity. We should start laying broad claims and proclaiming IE as one right way to ensure sustainable industrialization of Space. Let us join hands and do it, starting now.

HUMAN FACTORS IN AEROSPACE

For the sake of definitions, purpose, and scope, it is important to mention that in most of the world, ergonomics is synonymous with human factors. In the United States, the primary focus and definitions are on human factors. For fuller expatriation, we often refer to the field as a compound pursuit of HFE (Human Factors and Ergonomics). Ergonomics refers to the tools that humans use in their work environment. Thus, the term fitting the tool to the human and work environment is appropriate in this context.

Ergonomics and Human Factors are really the same subjects although the term Human Factors is mainly used in North America and is dominated by the psychology perspective. There are also many slightly different definitions of the subject. The broader view will be addressed in this course and an operational definition will be "Design with people in mind". There is a long history of the subject going back to when people first adopted tools or weapons to extend their capabilities and devices to protect them from the environment. Subsequently, ergonomics applications and opportunities may be found in any context where people are involved in work or other activities. Nowadays, ergonomics is widely applied in manufacturing, consumer products, including smart technology, the military, and even space exploration.

   a. Complexity

     All ergonomics situations are by nature complex and involve interactions among and within people, technology, operations, and the context. This complexity is succinctly described by the Edwards SHEL model (Figure 7.1).

     The key to these complex system analyses is the investigation of the interfaces, interactions, intraactions, interdependencies, interferences, and integration of the human, technology, operations, and contextual factors.

     Another way of looking at complex systems is through the Donabedian (1988) classification of Structures, Processes, and Outcomes. This approach is similar to the SHEL model except that it does address context. One way of looking at the context is to consider those things that cannot be changed but may require protection. A good example is the physical environment,

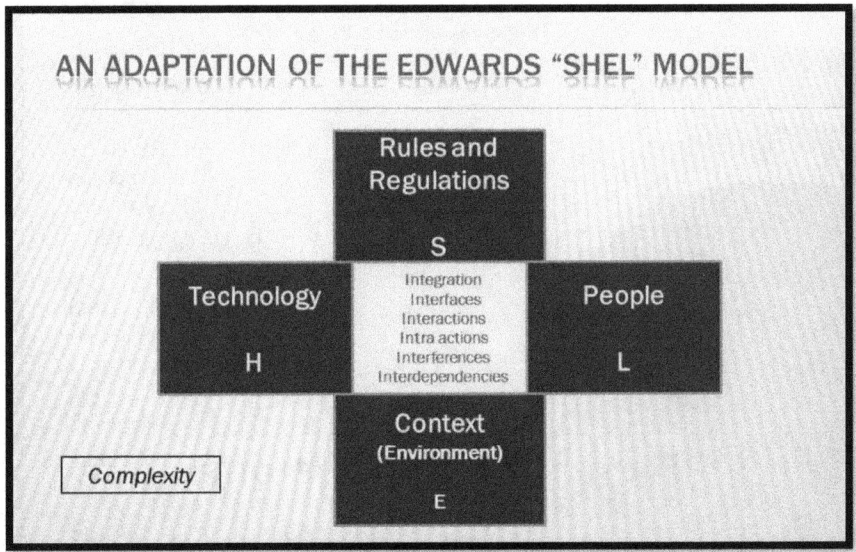

FIGURE 7.1   SHEL model for systems complexity.

such as the weather. Structures are generally considered to be things, such as technology that can be specified and designed; they are described by nouns and adjectives. For example, the design of workplaces for pilots, baggage handlers, and check-in counter staff can affect the effectiveness and efficiency of performance. Processes, on the other hand, are activities, operations, or tasks that are carried out by selected people and have a defined purpose. They are described by verbs and qualified or quantified by adjectives. Examples include flying an airplane or checking in for a flight.

b. **Outcomes and Purposes**

A simple view of the outcome or purpose of an operation is that of achieving a prescribed objective. For example, eating a meal, traveling to work, or using a computer booking utility have well-defined objectives – providing energy, earning money, or planning a holiday. But a deeper consideration of the operation reveals many more purposes and constraints. These purposes are described by the E4S4 model (Figure 7.2).

The basic purpose is process effectiveness – does the process achieve its intended objective – did you eat your food, get to work, or book a holiday? But these outcomes will be constrained by efficiency (use of resources), likeability, and ease of use. For example, going to work in a car may be costly, whereas taking the bus will be less expensive; but you may value comfort and convenience over cost or time spent on the activity may be very important. There will always be trade-offs!

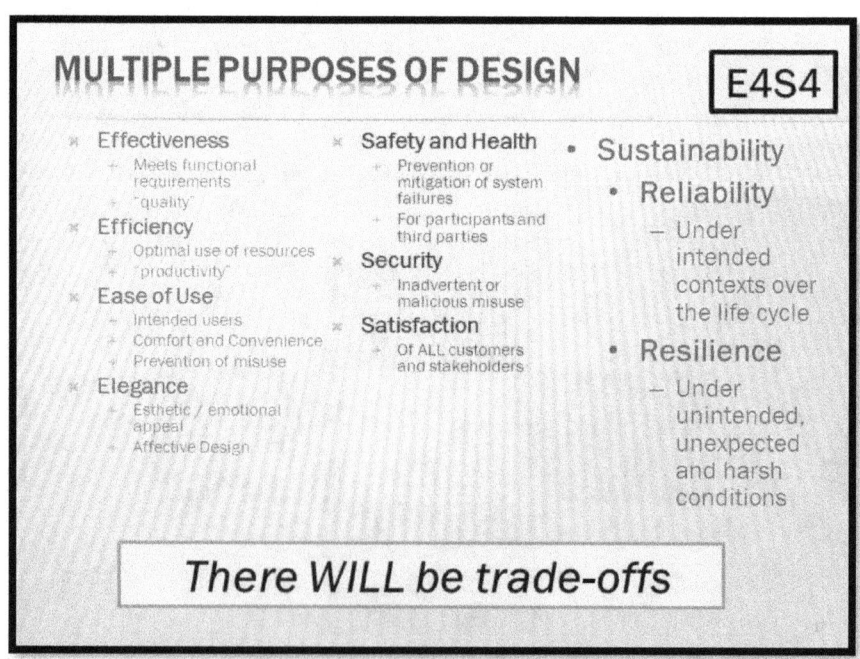

FIGURE 7.2   E4S4 for multiple purposes of design.

The 4Ss (Safety, Security, Satisfaction, and Sustainability) describe other constraints related to a process, activity, or operation. For example, getting to the gate to board an airplane requires hazardous luggage lifting, complex security checks, and satisfaction, both for you and the other people involved in the process.

It can be seen from the foregoing example that there will always be trade-offs in any process, activity, or operation.

The purposes of Ergonomics

- Contributing to the multiple purposes of design
- Helping with design tradeoffs

c. **Life Cycle Phases and Stakeholders**

There are multiple phases and multiple stakeholders in any operation or system design as shown in Figure 7.3.

All simple or complex operations involve multiple phases and multiple stakeholders who may have conflicting purposes and requirements. These phases include design, production, and use; in this "green" age, we may also have to address disposal. Consider the life cycle of a meal or a car. There is planning and design, cooking and manufacturing, and eating and driving. Many different people may be involved throughout this life cycle and they may all have different requirements. For example, the owner of a restaurant or car manufacturing company will probably want to make a profit, the cook

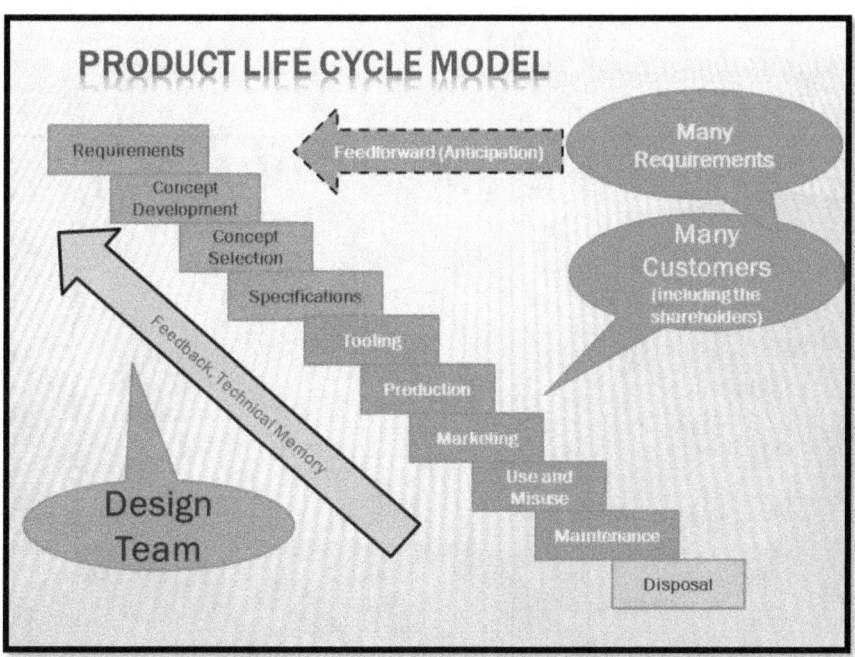

FIGURE 7.3 Product lifecycle model.

or assembly line worker will want a salary and safe workplace, and the end customer will want effectiveness, safety, and elegance (emotional appeal).

Sustainability implies that the system or process continues to operate to the satisfaction of all stakeholders. System reliability implies that the process outcomes are successful throughout the life cycle under normal intended conditions. Resilience, on the other hand, implies that the outcomes are achieved even in harsh or unpredictable contexts.

d. **Usability and Human Variability**

A major challenge of system design is that people vary on many dimensions. Therefore, systems must be designed to accommodate this variability. In particular, it is important to address intended use (users) and foreseeable misuse (misusers).

The fact of human variability is that people vary and people change

Consider the different capabilities and limitations of users in different age groups. Airlines must accommodate a wide range of people from babies to grandparents. Similarly, cell phones may be used by tech-savvy young people, inexperienced old people, and normal people while they are driving. Systems designers must anticipate this variability through the use of ergonomics data and methods. The 6Us and 2Ms model draws attention to the variety of users and usages of systems as well as the possible misusers and misusages. See Figure 7.4.

e. **Analyzing Complexity in System Design and Operations**

Systems can be large and complex. In each case, it is required to analyze the complexity of the system for the purpose of system design and required operations. The human in the loop of system complexity makes it imperative for the process of ergonomics to be understood and embraced.

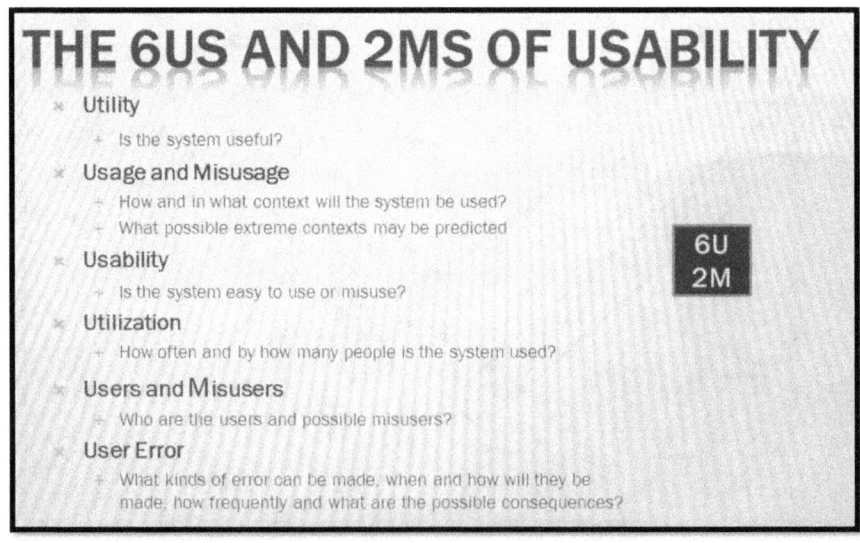

FIGURE 7.4   The 6US and 2MS of usability.

f. **The Process of Ergonomics**

As seen in the foregoing sections, ergonomics must deal with complexity, particularly with regard to user requirements and variable contexts of use. In any situation, there may be many different requirements and outcomes, various uncontrollable contexts, a limited set of design or intervention opportunities, and a wide range of possible users. Some interventions or designs may be easier or less expensive than others, some outcomes may be more important than others, and some contexts more demanding than others (Figure 7.5).

One temptation for ergonomists is to analyze and address those features of a system that are easy and inexpensive to change without due regard to the effect of these changes on the multiple outcomes. Another temptation may be to use overly sophisticated analyses where a simple checklist may suffice, and vice versa. Eventually, any decision about change or intervention will be the prerogative of management, who will certainly consider costs and benefits to the various stakeholders.

g. **The Activities of Ergonomists**

Ergonomists participate in various stages of a problem or system design. They may be involved in basic or applied research on human capabilities, limitations, and requirements in the context of a process design.

For example, in flight reservations, in flight meal provision, or airplane maintenance, the ergonomist may need to explore in detail the characteristics of the various users and the nature of the specific tasks. Next, they may make use of simulations or simulators to explore the solution possibilities.

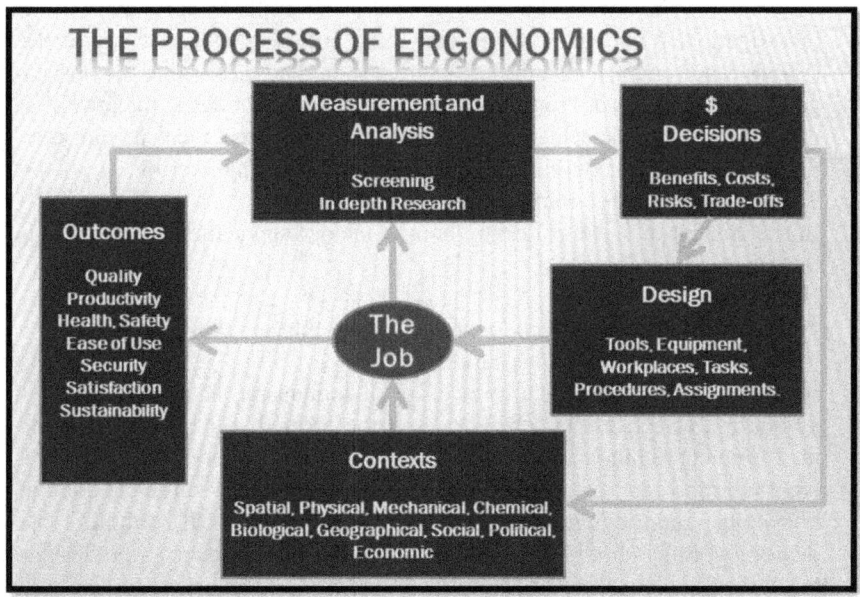

FIGURE 7.5   The process of ergonomics.

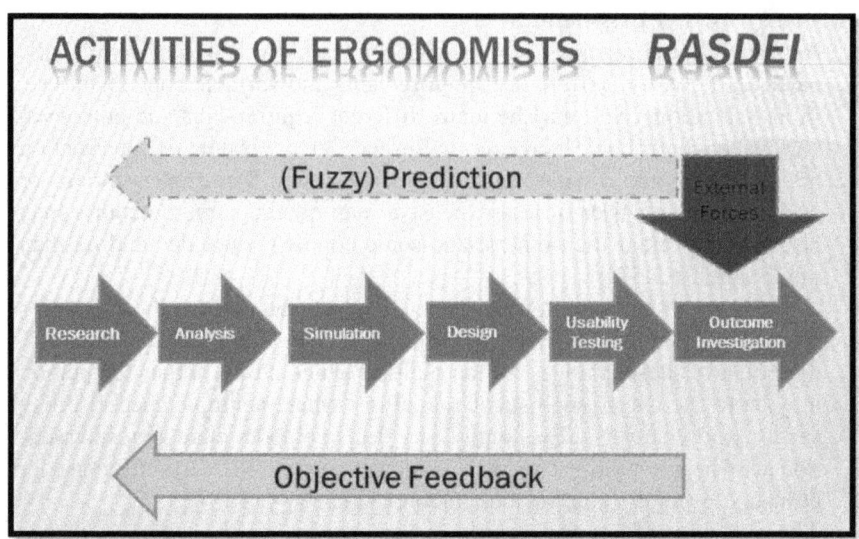

FIGURE 7.6   RASDEI framework for activities of ergonomists.

Then they may have the opportunity to work with the designer or engineer to map the user requirements into design specifications. As the design progresses, the ergonomist may be involved in system evaluation and usability testing, again in various operational contexts. See Figure 7.6.

Finally, when some unwanted outcome or accident occurs, the ergonomist may use his or her knowledge, tools, and skills to participate in an investigation. It should be noted that this "failure" may be of any of the (E4S4) requirements, also the ergonomist should be involved in the investigation and analysis of "success". A key role of ergonomics is the provision of documentation and feedback, sometimes called technical memory, so that designs can be continuously improved (Kansei Engineering).

h. **The Practice of Ergonomics**

In the context of systems design, human factors applications need the following characteristics:
- Domain knowledge
- Knowledge of human characteristics, capabilities, and limitations
- Measurement and analysis tools

An ergonomist needs basic knowledge of human physical, sensory, cognitive, and affective characteristics and the tools that are needed for the measurement of human performance, behavior, and preference (Figure 7.7). Another important requirement is that the ergonomist has sufficient domain knowledge and the ability to work with domain experts. It is unthinkable that ergonomists should offer gratuitous advice to airplane cockpit designers without understanding the tasks of pilots or maintenance operations and how airframes and engines function.

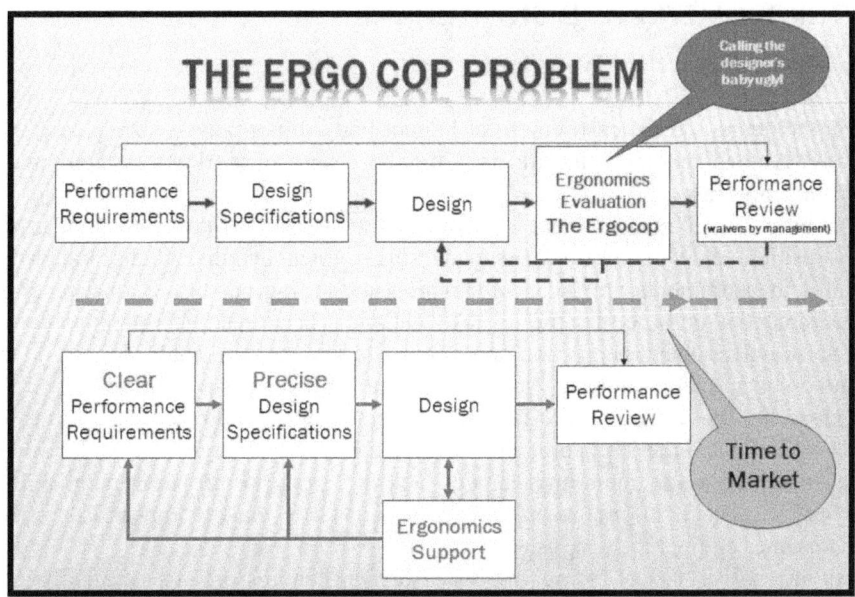

FIGURE 7.7 The ERGOCOP problem.

In this same context, it is important to not fall into the "ERGOCOP" (i.e., ergonomics police) trap. It is better to work with engineers, designers, and operations managers than walk in after the event and "call their baby ugly".

i. **The Scope of Ergonomics**
People vary on many dimensions – physical, sensory, cognitive, social, and affective. They also must function in many environmental and operational contexts. People, like technology, vary over time – they exhibit many individual differences, they learn, and they fluctuate in their capabilities and preferences.

The Broad Scope of Ergonomics encompasses the following:
Physical
Sensory
Cognitive
Social
Affective
Environmental
Temporal

These represent human, technological, operational, contextual, and temporal variability. IE tools and techniques are effective in addressing these systems' characteristics.

The later parts of this course will address many of these different human variables and the associated design options and process outcomes.

ANTHROPOMETRY AND BIOMECHANICS

a. Anthropometry and Workplace Design

Anthropometry is the measurement of human body dimensions – size, shape, and posture. Anthropometric data are used for the design of workplaces, equipment, tools, and clothing.

Anthropometric measurement is carried out with various fixtures to standardize body position, and tapes and calipers to measure segment lengths, widths, and girths. Usually, measurements are taken in standing and sitting postures. Consistency in measurement is obtained by the use of standard procedures and reference points that are usually subcutaneous bony prominences. The data are analyzed to produce population (sample) parameters, such as the mean and various percentiles.

Generally, the 5th percentile is used to accommodate the reach capabilities of smaller people and the 95th percentile is the fit capability of larger people. However, because of multiple segment involvement, clothing, work posture, movement ability, and various task functional requirements, it is usual to make compromises in the workspace or equipment designs.

An alternative way of estimating body segment lengths is to use the Dillis and Contini segment proportions model. Contemporary anthropometric methods use body scanning methods. Data for the Singapore population are contained in the attached hand-out.

A typical application of anthropometry is in the design of passenger accommodations in public transport, pilot or car driver workspaces, and auditorium seating.

Biomechanics and Manual Materials Handling

Mechanically speaking, the body consists of a bony framework with joints, muscles to control movement, a circulatory system to supply energy, and a nervous system to precisely control the movements. In a given static posture with the various limb and trunk segments held still by muscle action, the forces involved may be analyzed by drawing free body diagrams and analyzing the moments around various joints. In addition, especially in the spinal joints, there are various tension, compression, and shear forces.

Students are referred to the University of Michigan Three Dimensional Static Strength Prediction Program (UofM3DSSPP) for a detailed description of static biomechanical analysis.

Although the static analysis does provide useful evidence regarding the biomechanical stresses in manual materials handling, in reality, most activities are dynamic and the instantaneous and peak forces are difficult to measure. This complication along with the large variability between people in size, strength, postures, and movements somewhat limits the utility and practicality of static analyses.

A major outcome and problem of manual materials handling is the high prevalence of back injuries. Consequently, an analytic tool – the NIOSH Lift Equation – was developed to assess the contribution of key factors (Load, Horizontal Distance, Vertical Location and Movement, Asymmetry, Coupling, Frequency, and Duration) to the overall stress in a Lifting Task and to provide recommendations for task parameter limits. The tool is available as a free smartphone/tablet app.

A second tool – The Liberty Mutual (Snook) Manual Materials Handling Tables – was developed using psychophysical techniques. Psychophysical methods are a way of obtaining estimates of human capabilities based on the subjective reports of experimental subjects carrying out various (both cognitive and physical) tasks. Tables are available for Lifting Carrying, Pulling, and Pushing.

WORK PHYSIOLOGY, THERMAL ENVIRONMENT, AND CIRCADIAN RHYTHMS

Work physiology is concerned with the conversion of food energy into mechanical energy through muscle contraction. Those students interested in the details of this conversion process should study Krebs (Citric Acid) Cycle. A large component of the energy we use is to enable the vertical movement of our body, and carried objects, against gravity. It should also be noted that only about 20% of the energy we convert is into mechanical energy; the rest is the production of heat. Another key issue is that if the food energy we consume is not converted into mechanical and heat forms, it is stored as fat.

Ergonomists aspire to measure key indicators of this process in order to assess the physical demands of heavy work and work in hot environments. Apart from measuring the Calorific content of food consumed, we may also estimate the physical work we do; however, these methods are somewhat cumbersome and unreliable except in carefully controlled laboratory conditions. Consequently, we use the indirect methods of measuring oxygen uptake, circulation, and body temperature to indicate physical workload. Less accurate methods involve the use of look-up tables of food composition and activity demands. For example, brisk walking uses about 3 or 4 kilocalories (Physiological Calories) per minute. In terms of energy conversion template, we have the following units:

   1 kcal/min
   = 0.8 mets
   = 70 watts
   = 0.2 liters of oxygen/min
   = 3.6 ml/kg/min

1 met is equivalent to the resting metabolic rate of an average man or woman.

A widely used subjective method – the Borg Scale – is easy to obtain, relatively consistent and quite useful.

A general classification of physical work demands is shown below; however, these estimates are affected by such things as body weight, age, sex, level of conditioning, motivation, and fatigue (Figure 7.8):

   Sleeping: 1 kcal/min
   Standing: 2 kcal/min
   Walking: 3 kcal/min
   Running: 4–20 kcal/min
   Materials handling: 5–10 kcal/min
   Football: 5–20 kcal/min

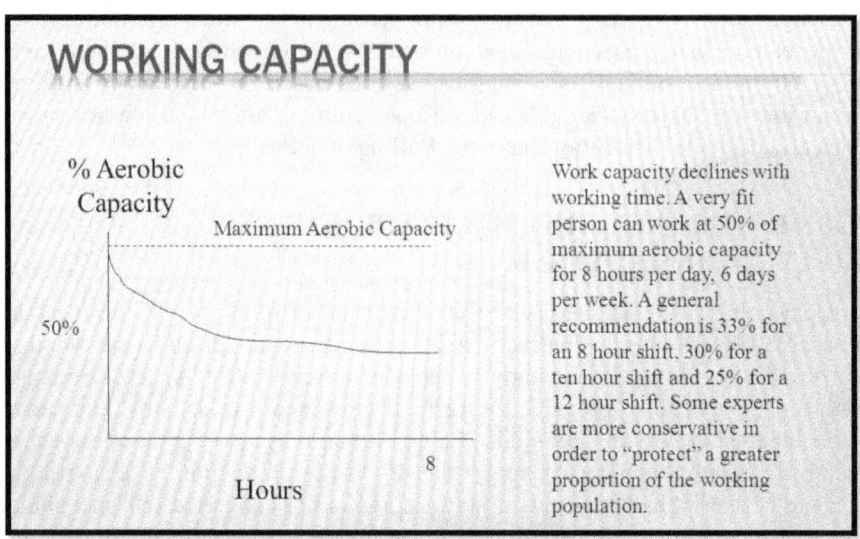

FIGURE 7.8 Human working capacity.

People's capacity for energy expenditure is also affected by the thermal environment and fatigue.

a. **The Thermal Environment**

Because the majority of the energy we use is converted into heat energy, we need a sophisticated thermal balance mechanism to maintain our body temperature at relatively consistent levels. This is particularly important in hot and humid climates where the possibility of heat disorders is very real, especially where the physical workload demands are high and fluid (and nutrient) replacement is not sufficient. Heat is transferred between the body and environment by conduction, convection, radiation, and evaporation. All of these mechanisms are potentially available for body temperature management. The biological process of thermal management involves the automatic control of heart output, cutaneous blood flow, and sweating. Behaviorally, we increase our heat production by activity and shivering in cold conditions and a reduction of physiological heat production by rest in hot conditions. Under normal thermal conditions, the core body temperature varies slightly (around 37°C) over the day and night, principally due to the varied physical work demands.

Measurement of the thermal environment includes temperature, humidity, air movement, and various combined measures, such as Effective Temperature and WBGT (Wet Bulb Globe Temperature). Humidity is measured by comparing dry and wet bulb temperatures. The (engineering) control of the thermal environment is by barriers to radiant heat, evaporative air conditioning, and increased air movement – fans. Administrative controls include reduced physical work demands, reduced exposure durations, acclimatization programs, increased fluids, and surveillance.

b. **Hypoxia**
- Lack of oxygen in the body tissues due to decreased quantity and molecular concentration.
- In aviation, it occurs due to a fall in partial pressure of oxygen in the inspired air with increasing altitudes.

Hypoxia Types
- Hypoxic Hypoxia – reduction of partial pressure of oxygen in the inspired air
- Anemic Hypoxia – reduction in oxygen-carrying capacity of the blood due to decreased hemoglobin content (carbon monoxide poisoning)
- Hypokinetic Hypoxia – malfunction of the circulatory system where there is an inadequate circulation of the blood
- Histotoxic Hypoxia – utilization of oxygen by the body tissues is interfered with by alcohol, narcotics, etc.

Signs and Symptoms of Hypoxia
- Breathlessness
- Excessive yawning
- Tiredness and fatigue
- Euphoria
- Impairment of psychomotor performance
- Decreased (night) visual acuity
- Impairment of mental performance (e.g., memory)
- Diminished alertness (maybe unconsciousness)

Predisposing Causes of Hypoxia
- Altitude – higher altitude results in lower partial pressure of alveolar oxygen
- Rate of ascent – the greater the rate of ascent, the more rapid the onset
- Duration at altitude – effects more severe if the duration at altitude is prolonged
- Ambient temperature – high or low environmental temperature
- Physical activity – physical effort at altitude raises the demand for Oxygen
- Individual susceptibility
- Physical fitness – regular physical training improves tolerance levels
- Smoking – smoking makes an individual more liable to suffer from hypoxia
- Organic diseases – diseases of the heart, lungs, or blood, which interfere with the normal oxygenation and circulation
- Emotional state – apprehension and anxiety make an individual more susceptible
- Acclimatization at high altitudes raises the individual's ability to withstand hypoxia

Altitude and Other Effects

- < 10,000 No/minimal effect
- 10,000–15,000 Compensatory behavior
- 15,000–20,000 Significant effects
- >20,000 Critical
- Less Than 12,000
  - Vision especially night vision is known to deteriorate even at altitudes less than 5000 ft.
  - Night vision may be lost by 5%–10% at 5000 ft and may be compromised by 25% at 12,000 ft.
  - Symptoms of hypoxia may be increased due to anemia, cigarette smoking, or exposure to carbon monoxide due to fumes in the cockpit.
  - Aircrew may not be aware of the hypoxia.
- 12,000–15,000
  - The effects of hypoxia on the nervous system become increasingly incapacitating.
  - Impairment of psychomotor skills
  - False sense of well-being and euphoria
  - Drowsiness
  - Judgment errors
  - Compensatory actions by the circulatory and respiratory systems can provide some defense against hypoxia.
  - Increase in heart rate
  - Increased and more forceful pumping of blood by the heart to improve circulation.
  - Increased rate and depth of respiration.
  - Compensatory responses occur spontaneously but once aware the pilot must take conscious corrective actions promptly.
- 15,000–20,000
  - Physiological compensations do not suffice to provide adequate oxygen to the tissues.
  - Subjective symptoms may include fatigue, lassitude, sleepiness, dizziness, headache, breathlessness, and euphoria.
  - There are no subjective sensations up to the time of loss of consciousness.
- Greater than 20,000
  - Lost consciousness
  - Circulatory or central nervous system failure
  - Convulsions
  - Failure of the respiratory center in the brain, can lead to death
- Other effects
  - Visual acuity is diminished.
  - Peripheral and central vision is impaired.
  - Range of accommodation for near vision is decreased.
  - Touch and pain are diminished or lost.
  - Hearing is one of the last senses to be lost.

- Intellectual impairment makes it impossible for the individual to comprehend his own disability.
- Thinking is slow and calculations are unreliable.
- Memory is faulty, particularly for recent events.
- Judgment is poor.
- Reaction time is delayed.
- Release of basic personality traits and emotions similar to intoxication.
- Euphoria, overconfidence, or an individual may become morose.
- Poor muscular coordination.
- Fine muscular movements may be impossible.
- Stammering, illegible handwriting.
- Poor coordination in aerobatics and formation flying.

c. **Circadian Rhythms and Shift Work**

In many production operations, many workers are subjected to shift work. It is important to understand and appreciate how work performance is affected by the circadian rhythms of workers. Circadian rhythms are the body's natural clocks, regulating sleep, hunger, hormones, and varied bodily functions. If the circadian rhythms are out of phase with the required work functions, performance may be adversely affected.

HUMAN INFORMATION PROCESSING

A traditional model of Human Information Processing includes sensing, attention, perception, cognition, memory, and effectors management. As with physical characteristics, people vary considerably in these processes and "human error" may occur at any stage (Figure 7.9).

a. **Attention**

A particularly vulnerable stage is the attention mechanism and cognition (Figures 7.10 and 7.11).

b. **Cognition**

Cognition is a term referring to the mental processes involved in gaining knowledge and comprehension. Some of the many different cognitive processes include thinking, knowing, remembering, judging, and problem-solving. Thus, cognition is an essential consideration in assessing ergonomics and work capabilities and performance.

c. **Memory**

Operational memory is vulnerable.

Factors affecting operational memory:

- Interference
  - Retroactive – information interpolated during the retention period
  - Proactive – information presented before the item to be remembered
- Primacy and recency
  - Items at the beginning and end of a list are more likely to be remembered than those in the middle
- Rehearsal
  - It is possible to maintain retention by rehearsing it.

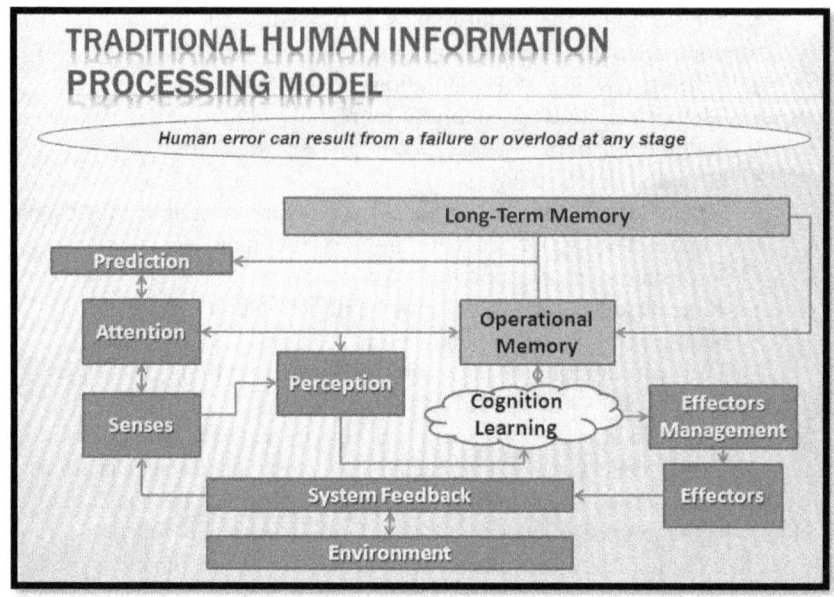

FIGURE 7.9   Traditional human information processing model.

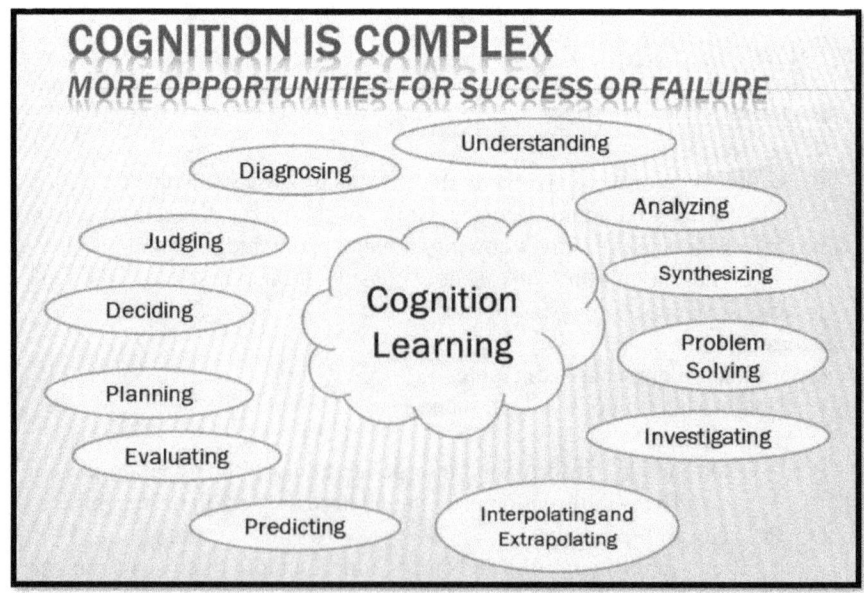

FIGURE 7.10   Patterns of human cognition.

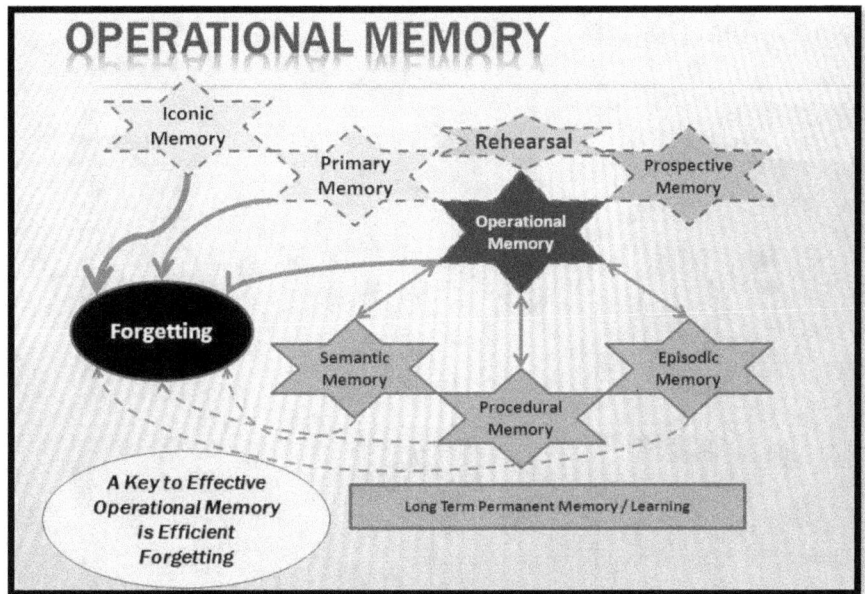

FIGURE 7.11   Operational memory in ergonomics.

HUMAN-MACHINE SYSTEMS AND CONTROL

One perspective of human factors is to observe the activities and information flow in a "Human-Machine System". In a simple system, like a hand tool, the user has a direct vision of the task and interacts directly with the tool (although handle length does allow some force gain and control adjustability). With a more complex system, like a car, displays are added to provide a system (e.g., engine temperature) and operational (e.g., speed) performance. Also, the steering wheel and foot pedals are connected to the front wheels, engine, and brakes via various linkages and power amplifiers. These additions of sophisticated controls and displays make control much more precise.

a. Interface Design

An alternative model of human-machine interaction is based on feedback control theory. In this model, the operator has an objective, anticipates the effects of uncontrollable external forces, and provides an appropriate system input. He/She then observes the output (feedback) and modifies the input accordingly.

CONTROL

b. **Communication**

Communication (Figure 7.12) is a generally familiar concept but breakdown of communication is probably the most frequent cause of system failure. Analysis and prevention of communication elements are therefore important

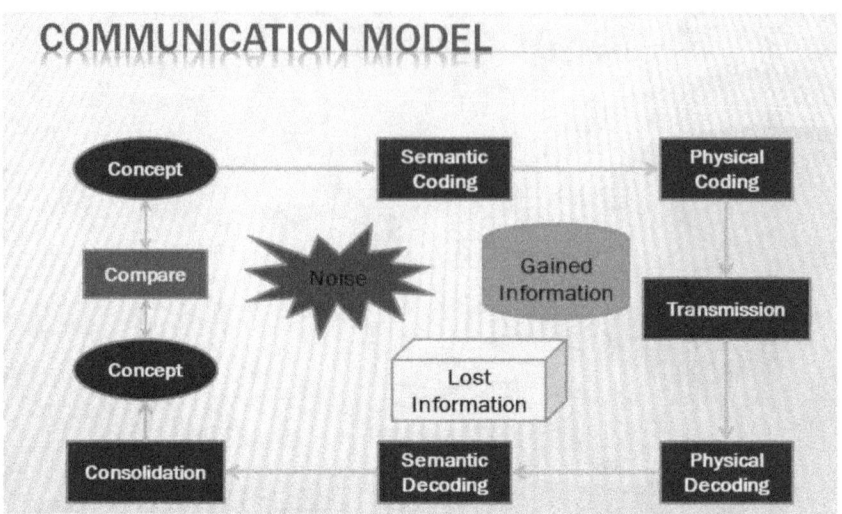

FIGURE 7.12   Systems communication model.

in system safety. Communication starts with an idea, concept, or objective that is then translated into some semantic form; the next step involves the choice of a physical mode of communication. The information is then transmitted to the receiver who has physical reception capabilities (eyes and ears); next, the receiver interprets the transmitted signal and translates it into some memory or action. The communication pathway is very dependent on the prior knowledge (added information) of the sender and receiver; also all of the stages are very vulnerable to noise and lost information.

SENSORY PROCESSES – VISION AND HEARING

VISION

The retina of the eye converts light energy into an electrical signal that is interpreted in the visual cortex of the brain. The pupil and lens are responsible for controlling and focusing the incoming light on the retina. The retina contains cones that are concentrated around the fovea and are sensitive to color, especially under high light levels. The rods are distributed around the outer parts of the retina, have a higher sensitivity but are not sensitive to color.

Functions of the Different Eye Structures

- Sclera – the eye's white outer protective coat, normally seen as the "white of the eye".
- Cornea – the transparent, curved structure at the front of the eye.
- Iris – the colored part of the eye – blue, brown, green, gray, etc. – that can be seen through the cornea.
- Pupil – the black part of the eye in the middle of the iris. It constricts or dilates according to the amount of light passing through it.

- Lens – the transparent disc (with both sides being convex) immediately behind the iris and pupil.
- Aqueous humor – the transparent fluid (with consistency similar to water) that circulates behind the cornea and in front of the lens.
- Vitreous humor – the material (like transparent jelly) that fills the eyeball between the lens and the retina.
- Retina – the light-sensitive layer of millions of nerve cells that line the back of the eyeball. The cells consist of two main groups, called rods and cones due to their appearance under the microscope.
- Rods – more numerous, spread out over the entire retina with more toward the outer edge, and respond to low levels of light.
- Cones – far fewer, concentrated around the fovea, respond to color and details.

Optic nerve and beyond – the "cord" of nerve cell connections that passes from the eyeball to destinations throughout the brain.

Visual Acuity

Visual acuity is dependent on the amount of light (especially that falling on the fovea where the cones are concentrated) and the visual angle subtended by the target. Visual acuity may be assessed by familiar eye charts or by more elaborate devices.

HEARING

Sound waves are concentrated along the outer ear canal and cause vibrations in the tympanic membrane that are transmitted through the middle ear by the ossicles to the nerve endings in the inner ear or cochlea.

Functions of the Different Ear Structures

The ear is differentially sensitive to different frequencies being most sensitive in the human speech range – from 300 Hz to 4000 Hz.

FITTS LAW

Fitts Law demonstrates that the time taken for a particular (reciprocal tapping) task is affected by both the amplitude and target width. This relationship can be extrapolated to a wide variety of control tasks.

$$MT = a + b\log_2(2A/W)$$

Where:
MT = Movement Time
a, b = Individual and situational constants
A = Amplitude of Movement
W = Width of Target
2A/W = Index of Difficulty (ID)

HICKS LAW

Hicks law demonstrates that more difficult decisions, in terms of the amount of information that is needed to be processed, require longer times.

$$MT = a + b\log_2 n$$

Where:
N = number of choices

For Hick's Law exercise for a complex task, measure the performance times and observe the errors and hesitations.

This relationship between the log of the number of choices and decision time can be extrapolated to many complex information-processing tasks. It must be noted, however, that the decision itself is significantly affected by the value of the outcome and the time available to make the decision.

DE JONG'S LAW

Motor skills and decision-making improve with practice as described by de Jong's Law.

MENTAL WORK LOAD

Mental work load is an assessment of the amount of mental work that a worker is expected to deal with in the process of performing his or her job functions. Mental workload reflects the relevant and irrelevant information available for decisions at any point in time. Sustained high mental workload will cause mental fatigue.

SITUATION AWARENESS

Situation awareness is the capacity to comprehend, assess, and project information in order to control a situation in a timely manner. This is a characteristic of individuals or a team and is aided by the appropriate use of technology and training. This is a widely used method of describing human performance in complex situations, such as aviation and process plant operations. The SA Model goes through Context, Responses, and Decisions with a feedback loop of Perception (Stage 1), Understanding (Stage 2), and Projection (Stage 3).

INTRODUCTION TO CRM

Crew Resource Management (CRM) is the application of human factors knowledge and skills to the conduct of flight (and other) operations with the objective of efficiently using all available resources (equipment, systems, and people) to achieve effective, efficient, and safe operations.

Commercial air transport remains one of the safest methods of moving people and goods from one point to another and the number of fatal incidents per mile traveled is extraordinarily low. But a single accident may have a disastrous outcome and the image of the safety practices in the industry gets negative publicity. It has been widely estimated that around 75% of accidents are caused by human error but what this term fails to recognize is that humans are one part of the wider environment – they must interact with many components including weather, technology, and social systems. Despite this, humans are at the most basic level of the root cause of almost every incident because humans ultimately design and/or interact with all elements of the wider environment. In most complex operations even though individuals may make errors, good CRM will provide redundancy to avoid adverse outcomes.

Modern crew resource management focuses on the management of all available resources to reduce error among all groups of aviation specialists (e.g., air traffic controllers, pilots, cabin crewmembers, mechanics, and dispatchers) through goal setting, teamwork, awareness, and both proactive and reactive feedback. Training of CRM for commercial aircrew has become a mandatory practice under the majority of the world aviation regulatory environments and the practice of Crew Resource Management is an integral part of commercial airline operations.

The successful application of CRM in aviation has been recognized and equivalent training methods are now widely applied in a range of other high-risk industries including, e.g., medicine, fire services, and maritime operations.

LEADERSHIP

- Leadership depends on the situation, not the rank
- There are important cultural differences
- A prerequisite of leadership is a team with the appropriate knowledge, skills, and abilities
- The primary purpose of leadership is to motivate others to contribute, collaborate, and communicate and to make hard decisions
- Leaders need to understand the motivations of the other team members
- Important components of motivation are goal setting and feedback
- Leadership is about continuous improvement in the processes and team capabilities

a. **Human Factors Analysis and Classification System**
   Shappel and Weigman converted this model into a formal accident analysis method – the Human Factors Analysis and Classification System (Figure 7.13) – which is now used widely in aviation, medicine, and maritime operations.

UNSAFE ACTS IN SYSTEMS IMPLEMENTATIONS

Unsafe acts may be errors or violations (see Figure 7.14) – unintentional or intentional. Errors may be due to any failure of the Human Information Processing System, particularly perception, decisions, and skills. Memory failures are another common cause of errors. Violations may be occasional or routine.

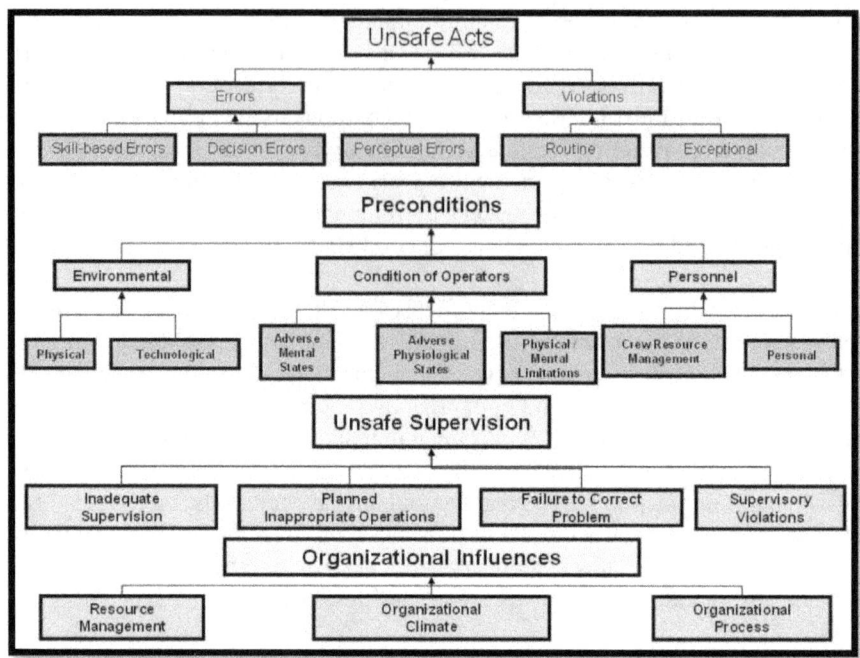

FIGURE 7.13   Human Factors Analysis and Classification System.

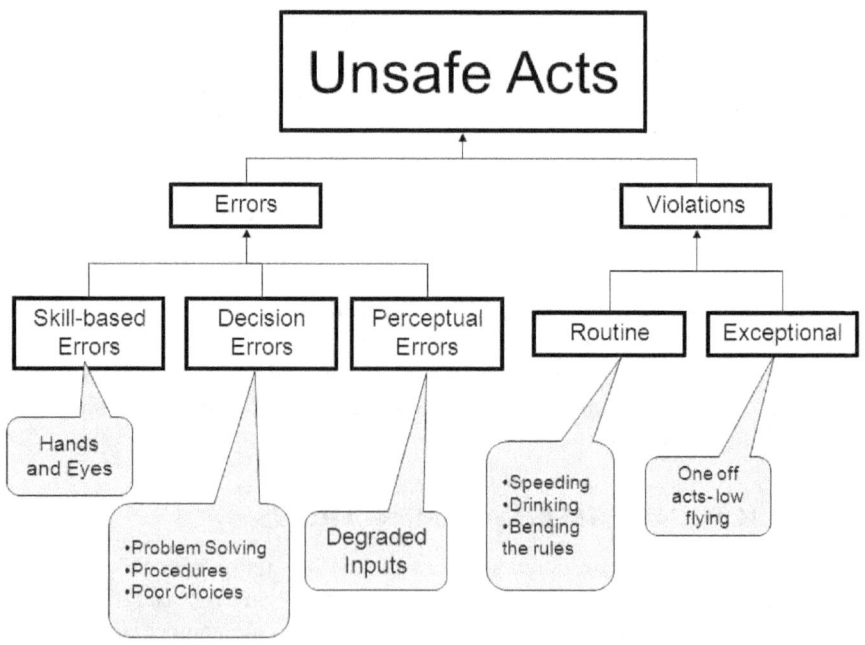

FIGURE 7.14   Elements of unsafe acts.

PRECONDITIONS

In almost all accidents, there are latent preconditions – "accidents waiting to happen" – such as a slippery floor, a staircase without a handrail, a tired operator, or routine risky behaviors. Commonly, these preconditions may be physical as with the design of equipment or the environmental context. On other occasions, they may be due to human, procedural, or organizational limitations. As discussed earlier with the SHEL model, these technological, human, contextual, and operational factors may interact to create preconditions conducive to system failure.

UNSAFE SUPERVISION

Many forms of inappropriate behavior occur because of lack of supervision – some authority makes sure that operators follow safe procedures. The "authority" may be anyone – a parent, colleague, instructor, supervisor, police, or security officer. The unsafe supervision opportunities may be occasional, planned, or routine.

ORGANIZATIONAL INFLUENCES

This is the most insidious form of latent system failure causes. They include a lack of or nonadherence to safe working policies and procedures, perhaps for productivity or cost-cutting reasons. Other inadequate policies may include inappropriate assignment of undertrained personnel. See Figure 7.15.

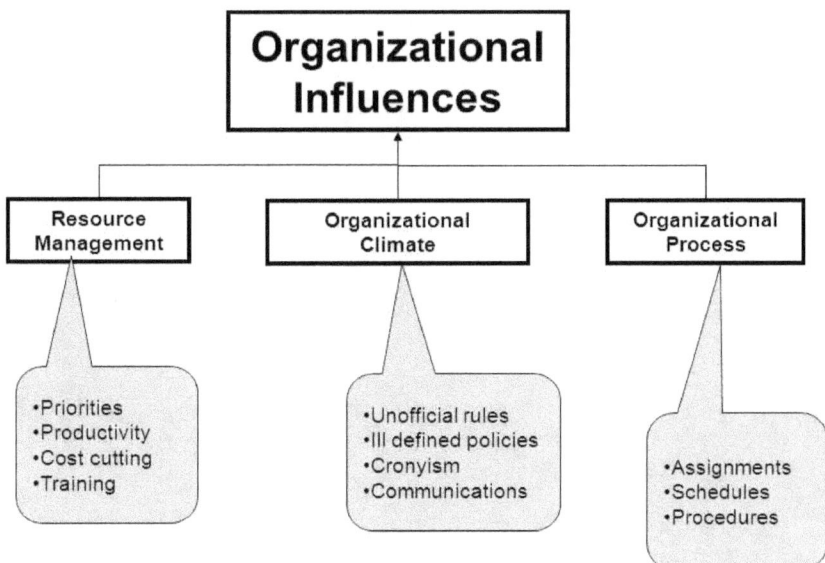

FIGURE 7.15   Scopes of organizational influences.

REFERENCES

Badiru, Adedeji B. (2022). The Next Frontier for Industrial Engineering. *ISE Magazine*, September 2022, p. 24.

Donabedian, A. (1988). The quality of care: How can it be assessed? *Journal of American Medical Association*, 260 (12): 1743–1748. https://doi.org/10.1001/jama.1988.03410120089033. PMID 3045356

Peacock, B. (2019). *Human Systems Integration*, Self-published manuscript, Fernandina Beach, FL.

Peacock, B. (2020). *How Ergonomics Works*, Self-published manuscript, Fernandina Beach, FL.

Peacock, B. (2021). *Ergonomics Tools and Applications*, Self-published manuscript, Fernandina Beach, FL.

8 NASA Case Example of Aerospace Systems Design

INTRODUCTION

Currie and Peacock (2002) present a good example of leveraging industrial engineering and systems modeling for the assembly and maintenance operations at the International Space Station (ISS), where scientists and engineers rely heavily on the use of extravehicular robotic systems. The diversity, flexibility, and versatility of industrial and systems engineering are demonstrated in complex systems (Peacock, 2019, 2020, 2021). After being fully assembled, the ISS robotics complement includes three main manipulators, two small dexterous arms, and a mobile base and transporter system. The complexity and mobility of the systems and limited opportunities for direct viewing of the Space Station's exterior make telerobotic operations an especially challenging task. Although fundamental manipulator design, control systems, and strategies for autonomous versus manual control vary greatly between the systems, commonality in the design of workstation controls and displays is considered essential to enhance operator performance and reduce the possibility of errors. Principal human factors opportunities are associated with workstation layout, human-computer interface considerations, adequacy of alignment cues for maintenance of safe approach corridors during mating tasks, spatial awareness challenges, integration of supplemental computer graphic displays to enhance operator global situational awareness, and training methodologies for the preservation of critical skills during long-duration missions. ISS is the largest and most complicated spacecraft ever assembled. This chapter is a retrospective case study of the construction of ISS. Five international space agencies – National Aeronautics and Space Administration (NASA), Russian Space Agency (RSA), Canadian Space Agency (CSA), National Space Development Agency of Japan (NASDA), and European Space Agency (ESA) – have developed the suite of robotic systems. Control of robotic systems in microgravity is a delicate and risky activity. The complexities are related to large masses (>100,000 kg), multi (6 and 7) degree-of-freedom systems, distinctive end effectors, limited direct visual information, diverse manual and automatic control conditions, Extravehicular activity (EVA) and Intravehicular activity (IVA) interfaces, and very high potential costs of error. Safe and efficient robotic operations require crewmembers to possess unique coordination and manipulation skills and detailed knowledge of the system's design and operation. Ensuring commonality between the systems will contribute to increases in mission success probabilities, expand safety margins of critical operations, and minimize system-specific training.

DOI: 10.1201/9781003328445-8

Although commonality considerations were paramount during the design phase of these systems, they were also weighted with respect to the flexibility and adaptability of individual system designs, design costs, and schedules.

OVERVIEW OF MANIPULATOR SYSTEMS

ISS robotic systems comprised a mix of manipulators with unique control systems and algorithms, capabilities, and flying characteristics. Space robotic system design features intended to optimize human and automatic control activity include: trajectory and motion limitations; collision avoidance algorithms; automatic safing schemes; force moment accommodation; and integrated procedures involving automated, ground, and crew control.

Mobile Servicing System (MSS). Jointly developed by Canada and NASA, the MSS will primarily function on the U.S. segments and truss assembly. It comprised five subsystems – the Space Station Remote Manipulator System (SSRMS), the Mobile Base System (MBS), the Mobile Transporter (MT), the Special Purpose Dexterous Manipulator (SPDM), and the Robotic Workstation (RWS). The SSRMS is a 17-foot-long manipulator consisting of two booms, seven joints, each with a range of $\pm270°$, and two latching end effectors. Power, data, and video are provided to the payloads via the latching end effectors. During assembly, the SSRMS will primarily be used to install pressurized modules and truss elements. The SSRMS can operate from any power and data grapple fixture on the ISS, giving the SSRMS the capability of "walking" or repositioning itself to a new base point. Control and monitoring of the SSRMS are from one of two modular workstations, both initially deployed in the U.S. Laboratory module. The RWS components are portable and include three video monitors, two hand controllers – one to effect translation and one for rotational inputs, a Display and Control panel, a Portable Computer System (PCS), and a cursor control device.

The SPDM is a dexterous manipulator with two symmetrical seven-joint arms attached to a central body structure. Its primary function is the changeout of robotically compatible small equipment on the Space Station's exterior. Additional SPDM tasks include scientific payload servicing and inspection and monitoring in support of EVA. The SPDM can either be operated from the end of the SSRMS or as a stand-alone manipulator system.

The MT provides transportation of the SSRMS along the exterior ISS truss. At its maximum velocity of 2.54 cm/sec, it will take 50 minutes to traverse the entire length of the truss when the ISS is fully assembled. The high mass/inertia, costly payloads, and great vulnerability of space vehicles to mechanical damage associated with space operations require very slow translation speeds to avoid overruns, reduce oscillations, and prevent collisions. These slow translation speeds require significant levels of operator vigilance.

The MBS provides a mobile base of operations for the MSS and serves as an interface between the MT and the SSRMS. It functions both as a work platform and as a base for the manipulators. The MBS provides four interfaces to support the attachment of the SSRMS and the SPDM and provisions for power and temporary storage of payloads and Orbital Replacement Units (ORUs).

The Japanese Experiment Module Robotics System (JEMRMS). The JEMRMS, developed by the NASDA of Japan, is primarily intended for use on the JEM Exposed

Facility (EF). This system comprised two manipulator devices, the Main Arm and the Small Fine Arm, and the JEMRMS console. The Main Arm is a 10-meter long, fixed-base, six-joint robotic arm with two main booms. The Small Fine Arm is a 2-meter long dexterous manipulator consisting of six joints, two booms, and an end effector mechanism. The Small Fine Arm can only be operated from and relocated by the Main Arm. The JEMRMS Console is located inside the JEM Pressurized Module and provides manual augmented, autotrajectory, and single-joint modes. Two different operational schemes are planned for the two JEMRMS manipulators. Control of the Main Arm will be primarily through the use of semi-autonomous autotrajectories designed to reduce crew workload since planned tasks include long and tedious operations (payload transfer). A manual control mode will be used most frequently for Small Fine Arm operations, which are primarily dexterous tasks requiring high positioning and trajectory accuracy.

The European Robotic Arm (ERA). The ERA, built by the ESA under contract from the RSA, is an 11-meter manipulator with seven joints and two booms. It will primarily perform automated operations. The ERA manipulator also has "walking" capability but can reposition only to base points located along the Russian Segment. Operational control is either through an EVA Man-Machine Interface (EMMI) or through an IVA Man-Machine Interface (IMMI) when the operator is inside the Russian Service Module. It is the only ISS manipulator that does not include hand controllers.

Control will primarily be through an autotrajectory mode but manually selectable single-joint modes are also available. Table 8.1 provides a comparative analysis of the respective characteristics of the ISS robotic manipulator systems.

TABLE 8.1
Manipulator System Characteristics

| System | Mobile Servicing System | | JEM RMS | | |
	SSRMS	SPDM	Main Arm	Small Fine Arm	ERA
Primary Function	ISS assembly/ maintenance, payload retrieval, and handling, EVA Support	ISS maintenance, small payload handling	JEM Exposure Facility payload handling	JEM Exposure Facility payload and ORU handling	Russian segment maintenance, EVA support
Joints	7	15 (7/arm + 1 body)	6	6	7
Manipulator Length	17 m	3.5 m	10 m	1.7 m	11.2 m
Max Payload Mass	116,000 kg	600 kg	7000 kg	300 kg	8000 kg
Operation Modes	Manual, Automatic, Single Joint	Manual, Automatic, Single Joint	Manual, Automatic, Single Joint	Manual, Automatic, Single Joint	Automatic, Single Joint
Max Arm Speed	0.36 m/s 4.0 deg/s	0.075 m/s 2.5 deg/s	0.06 m/s 2.5 deg/s	0.05 m/s 7.5 deg/s	0.2 m/s 2.9 deg/s

SYSTEMS FACTORS CONSIDERATIONS

Typically we talk about human factors considerations but in a systems-oriented environment, it makes sense to expand to systems factors considerations. Designs that optimize operator interfaces to the ISS robotic systems are key to enhancing operator performance and decreasing potential human errors during critical assembly and maintenance operations. Principal considerations include workstation topography and design, including mechanisms to assist with operator orientation and stabilization; graphical user interface commonality within and between systems; adequacy of alignment cues for maintenance of safe approach and mating corridors during berthing tasks; spatial awareness challenges; integration of supplemental computer graphic displays to enhance operator global situational awareness; and methodologies to preserve critical skills during long-duration missions.

CONTROL INTERFACES

Workstations should be designed to allow for emergency intervention by operators through hardware rather than software interfaces. It is clumsy and potentially hazardous to require an operator to negotiate through several layers of software to effect an immediate action to stop the manipulator. Further, numerous malfunction scenarios can render the PCS inoperative. Common hand controller designs will reduce operator training, enhance positive habit formation, and reduce the potential for inadvertent or errant commands during manual manipulator control. Although it is ideal to have two crewmembers available to operate the system, provisions should be made for both single and multi-crew access for manipulator, camera, and support equipment operation. Postural stability and comfort are essential for extended-duration operations (more than 7 hours) and delicate and high-stress activities – this is achieved by the design of restraint systems that accommodate a full spectrum of expected users and activities.

The MSS RWS and JEMRMS Workstation have very similar topography. For each system, the central operator interface is a PCS, located in the center of the workstation. Two hand controllers, each providing 3 degree-of-freedom manipulations, are positioned adjacent to the computer interface. The hand controller located on the left side of the workstation provides translational manipulator control (fore/aft, left/right, up/down). The hand controller on the right side provides rotational control of the manipulator's end effector (pitch, yaw, roll).

Each of the workstations provides multiple television views and includes hardware switches/buttons for controlling the cameras associated with each system. When fully assembled, the ISS will have 14 different locations where cameras can be installed. There are also cameras located on each of the manipulators.

GRAPHICAL USER INTERFACE COMMONALITY

The graphical user interfaces, icons, and procedures have to be readily understood by multinational crewmembers. The ISS Program Display and Graphics Commonality Standard documents the standards and guidelines utilized in the design and implementation of displays and graphical products used by both the onboard crew as well as ground control centers.

These standards are used in the development of onboard crew displays, reference drawings, and other graphical reference material. It is imperative for nomenclature and graphics used in robotics procedures to exactly match those used on the respective displays.

The presentation of joints and segment position and attitude information on robotics displays is consistent. The joints are arranged in order of their physical location in the manipulator, starting with the base. Starting from the left side of the graphic, each major component appears in the same order as on the actual manipulator hardware. The arm booms are further labeled with "Base" and "Tip", particularly important for manipulators that have the capability to swap base ends during "walking" maneuvers. Operation-critical component feedback on the displays includes base location (if applicable), joint angle data, end effector/tool status, payload identification and/or status, and command and display coordinate systems.

Specific display and graphic colors were developed in accordance with industry and international standards and each color has a specific operational meaning. Some key colors used on robotics systems displays are red, yellow, and orange. A red color is used to alert the operator to pay immediate attention to the robot motion in order to avoid a potentially catastrophic event. Yellow indicates an out-of-limits condition or warrants operator attention to the robot motion in order to avoid loss of time. Orange denotes that the robot arm or hardware is in motion, or ready to be commanded into motion. It is also used to notify the operator that the corresponding hardware switch will cause motion when that switch is selected (a switch in this state is referred to as "hot").

Colors are not used as the sole means for identification of the status of a component or subsystem. Alternate cues include labels, telemetry, or other graphical changes to acquire the attention of the operator. In some cases, the background color of each robotics system is tinted to differentiate systems that are similar in content. For example, since the SPDM has two identical manipulator arms, the "general" SPDM pages have a gray background while the Arm1 and Arm2-specific manipulator pages are pink and green, respectively.

Robotics-unique attention indicators (annunciators) are used on displays and appear only when the operator's immediate attention to the system is required. To facilitate proper operator scan techniques, these indicators are replaced by "place holders" when the attention-required event is not occurring. For example, one of the SSRMS attention indicators is used to alert the operator of an impending self-collision (one part of the robot is in close proximity to another part, and a collision would occur if motion in the same direction is not stopped). Although the manipulator control software, if enabled, should prevent actual contact of the manipulator components, it is important for the operator to note when they are approaching regions in which this may occur. Even if a self-collision event is not imminent, the rectangle and text containing the attention indicator are still distinguishable to the operator. Both color changes and flashing of the indicator are used as mechanisms to attain the operator's attention in the self-collision warning region. If the region is further encroached upon, the software will automatically transition the manipulator into a "hold" mode and will accept no further operator inputs. At this point, the indicator will transition to steady yellow and a black border appears around the rectangle.

ALIGNMENT CUES

Many of the mechanical systems used on the ISS for assembly and maintenance operations require precise alignment. During mating and berthing operations, a safe corridor must be maintained by operators to prevent contact of the element, payload, or manipulator with surrounding structures. More important than the physically constrained "hardware" corridor is the narrower "operational" corridor that the operator must remain within for safe operations.

Operational corridors are derived from actual hardware corridors and further modified or biased for ancillary items that can affect the accuracy of the target or alignment system. Allowances must be made for camera/target location and mounting precision complicated further by thermal, vibration, and pressure influences. Target design characteristics and imperfections and the operator's viewing angle of the scene can further impact total system accuracy. The familiar Fitts law challenges of target characteristics are paramount and the arm operators are always conservative in their speed/accuracy tradeoff – accuracy requirements dominate.

Alignment cues are dependent on lighting and shadow conditions and operations timing is enhanced by support from earth-based lighting models. Lighting (natural and supplemental) and shadow conditions are sometimes so extreme to cause interference with human and automatic sensing systems. Ground support to model the anticipated lighting conditions has proven extremely useful in the planning of robotic arm operations. Key status indicators, such as "ready to latch" micro-switches, provide integrated mechanical system support and can be used to assist in achieving final mechanism alignment.

SPATIAL AWARENESS CHALLENGES

The understanding and application of coordinate frames are essential for manual modes of operation and sustained attention during automatic modes of operation. During manual operations, the operator must understand along which axis the arm tip or payload will move and around which point in space it will rotate. Several coordinate frames are used to support the manual operations of ISS manipulators and to generate digital position and attitude displays. Fundamental elements are the Frame of Resolution (FOR), the Display Frame, and the Command Frame. The FOR defines the manipulator or attached payload's multi-dimensional position (x, y, z) and attitude (pitch, yaw, roll). The Display Frame is the reference coordinate frame for the FOR to compute and display the position and attitude. The Command Frame determines the direction of motion of the arm/attached payload when hand controllers are used in a manual mode.

The selection of desired FOR, Display Frame, and Command Frame is a contributing factor in determining the degree of difficulty of a robotics task. The position of the manipulator or attached payload with respect to a base structure, vector of arm maneuver, and available visual cues are the major factors in determining the optimum combination of these coordinate frames.

SUPPLEMENTAL DISPLAYS

The operator's situational awareness is dependent on multiple cameras and derived digital information sources. Field of view, reference frames, and dynamically changing conditions make high demands on the operator's ability to comprehend the current status and determine the implications of the next control input. During the initial stages of ISS assembly, there are limited external cameras available and almost no opportunities for direct viewing of the work site. Camera sources include zoom features that provide both global views and precise local information. Operators frequently must rely on cameras that are mounted on the manipulator itself resulting in a constantly changing point of reference as the task progresses.

Artificial or augmented reality cues, particularly bird's eye views, can be useful assist devices to enhance the operator's situational awareness. While supplemental views can be helpful, they also add complexity to the operator's mental model of the progress of a robotic arm task.

PRESERVATION OF CRITICAL SKILLS DURING LONG-DURATION MISSIONS

Organizational design and operator training are key to mission success. One-gravity simulations are very helpful training aids but will never substitute completely for an on-orbit experience. ISS crews are trained to perform the major assembly tasks that are scheduled to occur during their increment and are also trained in generic robotics skills as well.

Allocation of functions among long-duration ISS crewmembers, short-duration visiting Space Shuttle crews, and ground control is a developing science based on the limited mission experience available. There are relatively few simulators available outside the United States and limited skill-based training opportunities for crews in space. Training efforts are continually being improved through the use of video and computer-based techniques, simulators, and experienced expert advisors. Of particular importance in this organizational context is the provision of "just-in-time training" for tasks to supplement the robotics training that is only one part of a much more extensive mission training program, much of which is currently undertaken in Russia, away from the U.S. and Canadian robotic arm training facilities.

On-orbit training is necessary to maintain the high level of proficiency necessary for safe and effective robotics operations. This real-time training has recently included video teleconferences with Mission Control to discuss task procedures, planned operations, and operational impacts and workarounds in response to systems failures. Training videos and computer-based training systems have been uplinked or provided to ISS crews by visiting Space Shuttle crews.

The design of systems for future long-duration space missions should include provisions for operating the actual system in a "simulation" mode. Although there are drawbacks to this approach – notably the increased use of associated systems that could, in turn, impact failure incidence – stowage concerns for these extremely long-duration missions will dictate the optimum use of all available onboard equipment.

CONCLUSION

In conclusion, robotic arm operations are key to the assembly and maintenance of the ISS; to the transfer of materials from the Space Shuttle; and to the deployment, capture, and maintenance of satellites. When fully assembled, the ISS will have multiple extravehicular robotic systems, with different characteristics and a vast array of objects to be transferred, including EVA astronauts. The work is difficult, delicate, and dangerous. Human factors contributions, including training, are key to mission success. The diverse topics related to the theme of this chapter can be found in the pertinent references, including Bickford (1997), Fowler (1998), National Aeronautics and Space Administration (1997), National Aeronautics and Space Administration (2001), National Space Development Agency of Japan (2000), and Peacock (2019, 2020, 2021).

REFERENCES

Bickford, P. (1997). *Interface Design*. Academic Press, Boston, MA.
Currie, Nancy J., and B. Peacock (2002). "International Space Station Robotic Systems Operations – A Human Factors Perspective," *Proceedings of the Human Factors and Ergonomics Society 46th Annual Meeting*, September 29 – October 4, 2002, Baltimore, Maryland, USA.
Fowler, S. (1998). *GUI Design Handbook*. McGraw Hill, New York, NY.
National Aeronautics and Space Administration (1997*). Portable Computer System (PCS) Displays Software Requirements Specification* (JSC 26975). Space Station Program Office, Houston, TX.
National Aeronautics and Space Administration (2001). *Display and Graphics Commonality Standard, Rev. A* (SSP 50313). Space Station Program Office, Houston, TX.
National Aeronautics and Space Administration (2001). *International Space Station Robotic Overview Training Manual* (TD0102). Mission Operations Directorate, Houston, TX.
National Space Development Agency of Japan (2000). *Japanese Experiment Module Remote Manipulator System (JEMRMS) Training Manual*. NASDA, Japan.
Peacock, B. (2019). *Human Systems Integration*, Self-published manuscript, Fernandina Beach, FL.
Peacock, B. (2020). *How Ergonomics Works*, Self-published manuscript, Fernandina Beach, FL.
Peacock, B. (2021). *Ergonomics Tools and Applications*, Self-published manuscript, Fernandina Beach, FL.

9 Case Study of Cost Growth in Space Systems

INTRODUCTION

This case study, adapted from Rusnock et al. (2023), focuses on predicting cost growth for military and civil space systems. Although the context of the case study is within the national defense realm, the basic qualitative and quantitative analyses therein contained are generally applicable to engineering economic analysis scenarios.

Military and civil space acquisitions have received much criticism for their inability to produce realistic cost estimates. This research seeks to provide space systems cost estimators with a forecasting tool for space system cost growth by identifying factors contributing to growth, quantifying the relative impact of these factors, and establishing a set of models for predicting space system cost growth. The analysis considers data from both the U.S. Department of Defense (DoD) and National Aeronautics and Space Administration (NASA) space programs. The DoD dataset includes 21 space programs that submitted developmental Selected Acquisition Reports between 1969 and 2006. The analysis uses multiple regression to assess 22 predictor variables, finding that communications missions, ground equipment, firm-fixed-price (FFP) contracts, and increased program manager tenure are all predictive of lower cost growth for military space systems. The NASA analysis includes data sets from 71 satellites and spacecraft developed between 1964 and 2004. The analysis uses a two-stage logistic and multiple regression approach to analyze 31 predictor variables finding that smaller programs (by total cost), more massive spacecraft, microgravity missions, and space physics missions are predictive of higher cost growth.

The case study presented in this chapter is an adaptation of the award-winning conference paper from the 2015 Industrial and Systems Engineering Research Conference (ISERC) (Rusnock and White, 2015). In the profession of industrial engineering, case studies provide a sustainable platform for advancing the discipline's body of knowledge by linking past successes to future expectations. The methodology presented in the case study focuses on the military operating environment but it is generally applicable to nonmilitary systems in business and industry. This confirms the cliché that as the military goes, so does the nation. Lessons from the case study can find relevance in general industrial engineering applications in diverse areas. Based on data availability, the premise of the case study is cost growth. However, direct or indirect inferences can be made to schedule growth based on quantitative relationships implied in the triple constraints (or iron triangle) of cost, schedule, and performance (Badiru, 2019). Another desirable aspect of the case study is the emphasis on space systems, beyond typical aerial systems. The national importance of space, as the next technical frontier (Keller, 2014; Bridgeforth, 2015; Heron, 2015; Badiru, 2016), is confirmed by the 2020 formation of the U.S. Space Force as a new and focused arm of the U.S. military.

DOI: 10.1201/9781003328445-9

Over the past few decades, the United States has grown increasingly dependent on space systems in order to conduct military and civil operations. The combination of this dependence and recent difficulties in space systems acquisition has given cause for alarm among national leaders (Defense Science Board and Air Force Scientific Advisory Board Joint Task Force, 2003; Allard, 2005). Space acquisition programs, such as the Space Based Infrared System (SBIRS) High and the National Polar-orbiting Operational Environmental Satellite System (NPOESS) have received considerable national attention (and Congressional criticism) for their excessive cost growth. For example, the development of SBIRS High, originally estimated in 1996 to cost approximately $4 billion, experienced revised cost estimates in 2009 to over $10 billion, and as of 2015 was expected to eventually experience cost growth of 400% (Defense Science Board and Air Force Scientific Advisory Board Joint Task Force, 2003; Allard, 2005; Day, 2009; Defense Industry Daily, 2015). Similarly, NPOESS has almost doubled in cost growth, from an original estimate of approximately $6 billion to estimates of over $11 billion (Government Accountability Office, 2006). The extreme cost growth experienced by these and other military and civil space acquisition programs has led to the perception that the space acquisition process is "broken", ultimately eroding the credibility of the space acquisition community (Gourley, 2004; Lee, 2004; Allard, 2005).

Concerns over cost growth are not limited to the DoD; NASA has also received criticism for its inability to produce realistic cost estimates. While NASA points to technical problems and funding shortages as major contributors to cost growth, the Government Accountability Office (GAO) finds that NASA lacks a rigorous process for accurately estimating cost. NASA cost estimators lack access to sound financial and technical data, and thus are unable to produce reliable estimates (Government Accountability Office, 2004).

How can the military and civil space acquisition communities correct their problems of excessive cost growth? One critical step is to improve cost estimates for space systems acquisition. This study seeks to mitigate the impact of cost growth by providing space systems cost estimators with a forecasting tool for estimating growth. In so doing, this study identifies factors contributing to space system cost growth, quantifies the relative impact of these factors, and provides a set of models for predicting space system cost growth.

THE DATA ENVIRONMENT

This study engages in an exploratory analysis of DoD and NASA space system cost growth in order to identify a potential methodology for predicting growth as well as to identify predictors of growth (Rusnock, 2008). Due to the differences in available data for DoD and NASA space systems, this study analyzes military and civil systems separately.

Engineering economic analysis is a fundamental tool for justifying investments in highly technical endeavors (Sullivan et al., 2003; Newnan et al., 2004; Badiru and Omitaomu, 2007; White et al., 2014). In this regard, the triple constraints of cost, schedule, and performance are also in effect, albeit at different levels of criticality. It is, thus, essential to track cost growth, particularly in military operations, that

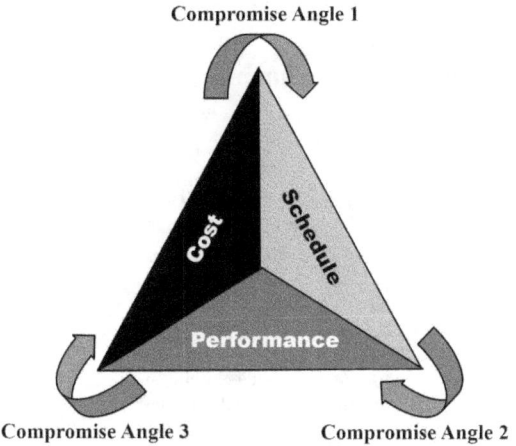

FIGURE 9.1 The triple constraints of cost, schedule, and performance.

are under pressure to be more cost-conscious. In many instances, trade-offs must be exercised among cost, schedule, and performance with respect to the urgency of national defense requirements. Figure 9.1 illustrates the typical interplay of the elements of the triple constraints, sometimes called the iron triangle (Badiru, 2019). If an analyst wants to be mathematically rigorous, he or she could do a combinatorial analysis of trade-offs with the constraints with two of the three taken at a time: Cost and Schedule, Cost and Performance, and Schedule and Performance.

DoD CASE STUDY

Using a specific DoD Dataset, the analysis of DoD space systems applies linear regression to identify predictors for military space system cost growth. The DoD dataset uses cost and programmatic information annually reported to Congress through Selected Acquisition Reports (SARs). The 21 space programs included in this dataset are satellites, launch vehicles, strategic missiles, and space-related ground equipment reported in SARs between 1969 and 2006. The dataset includes the total costs of all variance categories (with the adjustments described herein) for both development and procurement costs associated with the development phase of system acquisition. Like most studies using SAR data, this study uses a mix of completed and ongoing programs. To ensure enough cost data were available for each program, this study follows the example of McNichol by setting a minimum requirement that a program had to have reported SARs for at least three years in order to qualify for inclusion (McNichol, 2005). Table 9.1 presents summary data for the 21 space programs.

DoD RESPONSE VARIABLES

Military space programs often undergo adjustments in quantity over the course of their development cycle. Because decreases and increases in quantity can mask true cost growth or inadvertently display cost growth where none has actually occurred,

TABLE 9.1
DoD Program Summary Data

Description	Values
# of Programs	21
Commodity Type	Satellite: 10
	Launch Vehicle/Missile: 6
	Ground Equipment: 5
Mission Area	Communications: 7
	Navigation: 2
	Earth Observation: 4
	Space Support: 11
Prime Contractor	Lockheed Martin: 11
	Boeing: 7
	Northrop Grumman: 4
	Other Contractor: 2
Contract Type	CPAF: 13
	FFP: 7
	Neither: 4
Average PM tenure	Minimum: 1
	Median: 2
	Mean: 2.1
	Maximum: 3
% Cost Growth	Minimum: −0.24
	Median: 0.20
	Mean: 0.55
	Maximum: 4.34
% Unit Cost Growth	Minimum: −0.86
	Median: 0.22
	Mean: 0.42
	Maximum: 1.89

quantity adjustments are necessary. This study adjusts for quantity using two separate methodologies, thus, providing two response variables, *Total Cost Growth* and *Per Unit Cost Growth*.

Total Cost Growth. The *Total Cost Growth* response variable compares the actual or most current estimate (CE) to the original development estimate (DE) produced at program initiation. The CE is adjusted for quantity changes by subtracting the cost growth listed in the Quantity Variance category of the SAR. Both the DE and CE are adjusted for inflation by converting SAR data from 1969 to 2006 into Constant Year 2007 (CY07) dollars. Cost Growth is calculated as a percentage using Equation 9.1. Due to the unusually large cost growth factor for the Titan IV (over 400%) and the five strategic missile programs, this study omitted these data points from the analysis of Total Cost Growth in order to prevent misleading results and suggesting higher cost growth than the norm. The remaining 15 programs were used in the Total Cost Growth analysis.

$$\text{Total Cost Growth} = \frac{(CE - DE)}{DE} \qquad (9.1)$$

Per Unit Cost Growth. Similar to the *Total Cost Growth* response variable, the *Per Unit Cost Growth* response variable compares the actual or most current estimate (CE) to the original development estimate (DE), adjusted for inflation into CY07 dollars. The *Per Unit Cost Growth* is calculated using both the full dataset and the reduced (without strategic missiles) dataset. The *Per Unit Cost Growth* response variable accounts for changes in quantity by adjusting both the DE and the CE into a per unit cost, using Equation 9.2:

$$\text{Per Unit Cost Growth} = \frac{\dfrac{CE}{\#\ of\ units\ for\ CE} - \dfrac{DE}{\#\ of\ units\ for\ DE}}{\dfrac{DE}{\#\ of\ units\ for\ DE}} \qquad (9.2)$$

DoD PREDICTOR VARIABLES

Below are descriptions of the predictor variables used in this analysis for analyzing cost growth in DoD space systems. The predictor variables were selected based on available data and expected connection to cost growth from existing literature.

Commodity Type. This predictor variable captures the commodity classification of the particular program and is represented by three separate binary variables: *Satellite, Launch Vehicle/Missile,* and *Ground Equipment.*

Mission Area. This attribute captures the type of mission for the program and is represented by four separate binary variables: *Communications, Navigation, Earth Observation,* and *Space Support.* Note that a program may belong to more than one mission area. For multiple mission areas, all relevant variables are coded as "one".

Program Size. This attribute is a continuous variable measured in terms of actual system cost in CY07 dollars.

Development Duration. This is a continuous variable measuring the number of years between the first development estimate and the last development estimate.

Program Managers. Two Program Manager (PM) variables have been included: *# of PMs* and *PM Tenure. # of PMs* is a discrete variable measuring the number of Program Managers during the system's development phase, while *PM Tenure* is a continuous variable measuring the average tenure of the Program Managers, calculated by dividing the *Development Duration* by the *# of PMs.*

Baselines. Two baseline variables have been included to attempt to capture major programmatic changes. The first, *# of Baselines* is a discrete variable measuring the number of baselines for the system. A system baseline is the cost, schedule, technical performance, and quantity commitment established at the start of the program's development. It is used as the standard by which the program's subsequent performance is evaluated. As major changes occur in a program, the program may be "re-baselined", thus resulting in more than one baseline over the course of a program's development. The second, *Baselines/yr* is a continuous variable measuring the number of baselines adjusted for the length of the development, calculated by dividing the *# of Baselines* by the *Development Duration*.

Contract Type. This attribute captures the type of contracts that were used in the development of the program. It is measured by two separate binary variables: *Cost Plus Award Fee (CPAF)* and *Firm Fixed Price (FFP)*. Note that there are other contract types available, but these two are the most widely used for the programs in the DoD dataset. Many programs use multiple contract types for various portions of the development. Thus, it is feasible for a program to have both a CPAF and an FFP contract or to have neither. For multiple contract types, both binary variables are coded as "one".

Lead Service: Air Force. This is a binary variable capturing whether the system was developed by the Air Force. If this variable is predictive, then a positive co-efficient would be associated with cost growth.

Cost Breach. This is a binary variable capturing whether the system experienced a cost breach (cost exceeded 10% of the objective cost reported in the Acquisition Program Baseline) during development.

Schedule Breach. This is a binary variable capturing whether the system experienced a schedule breach (schedule exceeded six months from the objective schedule reported in the Acquisition Program Baseline) during development.

Prime Contractor. This attribute is measured by four separate binary variables: *Lockheed Martin*, *Boeing*, *Northrop Grumman*, and *Other Contractor*. Note that in some cases, a system may have multiple prime contractors. For multiple contract types, all relevant binary variables are coded as "one".

TOTAL COST GROWTH ANALYSIS

Since other military cost growth studies and current defense acquisition policies are inconsistent as to whether strategic missiles are treated as space systems, this study is interested in analyzing the defense data both with and without the inclusion of these systems. Equation 9.3 provides the final model for predicting *Total Cost Growth* (CG_T) using the reduced 15-program dataset that excludes strategic missiles. It was derived using step-wise multiple regression using all predictor variables. The adjusted R^2 was used to identify the preferred model, resulting in the preferred model being the simple linear regression provided in Equation 9.3. All results hereafter are significant at $\alpha = 0.05$.

$$CG_T = 0.715 - 0.577 * (Communications) \tag{9.3}$$

Thus, from this model, space systems with communications missions are predicted to experience a total cost growth of 13.8%, whereas those systems with other than communications missions are expected to incur a total cost growth of 71.5%. The R^2 for this model is 0.29 and the adjusted R^2 is 0.24. The relatively low adjusted R^2 values indicate that while the communications mission area is predictive, other unknown factors still play a sizable role in whether a program will experience cost growth. A number of diagnostics were performed to test the model for robustness. These diagnostics include Cook's Distance, a test for influential data points, the Shapiro-Wilk Test for the normality of the residuals, and the Breusch-Pagan Test for the constant variance of the residuals (Neter et al., 1996). All diagnostic tests for this model (as well as the remaining subsequent presented models) yielded satisfactory values.

Unfortunately, no models passed satisfactory diagnostics for Total Cost Growth for the full 21-program dataset when including strategic missiles. Therefore, including results from these might lead readers to deduce statistically invalid results. Consequently, we chose not to present results for the full 21-program dataset; too much uncertainty remained while modeling Total Cost Growth with missiles to present any credible results.

PER UNIT COST GROWTH

Equation 9.4 provides the model for *Per Unit Cost Growth* excluding strategic missile observations:

$$\text{CG}_{U3} = 0.945 - 1.153 * (Ground\ Equip) - 0.666 * (FFP) \qquad (9.4)$$

Thus, space systems other than ground equipment on FFP contracts are expected to experience 27.9% per unit cost growth and those not on FFP contracts are expected to have 94.5% per unit cost growth. Ground equipment systems are expected to come under budget, with those using FFP contracts anticipated to experience −87.4% per unit cost growth and those with other than FFP contracts experiencing −20.8% per unit cost growth. The R^2 for this model is 0.56 and the adjusted R^2 is 0.49. All diagnostic tests for this model yielded satisfactory values.

The entire DoD dataset (including strategic missiles) yielded two potential models for predicting *Per Unit Cost Growth*. These models are equally predictive, and thus both are presented. Equations 9.5 and 9.6 provide these models:

$$\text{CG}_{U1} = 0.869 - 0.941 * (Ground\ Equip) - 0.661 * (FFP) \qquad (9.5)$$

Note that this model includes the same factors as Equation 9.5, which excluded strategic missiles. As with Equation 9.5, this model predicts lower per unit cost growth for ground equipment systems and systems with FFP contracts. The R^2 for this model is 0.50 and the adjusted R^2 is 0.44. All diagnostic tests for this model yielded satisfactory values.

The second model for predicting *Per Unit Cost Growth* is:

$$\text{CG}_{U2} = 2.069 - 1.178 * (Ground\ Equip) - 0.664 * (PM\ Tenure) \qquad (9.6)$$

From this model, each additional year of average Program Manager (PM) tenure is associated with a decrease in per unit cost growth of 66.4 percentage points. In the DoD dataset, average PM tenures range from one to three years. Thus, a space system, other than ground equipment, with an average PM tenure of one year is predicted to experience 140.5% per unit cost growth, whereas an equivalent system with an average PM tenure of three years is anticipated to experience 7.7% per unit cost growth. The R^2 for this model is 0.47 and the adjusted R^2 is 0.41. All diagnostic tests for this model yielded satisfactory values.

DoD COST GROWTH DISCUSSION

The models for DoD cost growth, summarized in Table 9.2, reveal that communications missions, ground equipment, FFP contracts, and increased program manager tenure are all predictive of lower cost growth. One possible explanation for reduced cost growth for communications missions and ground equipment is the prevalence of these technologies in the commercial sector. The widespread use and availability of these types of technologies in both public and private sectors may make these technologies more mature, and thus less risky, than other missions and commodity types. Ground equipment also benefits from the ability to test in an operational environment, a luxury that most space-based systems do not have.

The study also found FFP contracts (contracts with a specified payment amount) to be predictive of lower cost growth. This finding is consistent with Rossetti's finding that FFP contracts are predictive of reduced support cost growth for DoD weapon systems (Rossetti and White, 2004). However, it is important to remember that regression analysis identifies relationships but does not indicate cause and effect. It could be that FFP contracts provide contractors with an incentive to minimize cost growth since additional costs reduce their profit margin. An alternative explanation is that government programs use FFP contracts on programs that are relatively well-defined, have mature technologies, and are less risky. Thus, while the models indicate that FFP contracts are associated with reduced cost growth, the models do not reveal whether these types of contracts lead to lower cost growth or are deliberately chosen for the types of programs that would be expected to have lower cost growth.

Both the Young Task Force (Defense Science Board and Air Force Scientific Advisory Board Joint Task Force, 2003) and the Government Accountability Office (Day, 2009) studies identify high turnover of Program Managers as a factor that contributes to the

TABLE 9.2
DoD Cost Growth Regression Equations

Model Title	Model	Fit	Exclusions
Total Cost Growth	$CG_T = 0.715 - 0.577*(Communications)$	R^2 0.29 Adj. R^2 0.24	Strategic Missiles and Titan IV
Per Unit Cost Growth Model 1	$CG_U = 0.869 - 0.941*(Ground\ Equip)$ $- 0.661*(FFP)$	R^2 0.50 Adj. R^2 0.44	None
Per Unit Cost Growth Model 2	$CG_U = 2.069 - 1.178*(Ground\ Equip)$ $- 0.664*(PM\ Tenure)$	R^2 0.47 Adj. R^2 0.41	None
Per Unit Cost Growth Model 3	$CG_U = 0.945 - 1.153*(Ground\ Equip)$ $- 0.666*(FFP)$	R^2 0.56 Adj. R^2 0.49	Strategic Missiles

cost growth of space systems. This study supports this assessment, finding that longer Program Manager tenures are predictive of lower cost growth (and thus, shorter tenures are predictive of higher cost growth). Additionally, this study quantifies the impact of Program Manager tenure, finding that a one-year increase in Program Manager tenure is associated with a reduction in per unit cost growth of 66.4 percentage points.

NASA CASE STUDY

The NASA dataset includes data compiled from the 1992 Institute for Defense Analyses study (Tyson, 1992), the 2004 NASA Headquarters Cost Analysis Division study (Schaffer, 2004), publicly available online NASA sources including National Space Science Data Center (NSSDC Master Catalog, 2007), JPL Mission and Space Craft Library (MSL: Mission and Spacecraft Library, 2007), and NASA's Science Mission Directorate (Science Missions, 2014), as well as data collected through personal communications with program personnel. The NASA dataset includes cost, schedule, and descriptive data for 71 satellites and spacecraft from 1964 to 2004. The cost data includes total development costs through the launch of the spacecraft. Unlike the DoD dataset that includes both completed and ongoing programs, all 71 NASA programs have completed development and been launched.

NASA RESPONSE VARIABLES

NASA Cost Growth. Similar to the DoD cost growth response variables, the NASA *Cost Growth* response variable compares actual development costs to the initial estimate in terms of a percentage, using Equation 9.7. The estimate and actual costs are adjusted for inflation by converting both into Constant Year 2007 (CY07) dollars. Because NASA programs tend to be formulated around the development of a single system, with each spacecraft considered a separate program, quantity adjustments are not required.

$$\text{Cost Growth} = \frac{(\text{Actual-Estimate})}{\text{Estimate}} \qquad (9.7)$$

Preliminary analyses and model diagnostics revealed numerous problems, i.e., non-constant variance and nonnormality of model residuals when analyzing cost growth using the entire NASA dataset. These problems were due to a number of programs experiencing extremely high-cost growth (over 100%). In fact, the NASA *Cost Growth* response represents a bimodal distribution whereby each subgroup followed a normal distribution and suggested separately modeling low *Cost Growth* and high *Cost Growth* for NASA space systems.

This study employs logistic regression to determine whether a program is likely to experience high-cost growth using the binary variable *High-Cost Growth* with a value "1" for high-cost growth programs and a value of "0" for low-cost growth programs. Logistic regression uses maximum likelihood to estimate the parameters that best model the data, creating a likelihood function that expresses the probability (as a value between 0 and 1) that the independent variables predict the dependent variable. For our purposes, a probability greater than or equal to 0.5 (50%) predicts a program will experience high-cost growth, and a probability of less than 0.5 predicts the program will *not* experience high-cost growth.

Thus, the first response variable, *High-Cost Growth*, is a binary variable measuring the likelihood that a program will experience high-cost growth. The second response variable, *Cost Growth*, is a continuous variable that measures the percentage of cost growth that a program is likely to incur. Because the NASA *Cost Growth* response represents a bimodal distribution, the analysis models this variable twice, once for each distribution, thus, providing a High-Cost Growth Linear Regression Model and a Low-Cost Growth Linear Regression Model.

NASA PREDICTOR VARIABLES

Below are descriptions of the predictor variables used in this analysis for analyzing cost growth in NASA space systems. The predictor variables were selected based on available data and expected connection to cost growth from existing literature. Table 9.3 presents summary data for the 71 space programs.

> *Program Size.* There are two program size variables: *Initial Program Size* and *Final Program Size. Initial Program Size* is a continuous variable measured in terms of the original estimate of the system cost in CY07 dollars. *Final Program Size* is a continuous variable measured in terms of actual system cost in CY07 dollars.
>
> *Program Start.* This is a continuous variable measured as the number of years from 1964 (this year was chosen as the baseline since it represents the earliest start date in the dataset). The dataset includes programs initiated between 1964 and 2000.
>
> *Schedule Characteristics.* There are three scheduled predictor variables: *Estimated Time to Launch, Actual Time to Launch,* and *Schedule Growth. Estimated Time to Launch* is a continuous variable measuring the initially planned launch date in the number of months from program initiation. *Actual Time to Launch* is a continuous variable measuring the actual launch date in the number of months from program initiation. *Schedule Growth* is a continuous variable measuring the percent growth between the estimated and actual launch schedule: (actual date − planned date)/planned date.
>
> *Mission Area.* This attribute captures the type of mission for the program as cataloged in the National Space Science Data Center database (NSSDC Master Catalog, 2007). Mission area is measured by ten binary variables: *Space Physics, Engineering, Earth Science, Planetary Science, Astronomy, Solar Physics, Human Crew, Communications, Life Science,* and *Microgravity.* Note that a program may have more than one mission area. For multiple mission areas, all relevant variables are coded as "one".
>
> *International Participation.* This is a binary variable capturing whether countries other than the United States participated in the scientific, technical, or design elements of the spacecraft.
>
> *Developer.* This attribute measures the primary organization responsible for designing and manufacturing the spacecraft. It is measured by nine binary

TABLE 9.3
NASA Program Summary Data

Description	Values
# of Programs	71
Initial Program Size	Minimum: $9.9K
	Median: $182.2K
	Mean: $796.0K
	Maximum: $27,802K
Program Start	1964–2000
Mission Area	Space Physics: 24
	Engineering: 11
	Earth Science: 22
	Planetary Science: 21
	Astronomy: 22
	Solar Physics: 16
	Human Crew: 4
	Communications: 4
	Life Science: 3
	Microgravity: 2
Developer	NASA: 10
	Jet Propulsion Laboratory: 9
	Johns Hopkins University: 5
	Lockheed Martin: 19
	Boeing: 9
	Northrop Grumman: 6
	DoD: 2
	International Developer: 6
	Other Developer: 13
Mass	Total Mass:
	Minimum: 124 kg
	Median: 954.5 kg
	Mean: 4079.6 kg
	Maximum: 109,000 kg
	Dry Mass:
	Minimum: 117 kg
	Median: 794.5 kg
	Mean: 4520 kg
	Maximum: 90,607 kg

variables: *NASA, Jet Propulsion Laboratory, Johns Hopkins University, Lockheed Martin, Boeing, Northrop Grumman, DoD, International Developer*, and *Other Developer*.

Life Span. There are two life span predictor variables: *Design Life* and *Actual Life*. *Design Life* is a continuous variable measuring the intended design life of the spacecraft in months (the average design life was 43 months). *Actual Life* is a continuous variable measuring the actual life span, or current estimate of the life span for programs still in operation, in months.

Mass. There are two mass predictor variables: *Total Mass* and *Dry Mass.* *Total Mass* is a continuous variable measuring the total mass of the spacecraft in kilograms, including consumable propellants, at the time of launch. *Dry Mass* is a continuous variable measuring the mass of the spacecraft in kilograms, excluding consumable propellants, at the time of launch.

NASA DATASET ANALYSIS

Due to the bimodal nature of the distribution, modeling NASA *Cost Growth* consists of a two-stage process. The first stage includes a logistic regression model to determine whether a program is likely to experience high-cost growth. The second stage includes separate linear regression models for both high- and low-cost growth to determine the likely percentage of cost growth. Note that the low-cost growth model also includes zero and negative cost growth.

> *Logistic Regression Models.* The analysis results in two logistic models for predicting the likelihood of a program to experience high-cost growth; Equations 9.8 and 9.9 provide these models. These models are equally predictive, and thus both are presented.

$$L_{HCG1} = \frac{e^{2.140 - 0.058*(Initial\ Program\ Size) + 0.001*(Total\ Mass)}}{1 + e^{2.140 - 0.058*(Initial\ Program\ Size) + 0.001*(Total\ Mass)}} \qquad (9.8)$$

For our purposes, an L_{HCG} probability greater than or equal to 0.5 (50%) predicts a program will experience high-cost growth, and a probability of less than 0.5 predicts the program will not experience high-cost growth. Based on the coefficients, larger *Initial Program Sizes* decreases the likelihood of experiencing high-cost growth (data range, in 1000s: \$9.9–\$27,801.9); whereas more massive spacecraft increases the likelihood of experiencing high-cost growth (data range: 124–109,000 kg).

Although the customary diagnostics used for linear regression analysis are not available for logistic regression, the Receiver Operating Characteristic (ROC) curve can be used to assess the model's accuracy. The ROC curve distinguishes between false positives and true positives; an ROC curve area of 1.0 would be perfectly predictive. There were no issues associated with the fitting of the logistic regression models considering we avoided repeated measurements or matched data setup, we assured low to no multicollinearity of independent variables by assessing Pearson's correlation between candidate independent variables, and we had a moderate sample size where the ratio was between 7 and 10:1 of independent variables to the total number of data points. The ROC curve is and continues to be the general standard for assessing logistic regression models. The Hosmer Lemeshow test generally doesn't take overfitting into account and tends to have low power (test-wise). Additionally, the selection of the number of subgroups used is relatively arbitrary. The logistic model in Equation 9.7 has an ROC curve area of 0.95, indicating an estimated accuracy of 95%. Equation 9.8 provides the second model for predicting *High-Cost Growth*:

$$L_{HCG2} = \frac{e^{0.741 - 0.038*(Initial\ Program\ Size) + 0.001*(Dry\ Mass) + 38.705*(Microgravity)}}{1 + e^{0.741 - 0.038*(Initial\ Program\ Size) + 0.001*(Dry\ Mass) + 38.705*(Microgravity)}} \qquad (9.9)$$

Based on the coefficients, larger *Initial Program Sizes* decreases the likelihood of experiencing high-cost growth (data range, in 1000s: $9.9–$27,801.9), more massive spacecraft increases the likelihood of experiencing high-cost growth (data range: 117–90,607 kg), and a microgravity mission increases the likelihood of experiencing high-cost growth. The ROC curve area is 0.94.

> *Linear Regression Models.* The analysis developed separate linear regression models for high and low-cost growth programs. In the dataset, 9 of the 71 programs were categorized as high-cost growth. These models are designed to be used in conjunction with the logistic regression models provided in Equations 9.8 and 9.9. If the logistic regression models predict that high-cost growth is likely to occur, then the High-Cost Growth Linear Regression Model (Equation 9.10) can be used to predict the likely percent cost growth. Similarly, if the logistic regression models predict that high-cost growth is not likely to occur, the Low-Cost Growth Linear Regression Model (Equation 9.11) can be used to predict the likely percent cost growth.

The High-Cost Growth Linear Regression Model is:

$$CG_H = 1.232 + 1.037 * (Space\ Physics) \tag{9.10}$$

This model predicts high-cost growth programs with space physics missions will experience 226.9% cost growth, and those with other types of missions will experience 123.2% cost growth.

The Low-Cost Growth Linear Regression Model is:

$$CG_L = 0.509 - 0.014 * (Program\ Start) \tag{9.11}$$

This model predicts low-cost growth programs that were initiated in the base year of 1964 (*Program Start* = 0) will experience 50.9% cost growth. Programs initiated in subsequent years are expected to experience a reduction in cost growth of 1.4 percentage points for each additional year after 1964. Thus, those programs initiated in 1994 (*Program Start* = 30) are predicted to have 8.9% cost growth (2000 is the latest initiation year in the data set, *Program Start* = 36).

COST GROWTH INFERENCES

Due to the bimodal nature of the cost growth data for the NASA dataset, the inferential analysis began with dividing the dataset into high-cost growth and low-cost growth programs, and then used logistic regression to assess whether a program was likely to experience high- or low-cost growth (low-cost growth includes no cost growth as well as negative cost growth). Table 9.4 provides the NASA cost growth models.

From the logistic regression analysis, this study found that a larger program size (measured in total cost) decreased the likelihood of being a high-cost growth program, whereas more massive spacecraft and microgravity missions increased the likelihood of being a high-cost growth program. This finding of larger programs being associated

TABLE 9.4

NASA Cost Growth Regression Equations

Model Title	Model	Fit
High-Cost Growth? Logistic Model 1	$L_{HCG} = \dfrac{e^{2.140-0.058*(Initial\ Program\ Size)+0.001*(Total\ Mass)}}{1+e^{2.140-0.058*(Initial\ Program\ Size)+0.001*(Total\ Mass)}}$	R^2 (U) 0.57
High-Cost Growth? Logistic Model 2	$L_{HCG} = \dfrac{e^{0.741-0.038*(Initial\ Program\ Size)+0.001*(Dry\ Mass)+38.705*(Microgravity)}}{1+e^{0.741-0.038*(Initial\ Program\ Size)+0.001*(Dry\ Mass)+38.705*(Microgravity)}}$	R^2 (U) 0.50
High-Cost Growth Linear Model	$CG_H = 1.232 + 1.037*(Space\ Physics)$	R^2 0.77 Adj. R^2 0.74
Low-Cost Growth Linear Model	$CG_L = 0.509 - 0.014*(Program\ Start)$	R^2 0.18 Adj. R^2 0.16

with lower cost growth is consistent with many other cost growth studies (Drezner et al., 1993; Pannell, 1994; Dameron et al., 2002; McCrillis et al., 2003; Schaffer, 2004). Drezner et al. explain that smaller programs are more likely to experience high-cost growth due to minimal oversight and because equivalent costs and increases in costs represent proportionally greater amounts of the total cost for smaller programs (Drezner et al., 1993). Further study is recommended to determine the cause of the increased likelihood of high-cost growth for more massive spacecraft and microgravity missions. While this increased likelihood could be an indication of the increased technical complexity of these types of systems, it may also be an indication of other problems unique to these programs, such as inadequate cost-estimating procedures, deficient program acquisition processes, or other technical or scientific issues.

After using the logistic regression to determine the likelihood of high-cost growth, the linear regression models are then used for determining the quantity of cost growth. For those programs that are likely to experience high-cost growth, the amount of cost growth increases for those programs from a space physics mission. Again, further study is recommended to identify the root causes of this relationship.

For programs predicted to experience low-cost growth by the logistic model, the program start date is the best predictor of the amount of cost growth, with more recent programs associated with lower cost growth. Further study is recommended to determine if this relationship is an indication of improved program acquisition or cost-estimating processes.

CONCLUSION

This research consists of an exploratory analysis of space systems cost growth in order to provide defense and civil cost estimators and space system acquirers with a set of models to aid in predicting cost growth. Since many of the systems that the defense space acquisition community will be tasked to acquire will be systems other than ground-based equipment and communications systems, the analysis suggests that cost estimators and acquirers should anticipate that other systems are likely to experience higher cost growth and should plan accordingly. Additionally, this research indicates that longer Program Manager tenures are associated with decreased cost growth. The respective model predicts that increasing the average Program Manager tenure by one year will reduce the anticipated per unit cost growth by 66.4 percentage points. Thus, this research supports the recommendation of Young's Task Force to increase the length of Program Managers' tenure (Defense Science Board and Air Force Scientific Advisory Board Joint Task Force, 2003).

Similarly, while NASA will continue to procure a variety of systems, with wide ranges of program sizes, spacecraft sizes, and mission types, it would behoove cost estimators and acquirers to recognize that smaller programs, more massive spacecraft, and microgravity and space physics missions are more vulnerable to experiencing higher cost growth. Additionally, cost estimators and acquirers should also recognize that while larger programs are less vulnerable to cost growth, they are more vulnerable to schedule growth.

Using the template of this case study, potential future applications for consideration include testing the models provided herein with additional data from other NASA and DoD programs to validate models or establish more robust models; exploring data from commercial and classified systems to see if the models and methodology applied herein are applicable to these types of systems; exploring additional predictor variables not evaluated herein, such as requirements, systems engineering expertise, or technological maturity; augmenting NASA data with additional space programs to see if the bimodal distributions for cost growth hold; and/or further exploring relationships identified herein using a more in-depth qualitative analysis. It is often said that "as the military goes, so goes the nation". This case study provides an example of a military cost-growth methodology that can be adapted for civilian applications. Commercial and nonmilitary organizations can adapt the methodology of this case study for their own cost-growth analyses of interest.

REFERENCES

Allard, W., United States Senator (2005). Address to the National Defense Industrial Association, 2005 Space Policy and Architecture Symposium.

Badiru, A. B. (2016). Space: An Engineering Frontier, *ASEE PRISM*, October 2016, p. 56.

Badiru, A. B. (2019). *Project Management: Systems, Principles, and Applications.* (2nd ed.). Taylor & Francis CRC Press, Boca Raton, FL.

Badiru, A. B., and O. A. Omitaomu (2007). *Computational Economic Analysis for Engineering and Industry.* Taylor & Francis CRC Press, Boca Raton, FL.

Bridgeforth, S., J. D. Ritschel, E. White, and G. Keaton (2015). Using Earned Value Data to Forecast the Duration of Department of Defense Space Acquisition Programs, *Journal of Cost Analysis and Parametrics* 8(2): 92–107. https://doi.org/10.1080/1941658X.2015.1062817

Dameron, M. E., R. L. Coleman, J. R. Summerville, C. L. Pullen, D. M. Snead, and S. L. Van Drew, (2002). Modeling the Effect of Program Size on Cost Growth. *Society of Cost Estimating and Analysis (SCEA) 2002 National Conference*, June 12, 2002, Scottsdale, AZ.

Day, D. (2009). Things Are Rough All Over..., *The Space Review.* November 16, 2009. http://thespacereview.com/article/1511/1

Defense Industry Daily (2015). Budget Busters: The USA's SBIRS-High Missile Warning Satellites. *Defense Industry Daily.* March 3, 2015. https://www.defenseindustrydaily.com/Despite-Problems-SBIRS-High-Moves-Ahead-With-3rd-Satellite-Award-05467/

Defense Science Board and Air Force Scientific Advisory Board Joint Task Force (2003). *Acquisition of National Security Space Programs.* Office of the Under Secretary of Defense for Acquisition, Technology, and Logistics, Washington, DC.

Drezner, J. A., J. M. Jarvaise, R. W. Hess, P. G. Hough, and D. Norton (1993). *An Analysis of Weapon System Cost Growth.* RAND Corporation, Santa Monica, CA.

Gourley, Scott R, 2004. Space Warriors and Wizards, *Military Aerospace Technology,* 3(3): 1–4.

Government Accountability Office (2004). *NASA: Lack of Disciplined Cost-Estimating Processes Hinders Effective Program Management.* GAO-04-642 report to Subcommittee on Science, House of Representatives. Washington, DC.

Government Accountability Office (2006). *Space Acquisitions: DoD Needs to Take More Action to Address Unrealistic Initial Cost Estimates of Space Systems.* GAO-07-96 report to Subcommittee on Strategic Forces, Committee on Armed Services, House of Representatives. Washington, DC.

Heron, Rey A. (2015). Forecasting DoD mid-acquisition space program final costs, using WBS level 2 and 3 data, Master thesis, Air Force Institute of Technology, Dayton, OH.

Keller, S., P. Collopy, and P. Componation (2014). What Is Wrong With Space System Cost Models? A Survey and Assessment of Cost Estimating Approaches, *Acta Astronautica* 93: 345–351. https://doi.org/10.1016/j.actaastro.2013.07.014

Lee, D. E. (2004). Space Reform, *Air and Space Power Journal* 18(2): 103–112.

McCrillis, J., Office of the Secretary of Defense, Cost Analysis Improvement Group (2003). Cost Growth of Major Defense Programs, *36ᵗʰ Annual Department of Defense Cost Analysis Symposium*, January 30, 2003, Williamsburg, VA.

McNicol, D. L. (2005). *Cost Growth in Major Weapon Procurement Programs.* (2nd ed.). Institute for Defense Analyses, Alexandria, VA.

MSL (2007) "MSL: Mission and Spacecraft Library" in *JPL Mission and Spacecraft Library.* http://msl.jpl.nasa.gov. December 22, 2007.

Neter, J., M. H. Kutner, W. Wasserman, and C. J. Nachtsheim (1996). *Applied Linear Statistical Models.* (4th ed.). Irwin, Chicago, IL.

Newnan, D. G., T. G. Eschenbach, and J. P. Lavelle (2004). *Engineering Economic Analysis.* (9th ed.). Oxford University Press, New York, NY.

NSSDC (2007). "NSSDC Master Catalog" in *National Space Science Data Center.* October 30, 2007. http://nssdc.gsfc.nasa.gov

Pannell, B. J. (1994). *A Quantitative Analysis of Factors Affecting Weapon System Cost Growth.* MS thesis. Naval Postgraduate School, Monterey, CA (ADA280342).

Rossetti, M. B., and E. D. White (2004). A Two-Pronged Approach to Estimate Procurement Cost Growth in Major DoD Weapon Systems, *Journal of Cost Analysis and Management* 6(2), 11–21.

Rusnock, C. F. (2008). Predicting Cost and Schedule Growth for Military and Civil Space Systems, MS Thesis, Air Force Institute of Technology, Dayton, OH, 2008.

Rusnock, C. F., and E. D. White (2015). Predicting Cost Growth for Military and Civil Space Systems. *Proceedings of the 2015 Institute of Industrial Engineers (IIE) Industrial & Systems Engineering Research Conference*, Nashville, TN, May 30-June 2, 2015. *Recognition: Engineering Economy Track Best Paper Award.*

Rusnock, C. F., E. D. White, and A. B. Badiru (2023). Chapter 16. In: B. Bidanda (Ed.), *Maynard's Industrial & Systems Engineering Handbook*, (6th ed.). McGraw-Hill, New York, NY.

Schaffer, M. (2004). NASA Cost Growth: A Look at Recent Performance. Presentation to NASA Comptroller, NASA Headquarters. 5 February 2004.

"Science Missions" in *Science@NASA: The Science Mission Directorate.* December 22, 2007. http://science.hq.nasa.gov/missions/phase.html

Sullivan, W. G., E. M. Wicks, and J. T. Luxhoj (2003). *Engineering Economy.* (12th ed.). Pearson Education, Inc., Upper Saddle River, NJ.

Tyson, K. W., J. R. Nelson, and D. M. Utech (1992). *A Perspective on Acquisition of NASA Space Systems.* IDA Document D-1224. Institute for Defense Analyses (ADA263246), Alexandria, VA.

White, J. A., K. S. Grasman, K. E. Case, K. L. Needy, and D. B. Pratt (2014). *Fundamentals of Engineering Economic Analysis.* John Wiley, Hoboken, NJ.

10 People Skills in Systems Management

INTRODUCTION

The human side of industrial engineering is what sets the profession apart from other engineering disciplines (Badiru, 2023a). In this regard, soft skills for the bastion of the practice of industrial engineering. Using a systems framework, industrial engineers enmesh qualitative and quantitative tools and techniques to manage integrated systems of people, tools, and processes. The premise of this chapter is to combine the systems viewpoint and human cognitive reasoning to improve functional efficiency and productivity in the work environment. A key systems tool discussed in the chapter is the DEJI Systems Model®, which provides a structural pathway for human-based work design, work evaluation, work justification, and work integration (Badiru, 2019). The quote below demonstrates how soft skills facilitate building, actuating, and managing teams.

> Working together productively requires that the work be designed appropriately to permit teamwork.

Adedeji Badiru

DESIGN OF WORK

The planning, organizing, and coordination of work elements all fall under the category of design under the DEJI model. It is essential that a structured approach be applied to work design right at the outset. Retrofitting a work element only after problems develop not only impedes the overall progress of work in an organization but also leads to inefficient use of limited human and material resources. This stage of the model is expected to guide work designers onto the path of strategic thinking about work elements down the line rather than just the tactical manipulation of work for the present needs. In this regard, Badiru (2016) says, "Right next to innovation, structured methods for producing effective work results are a survival imperative for every large organization".

EVALUATION OF WORK

Following the design of a work element, the DEJI model calls for a formal evaluation of the intended purpose of the work vis-à-vis other work elements going on in the organization. Such an evaluation may lead to a need to go back and re-design the work element. Evaluation can be done as a combination of both qualitative and quantitative assessment of the work element, depending on the specific nature of the work, the main business of the organization, and the managerial capabilities of the organization.

DOI: 10.1201/9781003328445-10

JUSTIFICATION OF WORK

According to the concept of the DEJI model, not only should a work element be designed and evaluated, it should also be formally and rigorously justified. If this is not done, errant work elements will creep into the organizational pursuits. What is worth doing is worth doing well. Otherwise, it should not be done at all. The principles of Lean operations (Agustiady and Badiru, 2012) suggest weeding out functions that do not add value to the organizational goal. In this regard, each and every work element needs to be justified. But it should be realized that not all work elements are expected to generate physical products in the work environment. A work element may be justified on the basis of adding value to the well-being of the worker with respect to mental, emotional, spiritual, and physical characteristics. The point of this stage of the DEJI model is to ensure that the work element is needed at all. Or, the do-nothing alternative is always an option.

INTEGRATION OF WORK

This last stage of the DEJI model is of utmost importance, but it is often neglected. The model affirms that the most sustainable work elements are those that fit within the normal flow of operations, existing practices, or other expectations within an organization. Does the work fit in? Will a new work element under consideration be an extraneous pursuit or a detraction in the overall work plan? If a work element is not integrated with other normal pursuits, it cannot be sustained for the long haul. This is why many organizations suffer from repeated program starts and stops. For example, inasmuch as worker wellness programs are desirable pursuits in an organization, they cannot be sustained if not integrated into the culture and practices of the organization. Unintegrated flash-in-the-plan programs, activities, and work elements often fall by the wayside over time. For this stage of the DEJI model, work elements must be tied to the end goal of the person and the organization. Badiru (2016) remarks that "aimless work is so insidious because it tends to covertly masquerade as fruitful labor", which is not connected to real organizational goals. This is the premise of applying the multi-dimensional hierarchy of needs (of the worker and the organization) as discussed in a subsequent section of this chapter.

Work is the means to accomplishing a goal. For the purpose of the theme of this book, work is literally considered as an activity to which strength, mental acuity, and resources are applied to get something done. This may involve a sustained physical and/or mental effort to overcome impediments in the pursuit of an outcome, an objective, a result, or a product. In an operational context, work can be viewed as a process of performing a defined task or activity, such as research, development, operations, maintenance, repair, assembly, production, administration, sales, software development, inspection, data collection, data analysis, teaching, and so on. The opening quote in this chapter signifies the meaning of work as a means to an end goal, where workers thrive and the work effort succeeds. If you understand your work, from a system's viewpoint, you will enjoy the work and you will want to do more of it. In this book, we view work as a "work system" rather than work in isolation or disconnected from other human endeavors.

A systems view of work is essential because of the several factors and diverse people that may be involved in the performance of the work. There are systems and subsystems involved in each execution of work whether small or large, whether simple or complex, and whether localized or multi-locational. For an activity to be workable consistently, all the attendant factors and issues must be taken into account. If some crucial factors are neglected, the *workability* of the activity may be in jeopardy. For the purpose of a systems view of work, we define a system as a collection of interrelated elements (subsystems) working together synergistically to generate a collective and composite outcome (value) that is higher than the sum of the individual outcomes of the subsystems. Even simple tasks run the risk of failure if some minute subsystem is not accounted for. The following two specific systems engineering models are used for the purpose of the theme of this book.

1. The V-Model of Systems Engineering
2. The DEJI Model of Systems Quality Integration (Badiru, 2023b)

Figure 10.1 presents an illustration of the V-Model applied to a manufacturing enterprise consisting of a series of work elements. Although the model is most often used in the software development process, it is also applicable to hardware development as well as general work in systems engineering. In the model, instead of moving down in a linear fashion, the process steps are bent upward after the coding phase to form the typical V-shape. The V-model demonstrates the relationships between each phase of the development life cycle and its associated phase of testing. The horizontal and vertical axes represent time or project completeness (left-to-right) and the level of abstraction, respectively.

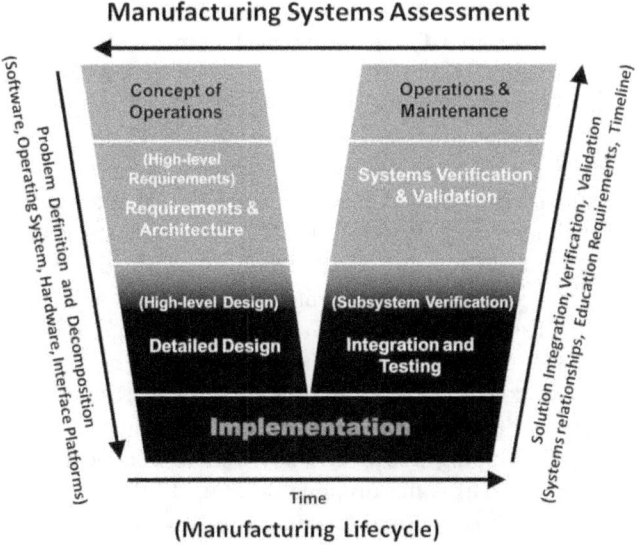

FIGURE 10.1   V-model of systems engineering applied to manufacturing system.

There are several different ways that a work system can be developed and delivered using the V-model. The best development strategy depends on how much the work analyst knows about the system for which work is being designed. Three basic design strategies can be used:

Once-Through Approach

In this case, we plan, specify, and implement the complete work system in one pass through the V-shape. This approach, also sometimes called the "waterfall" approach, works well if the vision is clear, the requirements are well understood and stable, and there is sufficient funding. The problem is that there isn't a lot of flexibility or opportunity for recovery if the vision, work environment, or requirements change substantially.

Incremental Approach

Here, we plan and specify the work system and then implement it in a series of well-defined increments or phases, where increment delivers a portion of the desired end goal. This is like moving through value-adding increments of the work. In this case, we are making one pass through the first part of the V-shape and then iterating through the latter part for each phased-in increment. This is a common strategy for field equipment deployment where system requirements and design can be incrementally implemented and deployed across a given area in several phases and several projects.

Evolutionary Approach

In this approach, we plan, specify, and implement an initial system capability, learn from the experience with the initial system, and then define the next iteration to address issues and extend capabilities (or add value). Thus, we refine the Concept of Operations, add and change system requirements, and revise the design as necessary. We will continue with successive iterative refinements until the work system is complete. This strategy can be shown as a series of "Vs" that are placed end to end since system operation on the right side of the "V" influences the next iteration. This strategy provides the most flexibility but also requires project management expertise and vigilance to make sure the development stays on track. It also requires patience from the stakeholders as the design moves along in incremental stages.

Figure 10.2 presents an illustration of the application of the DEJI model for systems integration of work factors. The key benefit of the DEJI model is that it moves the effort systematically through the stages of Design, Evaluation, Justification, and Integration. The approach pings the work analyst about what needs to be addressed in each stage so that the ball is not dropped on critical requirements. The greatest aspect of the DEJI model is the final stage that calls out the need to integrate the work with other efforts within the work environment. If there is a disconnect, then the work may end up being a misplaced effort. If a work effort is properly integrated, then it will be sustainable.

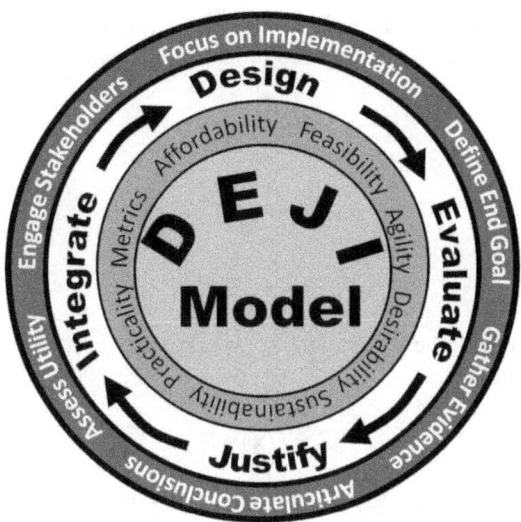

FIGURE 10.2 Framework for the application the DEJI model to systems integration.

SYSTEMS DEFINITION FOR SOFT SKILLS

A system is defined as a collection of interrelated elements working together synergistically to achieve a set of objectives. Any work is essentially a collection of interrelated activities, tasks, people, tools, resources, processes, and other assets brought together in the pursuit of a common goal. The goal may be in terms of generating a physical product, providing a service, or achieving a specific result. This makes it possible to view any work as a system that is amenable to all the classical and modern concepts of systems management.

Work is the foundation of everything we do. Having some knowledge is not enough. The knowledge must be applied to do something in the pursuit of objectives. Work management facilitates the application of knowledge and willingness to actually accomplish tasks. Where there is knowledge, willingness often follows. But it is work execution that actually gets things done. From the very basic tasks to the very complex endeavors, work management must be applied to get things done. It is, thus, essential that systems thinking to be a part of the core of every work pursuit in business, industry, education, government, and even at home. In this regard, a systems approach is of utmost importance because work accomplishment is a "team sport" that has several underlying factors as elements of the overall work system.

TECHNICAL SYSTEMS CONTROL

Classical technical system control focuses on control of the dynamics of mechanical objects, such as a pump, electrical motor, turbine, rotating wheel, and so on. The mathematical basis for such control systems can be adapted (albeit in iconical formats) for management systems, including work management. This is because both technical and managerial systems are characterized by inputs, variables, processing, control, feedback, and output. This is represented graphically by input-process-output relationship block diagrams. Mathematically, it can be represented as:

$$z = f(x) + \varepsilon,$$

where:

  z = output
  $f(.)$ = functional relationship
  ε = error component (noise, disturbance, etc.)

For multi-variable cases, the mathematical expression is represented as vector-matrix functions as shown below:

$$\mathbf{Z} = \mathbf{f}(\mathbf{x}) + \mathbf{E},$$

where each term is a matrix. $\mathbf{Z}$ is the output vector, $\mathbf{f}(\mathbf{x})$ is the input vector, and $\mathbf{E}$ is the error vector. Regardless of the level or form of mathematics used, all systems exhibit the same input-process-output characteristics, either quantitatively or qualitatively. The premise of this book is that there should be a cohesive coupling of quantitative and qualitative approaches in managing a work system. In fact, it is this unique blending of approaches that makes systems application for work management more robust than what one will find in mechanical control systems, where the focus is primarily on quantitative representations.

SOFT SKILLS AND ORGANIZATIONAL PERFORMANCE

Systems engineering efficiency and effectiveness are of interest across the spectrum of work management for the purpose of improving organizational performance. Managers, supervisors, and analysts should be interested in having systems engineering serve as the umbrella for improving work efforts throughout the organization. This will get everyone properly connected with the prevailing organizational goals as well as create collaborative avenues among the personnel. Systems application applies across the spectrum of any organization and encompasses the following elements:

- Technological systems (e.g., engineering control systems and mechanical systems)
- Organizational systems (e.g., work process design and operating structures)
- Human systems (e.g., interpersonal relationships and human-machine interfaces)

A systems view of the world makes everything work better and work efforts more likely to succeed. A systems view provides a disciplined process for the design, development, and execution of work both in technical and nontechnical organizations. One of the major advantages of a systems approach is the win-win benefit for everyone. A systems view also allows the full involvement of all stakeholders and constituents of a work center. This is very well articulated by the Chinese saying below:

Tell me and I forget;
Show me and I remember;
Involve me and I understand.

Confucius, Chinese Philosopher

For example, the pursuit of organizational or enterprise transformation is best achieved through the involvement of everyone, from a systems perspective. Every work environment is very complex because of the diversity of factors involved, including the following:

- The worker's overall health and general wellbeing
- The worker's physical attributes
- The worker's mental abilities
- The worker's emotional stability
- The worker's spiritual interests
- The worker's psychological profile

There are differing human personalities, technical requirements, expectations, and environmental factors. Each specific context and prevailing circumstances determine the specific flavor of what can and cannot be done in the work environment. The best approach for effective work management is to adapt to what each work requirements and specifications. This requires taking a systems view of the work. The work systems approach presented in this book is needed for "working across" organizations, countries, across cultures, and across unique nuances of each project. This is an essential requirement in today's globalized and intertwined personal and professional goals. A systems view requires a disciplined embrace of multidisciplinary execution of work in a way that each component complements other components in the work system. Formal work management represents an excellent platform for the implementation of a system approach. A comprehensive work management program requires control techniques, such as operations research, operations management, forecasting, quality control, and simulation to achieve goals. Traditional approaches to management use these techniques in a disjointed fashion, thus, ignoring the potential interplay among the techniques. The need for an integrated systems-based work management worldwide has been recognized for decades. As long ago as 1993, the World Bank reported that a lack of systems accountability led to several worldwide project failures. The bank, which has loaned more than $300 billion to developing countries over the last half-century, acknowledged that there has been a dramatic rise in the number of failed projects around the world. In other words, the work efforts failed. A lack of an integrated system approach to managing the projects was cited as one of the major causes of failure. More recent reports by other organizations point to the same flaws in managing global projects; and point to the need to apply better project management to major projects. Press headlines in April 2008 highlighted that "Defense needs better management of projects". This was in the wake of a government audit that reveals gross inefficiencies in managing large defense projects. In a national news release on April 1, 2008, it was reported that auditors at the Government Accountability Office (GAO) issued a scathing review of dozens of the Pentagon's biggest weapons systems, citing that ships, aircraft, and satellites are billions of dollars over budget and years behind schedule. According to the review, "95 major systems have exceeded their original budgets by a total of $295 billion; and are delivered almost two years late on average". Further, "none of the systems that the GAO looked at had met all of the standards for best management practices during their development stages". Among programs noted for increased development costs were the "Joint Strike Fighter and Future Combat Systems". The costs of those programs had risen "36 percent and 40 percent, respectively", while C-130 avionics modernization costs had risen 323%. And, while "Defense Department officials have tried to improve the procurement process, the GAO" added that "significant policy changes have not yet translated into best practices on individual programs". In the view of this book, a failed program is an indicator of failed work efforts. A summary of the report of the accounting office reads:

> Every dollar spent inefficiently in developing and procuring weapon systems is less money available for many other internal and external budget priorities, such as the global war on terror and growing entitlement programs. These inefficiencies also often result in the delivery of less capability than initially planned, either in the form of fewer quantities or delayed delivery to the warfighter.

In as much as the military represents the geo-political-economic landscape of a nation, the above assessment is representative of what every organization faces, whether public or private. In systems-based project management, it is essential that related techniques be employed in an integrated fashion so as to maximize the total project output. One definition of systems project management (Badiru, 2012) is stated as follows:

> Systems project management is the process of using systems approach to manage, allocate, and time resources to achieve systems-wide goals in an efficient and expeditious manner.

The above definition calls for a systematic integration of technology, human resources, and work process design to achieve goals and objectives. There should be a balance in the synergistic integration of humans and technology. There should not be an over-reliance on technology nor should there be an over-dependence on human processes. Similarly, there should not be too much emphasis on analytical models to the detriment of common-sense human-based decisions.

SOFT SKILLS IN SYSTEMS ENGINEERING

Systems engineering is growing in appeal as an avenue to achieve organizational goals and improve operational effectiveness and efficiency. Researchers and practitioners in business, industry, and government are all embrace systems engineering implementations. So, what is systems engineering? Several definitions exist. Below is one quite comprehensive definition:

> Systems engineering is the application of engineering to solutions of a multi-faceted problem through a systematic collection and integration of parts of the problem with respect to the lifecycle of the problem. It is the branch of engineering concerned with the development, implementation, and use of large or complex systems. It focuses on specific goals of a system considering the specifications, prevailing constraints, expected services, possible behaviors, and structure of the system. It also involves a consideration of the activities required to assure that the system's performance matches the stated goals. Systems engineering addresses the integration of tools, people, and processes required to achieve a cost-effective and timely operation of the system.

INCOSE (International Council on Systems Engineering) defines systems engineering follow:

> Systems Engineering is an interdisciplinary approach and means to enable the realization of successful systems. It focuses on defining customer needs and required functionality early in the development cycle, documenting requirements, then proceeding with design synthesis and system validation while considering the complete problem.

Systems Engineering integrates all the disciplines and specialty groups into a team effort forming a structured development process that proceeds from concept to production to operation. Systems Engineering considers both the business and the technical needs of all involved with the organizational goals.

HUMAN WORK SYSTEMS

Logistics can be defined as the planning and implementation of a complex task, the planning and control of the flow of goods and materials through an organization or manufacturing process, or the planning and organization of the movement of personnel, equipment, and supplies. Complex projects represent a hierarchical system of operations. Thus, we can view a project system as a collection of interrelated projects all serving a common end goal. Consequently, we present the following universal definition:

> Work systems logistics is the planning, implementation, movement, scheduling, and control of people, equipment, goods, materials, and supplies across the interfacing boundaries of several related projects.

Conventional organizational management must be modified and expanded to address the unique logistics of work systems.

SYSTEMS CONSTRAINTS IN HUMAN WORK

Systems management is the pursuit of organizational goals within the constraints of time, cost, and quality expectations. The iron triangle model shows that project accomplishments are constrained by the boundaries of quality, time, and cost. In this case, quality represents the composite collection of project requirements. In a situation where precise optimization is not possible, there will have to be trade-offs among these three factors of success. The concept of the iron triangle is that a rigid triangle of constraints encases the project. Everything must be accomplished within the boundaries of time, cost, and quality. If better quality is expected, a compromise along the axes of time and cost must be executed, thereby altering the shape of the triangle. The trade-off relationships are not linear and must be visualized in a multi-dimensional context. This is better articulated by a 3D-view of the systems constraints. Scope requirements determine the project boundary and trade-offs must be done within that boundary. If we label the eight corners of the box as (a), (b), (c),..., (h), we can iteratively assess the best operating point for the project. For example, we can address the following two operational questions:

1. From the point of view of the project sponsor, which corner is the most desired operating point in terms of the combination of requirements, time, and cost?
2. From the point of view of the project executor, which corner is the most desired operating point in terms of the combination of requirements, time, and cost?

Note that all the corners represent extreme operating points. We notice that point (e) is the do-nothing state, where there are no requirements, no time allocation, and no cost incurrence. This cannot be the desired operating state of any organization that seeks to remain productive. Point (a) represents an extreme case of meeting all requirements with no investment of time or cost allocation. This is an unrealistic extreme in any practical environment. It represents a case of getting something for nothing. Yet, it is the most desired operating point for the project sponsor. By comparison, point (c) provides the maximum possible for requirements, cost, and time. In other words, the highest levels of requirements can be met if the maximum possible time is allowed and the highest possible budget is allocated. This is an unrealistic expectation in any resource-conscious organization. You cannot get everything you ask for to execute a project. Yet, it is the most desired operating point for the project executor. Considering the two extreme points of (a) and (c), it is obvious that the project must be executed within some compromise region within the scope boundary. A graphical representation will review a compromised surface with peaks and valleys representing give-and-take trade-off points within a constrained rectangular box (see Badiru, 2023b). The challenge is to come up with some analytical modeling techniques to guide decision-making over the compromise region. If we could collect sets of data over several repetitions of identical projects, then we could model a decision surface that can guide future executions of similar projects. Such typical repetitions of an identical project are most readily apparent in construction projects, e.g., residential home development projects.

HUMAN WORK EFFECTIVENESS

Systems influence philosophy suggests the realization that you control the internal environment while only influencing the external environment. The inside (controllable) environment is represented as a black box in the typical input-process-output relationship. The outside (uncontrollable) environment is bounded by the cloud representation. In the comprehensive systems structure, inputs come from the global environment are moderated by the immediate outside environment and are delivered to the inside environment. In an unstructured inside environment, work functions occur as blobs. A "blobby" environment is characterized by intractable activities where everyone is busy but without a cohesive structure of input-output relationships. In such a case, the following disadvantages may be present:

- Lack of traceability
- Lack of process control
- Higher operating cost
- Inefficient personnel interfaces
- Unrealized technology potentials

Organizations often inadvertently fall into the blobs structure because it is simple, low-cost, and less time-consuming until a problem develops. A desired alternative is to model the project system using a systems value-stream structure that will require a coordinated linear flow of elements from left to right. This will use a proactive and problem-preempting approach to execute projects. This alternative has the following advantages:

- Problem diagnosis is easier
- Accountability is higher
- Operating waste is minimized
- Conflict resolution is faster
- Value points are traceable

QUANTITATIVE VALUE ASSESSMENT OF HUMAN WORK

A technique that can be used to assess the overall value-added components of a process improvement program is the systems value model (SVM), which is an adaptation of the manufacturing system value (MSV) model presented by Troxler and Blank (1989). The model provides an analytical decision aid for comparing process alternatives. Value is represented as a p-dimensional vector:

$$V = f\left(A_1, A_2, \ldots, A_p\right)$$

where $A = \left(A_1, \ldots, A_n\right)$ is a vector of quantitative measures of tangible and intangible attributes. Examples of process attributes are quality, throughput, capability, productivity, cost, and schedule. Attributes are considered to be a combined function of factors, x_1, expressed as:

$$A_k\left(x_1, x_2, \ldots, x_{m_k}\right) = \sum_{i=1}^{m_k} f_i\left(x_i\right)$$

where $\{x_i\}$ = set of m factors associated with attribute A_k $(k = 1, 2, \ldots, p)$ and f_i = contribution function of factor x_i to attribute A_k. Examples of factors include reliability, flexibility, user acceptance, capacity utilization, safety, and design functionality. Factors are themselves considered to be composed of indicators, v_i, expressed as

$$x_i(v_1, v_2, \ldots, v_n) = \sum_{j=1}^{n} z_i(v_i)$$

where $\{v_j\}$ = set of n indicators associated with factor $x_i (i = 1, 2, \ldots, m)$ and z_j = scaling function for each indicator variable v_j. Examples of indicators are project responsiveness, lead time, learning curve, and work rejects. By combining the above definitions, a composite measure of the value of a process can be modeled as:

$$V = f(A_1, A_2, \ldots, A_p)$$

$$= f \left\{ \left[\sum_{i=1}^{m_1} f_i \left(\sum_{j=1}^{n} z_j(v_j) \right) \right]_1, \left[\sum_{i=1}^{m_2} f_i \left(\sum_{j=1}^{n} z_j(v_j) \right) \right]_2, \ldots, \left[\sum_{i=1}^{m_k} f_i \left(\sum_{j=1}^{n} z_j(v_j) \right) \right]_p \right\}$$

where m and n may assume different values for each attribute. A subjective measure to indicate the utility of the decision maker may be included in the model by using an attribute weighting factor, w_i, to obtain a weighted PV:

$$PV_w = f(w_1 A_1, w_2 A_2, \ldots, w_p A_p)$$

where

$$\sum_{k=1}^{p} w_k = 1, \quad (0 \le w_k \le 1)$$

With this modeling approach, a set of process options can be compared on the basis of a set of attributes and factors. To illustrate the model above, suppose three IT options are to be evaluated based on the four attribute elements in Table 10.1: *capability, suitability, performance,* and *productivity* For this example, the value vector is defined as:

$$V = f(\text{capability, suitability, performance, productivity})$$

Capability: The term "capability" refers to the ability of IT equipment to satisfy multiple requirements. For example, a certain piece of IT equipment may only provide computational service. A different piece of equipment may be capable of generating reports in addition to computational analysis, thus, increasing the service variety that can be obtained. In the computational analysis, the levels of increase in service variety from the three competing equipment types are 38%, 40%, and 33%, respectively. *Suitability:* "Suitability" refers to the appropriateness of the IT equipment for current operations. For example, the respective percentages of operating scope for which the three options are suitable are 12%, 30%, and 53%. *Performance:*

TABLE 10.1
Comparison of IT Work Value Options

IT Equipment Options	Suitability ($k = 1$)	Capability ($k = 2$)	Performance ($k = 3$)	Productivity ($k = 4$)
Option A	0.12	0.38	0.18	0.02
Option B	0.30	0.40	0.28	−1.00
Option C	0.53	0.33	0.52	−1.10

"Performance", in this context, refers to the ability of the IT equipment to satisfy schedule and cost requirements. In the example, the three options can, respectively, satisfy requirements for 18%, 28%, and 52% of the typical set of jobs. ***Productivity:*** "Productivity" can be measured by an assessment of the performance of the proposed IT equipment to meet workload requirements in relation to the existing equipment. For the example, in Table 10.1, the three options, respectively, show normalized increases of 0.02, −1.0, and −1.1 on a uniform scale of productivity measurement. A plot of the histograms of the respective "values" of the three IT options was evaluated to find Option C as the best "value" alternative in terms of suitability and performance. Option B shows the best capability measure, but its productivity is too low to justify the needed investment. Option A offers the best productivity but its suitability measure is low. The analytical process can incorporate a lower control limit into the quantitative assessment such that any option providing value below that point will not be acceptable. Similarly, a minimum value target can be incorporated into the graphical plot such that each option is expected to exceed the target point on the value scale.

The relative weights used in many justification methodologies are based on the subjective propositions of decision-makers. Some of those subjective weights can be enhanced by the incorporation of utility models. For example, the weights shown in Table 10.1 could be obtained from utility functions. There is a risk of spending too much time maximizing inputs at "point-of-sale" levels with little time defining and refining outputs at the "wholesale" systems level. Without a systems view, we cannot be sure we are pursuing the right outputs.

SOFT SKILLS FOR PROJECT MANAGEMENT

Project management continues to grow as an effective means of managing functions in any organization. Project management should be an enterprise-wide systems-based endeavor. Enterprise-wide project management is the application of project management techniques and practices across the full scope of the enterprise. This concept is also referred to as management by project (MBP). Management by project is a contemporary concept that employs project management techniques in various functions within an organization. MBP recommends pursuing endeavors as project-oriented activities. It is an effective way to conduct any business activity. It represents a disciplined approach that defines any work assignment as a project. Under MBP, every undertaking is viewed as a project that must be managed just like a traditional project. The characteristics required of each project so defined are

1. An identified scope and a goal
2. A desired completion time
3. Availability of resources
4. A defined performance measure
5. A measurement scale for review of work

An MBP approach to operations helps in identifying unique entities within functional requirements. This identification helps to determine where functions overlap and how they are interrelated, thus paving the way for better planning, scheduling, and control. Enterprise-wide project management facilitates a unified view of organizational goals and provides a way for project teams to use the information generated by other departments to carry out their functions.

The use of project management continues to grow rapidly. The need to develop effective management tools increases with the increasing complexity of new technologies and processes. The life cycle of a new product to be introduced into a competitive market is a good example of a complex process that must be managed with integrative project management approaches. The product will encounter management functions as it goes from one stage to the next. Project management will be needed throughout the design and production stages of the product. Project management will be needed in developing marketing, transportation, and delivery strategies for the product. When the product finally gets to the customer, project management will be needed to integrate its use with those of other products within the customer's organization.

The need for a project management approach is established by the fact that a project will always tend to increase in size even if its scope is narrowing. The following three literary laws are applicable to any project environment:

Parkinson's law: work expands to fill the available time or space.
Peter's principle: people rise to the level of their incompetence.
Murphy's law: whatever can go wrong will.
Badiru's Rule: the grass is always greener where you most need it to be dead.

An integrated systems project management approach can help to diminish the adverse impacts of these laws through good project planning, organizing, scheduling, and control.

INTEGRATED SYSTEMS IMPLEMENTATION

Project management tools can be classified into three major categories:

1. Qualitative tools. These are the managerial tools that aid in the interpersonal and organizational processes required for project management.

2. Quantitative tools. These are analytical techniques that aid in the computational aspects of project management.
3. Computer tools. These are software and hardware tools that simplify the process of planning, organizing, scheduling, and controlling a project. Software tools can help in both the qualitative and quantitative analyses needed for project management.

Although individual books dealing with management principles, optimization models, and computer tools are available, there are few guidelines for the integration of the three areas for project management purposes. In this book, we integrate these three areas for a comprehensive guide to project management. The book introduces the *Triad Approach* to improve the effectiveness of project management with respect to schedule, cost, and performance constraints within the context of systems modeling. Figure 10.3 illustrates this emphasis from a work systems management perspective. The approach considers not only the management of the work itself but also the management of all the worker-related functions that support the work.

It is one thing to have a quantitative model but it is a different thing to be able to apply the model to real-world problems in a practical form. The systems approach presented in this book illustrates how to make the transition from model to practice.

A system approach helps to increase the intersection of the three categories of project management tools and, hence, improve overall management effectiveness. A crisis should not be the instigator for the use of project management techniques. Project management approaches should be used upfront to prevent avoidable problems rather than to fight them when they develop. What is worth doing is worth doing well, right from the beginning.

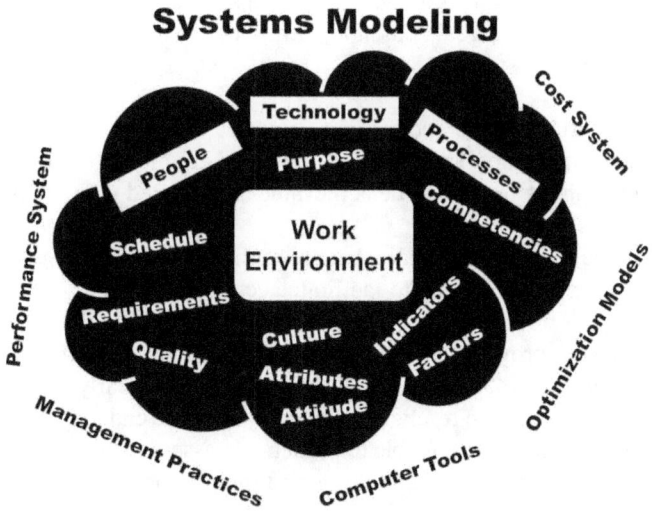

FIGURE 10.3 Project systems modeling in the work environment.

WORKER-BASED FACTORS FOR EFFICIENCY AND EFFECTIVENESS

The premise of this book is that the critical factors for systems success revolve around people and the personal commitment and dedication of each person. No matter how good a technology is and no matter how enhanced a process might be, it is ultimately the people involved that determine success. This makes it imperative to take care of people issues first in the overall systems approach to project management. Many organizations recognize this, but only a few have been able to actualize the ideals of managing people productively. Execution of operational strategies requires forthrightness, openness, and commitment to get things done. Lip service and arm waving are not sufficient. Tangible programs that cater to the needs of people must be implemented. It is essential to provide incentives, encouragement, and empowerment for people to be self-actuating in determining how best to accomplish their job functions. A summary of critical factors for systems success encompasses the following:

Total system management: hardware, software, and people.

- Operational effectiveness
- Operational efficiency
- System suitability
- System resilience
- System affordability
- System supportability
- System life cycle cost
- System performance
- System schedule
- System cost

Systems engineering tools, techniques, and processes are essential for project life-cycle management to make goals possible within the context of **SMART** principles, which are represented as follows:

1. Specific: pursue specific and explicit outputs.
2. Measurable: design of outputs that can be tracked, measured, and assessed.
3. Achievable: make outputs to be achievable and aligned with organizational goals.
4. Realistic: pursue only the goals that are realistic and result-oriented.
5. Timed: make outputs timed to facilitate accountability.

Systems engineering provides the technical foundation for executing a project successfully. A systems approach is particularly essential in the early stages of the project in order to avoid having to re-engineer the project at the end of its life cycle. Early systems engineering makes it possible to proactively assess the feasibility of meeting user needs, adaptability of new technology, and integration of solutions into regular operations.

SYSTEMS HIERARCHY FOR HUMAN WORK

The traditional concepts of systems analysis are applicable to the project process. The definitions of a project system and its components are presented next.

> **System.** A project system consists of interrelated elements organized for the purpose of achieving a common goal. The elements are organized to work synergistically to generate a unified output that is greater than the sum of the individual outputs of the components.
>
> **Program.** A program is a very large and prolonged undertaking. Such endeavors often span several years. Programs are usually associated with particular systems. For example, we may have a space exploration program within a national defense system.
>
> **Project.** A project is a time-phased effort of much smaller scope and duration than a program. Programs are sometimes viewed as consisting of a set of projects. Government projects are often called *programs* because of their broad and comprehensive nature. The industry tends to use the term *project* because of the short-term and focused nature of most industrial efforts.
>
> **Task.** A task is a functional element of a project. A project is composed of a sequence of tasks that all contribute to the overall project goal.
>
> **Activity.** An activity can be defined as a single element of a project. Activities are generally smaller in scope than tasks. In a detailed analysis of a project, an activity may be viewed as the smallest, practically indivisible work element of the project. For example, we can regard a manufacturing plant as a system. A plant-wide endeavor to improve productivity can be viewed as a program. The installation of a flexible manufacturing system is a project within the productivity improvement program. The process of identifying and selecting equipment vendors is a task, and the actual process of placing an order with a preferred vendor is an activity.

The emergence of systems development has had an extensive effect on project management in recent years. A system can be defined as a collection of interrelated elements brought together to achieve a specified objective. In a management context, the purposes of a system are to develop and manage operational procedures and to facilitate an effective decision-making process. Some of the common characteristics of a system include:

1. Interaction with the environment
2. Objective
3. Self-regulation
4. Self-adjustment

Representative components of a project system are the organizational subsystem, planning subsystem, scheduling subsystem, information management subsystem, control subsystem, and project delivery subsystem. The primary responsibilities of project analysts involve ensuring the proper flow of information throughout the project system. The classical approach to the decision process follows rigid lines of organizational charts. By contrast, the systems approach considers all the interactions necessary among the various elements of an organization in the decision process.

The various elements (or subsystems) of the organization act simultaneously in a separate but interrelated fashion to achieve a common goal. This synergism helps to expedite the decision process and enhance the effectiveness of decisions. The supporting commitments from other subsystems of the organization serve to counterbalance the weaknesses of a given subsystem. Thus, the overall effectiveness of the system is greater than the sum of the individual results from the subsystems.

The increasing complexity of organizations and projects makes the systems approach essential in today's management environment. As the number of complex projects increase, there will be an increasing need for project management professionals who can function as systems integrators. Project management techniques can be applied to the various stages of implementing a system as shown in the following guidelines:

1. Systems definition. Define the system and associated problems using keywords that signify the importance of the problem to the overall organization. Locate experts in this area who are willing to contribute to the effort. Prepare and announce the development plan.
2. Personnel assignment. The project group and the respective tasks should be announced, a qualified project manager should be appointed, and a solid line of command should be established and enforced.
3. Project initiation. Arrange an organizational meeting during which a general approach to the problem should be discussed. Prepare a specific development plan and arrange for the installation of needed hardware and tools.
4. System prototype. Develop a prototype system, test it, and learn more about the problem from the test results.
5. Full system development. Expand the prototype to a full system, evaluate the user interface structure, and incorporate user training facilities and documentation.
6. System verification. Get experts and potential users involved, ensure that the system performs as designed, and debug the system as needed.
7. System validation. Ensure that the system yields expected outputs. Validate the system by evaluating performance levels, such as the percentage of success in so many trials, measuring the level of deviation from expected outputs, and measuring the effectiveness of the system output in solving the problem.
8. System integration. Implement the full system as planned, ensure the system can coexist with systems already in operation, and arrange for technology transfer to other projects.
9. System maintenance. Arrange for continuing maintenance of the system. Update solution procedures as new pieces of information become available. Retain responsibility for system performance or delegate to well-trained and authorized personnel.
10. Documentation. Prepare full documentation of the system, prepare a user's guide, and appoint a user consultant.

Systems integration permits sharing of resources. Physical equipment, concepts, information, and skills may be shared as resources. Systems integration is now a major concern for many organizations. Even some of the organizations that traditionally compete and typically shun cooperative efforts are beginning to appreciate the value of integrating their operations. For these reasons, systems integration has emerged as a major interest in the business. Systems integration may involve the physical integration of technical components, objective integration of operations, conceptual integration of management processes, or a combination of any of these.

Systems integration involves the linking of components to form subsystems and the linking of subsystems to form composite systems within a single department and/or across departments. It facilitates the coordination of technical and managerial efforts to enhance organizational functions, reduce cost, save energy, improve productivity, and increase the utilization of resources. Systems integration emphasizes the identification and coordination of the interface requirements among the components in an integrated system. The components and subsystems operate synergistically to optimize the performance of the total system. Systems integration ensures that all performance goals are satisfied with a minimum expenditure of time and resources. Integration can be achieved in several forms including the following:

1. Dual-use integration: this involves the use of a single component by separate subsystems to reduce both the initial cost and the operating cost during the project life cycle.
2. Dynamic resource integration: this involves integrating the resource flows of two normally separate subsystems so that the resource flow from one to or through the other minimizes the total resource requirements in a project.
3. Restructuring of functions: this involves the restructuring of functions and reintegration of subsystems to optimize costs when a new subsystem is introduced into the project environment.

Systems integration is particularly important when introducing a new work into an existing system. It involves coordinating new operations to coexist with existing operations. It may require the adjustment of functions to permit the sharing of resources, development of new policies to accommodate product integration, or realignment of managerial responsibilities. It can affect both the hardware and software components of an organization. Presented below are guidelines and important questions relevant to work systems integration.

- What are the unique characteristics of each component in the integrated system?
- How do the characteristics complement one another?
- What physical interfaces exist among the components?
- What data/information interfaces exist among the components?
- What ideological differences exist among the components?
- What are the data flow requirements for the components?

- Are there similar integrated systems operating elsewhere?
- What are the reporting requirements in the integrated system?
- Are there any hierarchical restrictions on the operations of the components of the integrated system?
- What internal and external factors are expected to influence the integrated system?
- How can the performance of the integrated system be measured?
- What benefit/cost documentation are required for the integrated system?
- What is the cost of designing and implementing the integrated system?
- What are the relative priorities assigned to each component of the integrated system?
- What are the strengths of the integrated system?
- What are the weaknesses of the integrated system?
- What resources are needed to keep the integrated system operating satisfactorily?
- Which section of the organization will have primary responsibility for the operation of the integrated system?
- What are the quality specifications and requirements for the integrated systems?

HIERARCHY OF NEEDS OF WORKERS

Maslow's Hierarchy of Needs is very much applicable to any work environment. According to Maslow's theory, the five different orders of human needs are:

1. *Basic Physiological Needs*: includes food, water, shelter, and the like. In modern society, the basic drives of human existence cause individuals to become involved in organizational life. People become participants in the organization that employs them. Thus, at the simplest level of human needs, people are motivated to join organizations, remain in them, and contribute to their objectives.
2. *Security and Safety*: security means many things to different people in different circumstances. For some, it means earning a higher income to assure freedom from what might happen in case of sickness or during old age. Thus, many people are motivated to work harder to seek success that is measured in terms of income. It can also be interpreted as job security. To some people, such as civil servants and teachers, the assurance of life tenure and a guaranteed pension may be strong motivators in their participation in employing organizations.
3. *Social affiliation*: an employee with a reasonable well-paying and secure job will begin to feel that belonging and approval are important motivators in his/her organizational behavior.
4. *Esteem*: the need to be recognized, to be respected, and to have prestige (self-image and the view that one holds of oneself). There is a dynamic interplay between one's own sense of satisfaction and self-confidence on

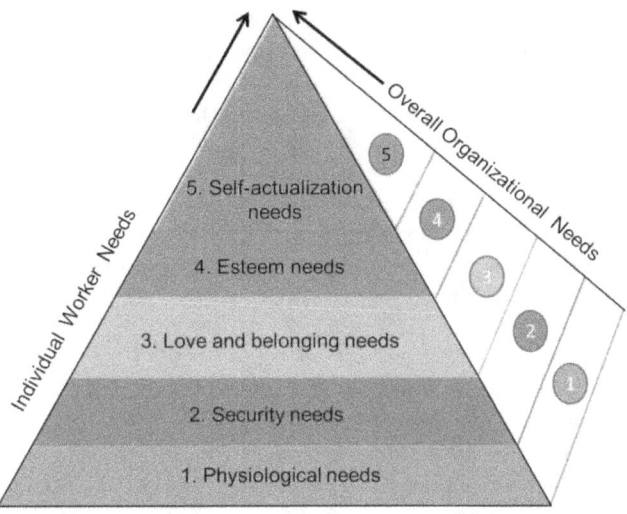

FIGURE 10.4   Multidimensional pyramid of needs of workers and the organization.

one hand, and feedback from others in such diverse forms as being asked for advice on the other hand.

5. *Self-actualization*: the desire to become more and more of what one is, to become everything that one is capable of becoming. The self-actualized person is strongly inner-directed, seeks self-growth, and is highly motivated by loyalty to cherished values, ethics, and beliefs. Not everyone reaches the self-actualized state. It is estimated that these higher level needs are met about 10% of the time.

In any organization, the prevailing hierarchy of needs of the worker must be evaluated in the context of the organization's own hierarchy of needs. In this regard, Badiru (2008) developed an adaptation of the conventional triangle of the hierarchy into a multidimensional pyramid of needs as illustrated in Figure 10.4.

SOFT SKILLS FOR ECONOMIC DEVELOPMENT

Not only is work essential for personal and organizational advancement, it is also essential, from a synergistic systems perspective, for national social and economic development. The gross domestic product (GDP) is the eventual coalescing of work done at various levels of the nation. GDP is the monetary value of all the finished goods and services produced within a country in a defined period of time. Though GDP is usually calculated on an annual basis, it can be calculated on a quarterly basis as well in order to increase the granularity of management policies and practices to increase the national output. GDP includes all private and public consumption, government spending, investments, and exports (minus imports) that occur within a national boundary. In other words, GDP is a broad and composite measurement of a nation's overall economic activity. It can be calculated as follows:

$$GDP = C + G + I + NX$$

where
  C = all private consumption, or consumer spending
  G = the sum of government spending
  I = sum of all the country's investment, including corporate capital expenditures
  NX = the nation's total net exports, calculated as total exports minus total imports

GDP is commonly used as an indicator of the economic health of a country, as well as a gauge of a country's standard of living. If a country's standard of living is high, the workers in the country do well for themselves. Since the mode of measuring GDP is uniform from country to country, GDP can be used to compare the productivity of various countries with a high degree of accuracy. Adjusting for inflation from year to year allows for an objective comparison of current GDP measurement trends. Thus, a country's GDP from any period can be measured as a percentage relative to previous years or quarters. Consequently, GDP trends can be used in measuring a nation's economic growth or decline as well as in determining if an economy is in recession, which has a direct impact on workers and the work environment.

CONCLUSION

In summary for this chapter, we reaffirm that soft skills can impact Total Worker Health® (Schill and Chosewood, 2013; Schill, 2016). A healthy workforce is a more productive workforce. It is through the application of soft skills that organizations can guide employees toward benchmarks and expectations for total work health, which then has an impact on organizational efficiency, effectiveness, and productivity. Work, wellness, and wealth can go hand-in-hand. Workers' health directly affects GDP, based on a systems view of work, cascading from one person's level all the way (collectively) to the national level. Good health is related to good work performance. Health is an individual attribute that compliments each worker's hierarchy of needs. Without good health, even the best worker cannot perform. Without good health, even the best athlete cannot succeed. Without good health, even the most proficient expert cannot manifest his or her expertise. Without total health, even the most dedicated and experienced employee cannot contribute to the accomplishment of the organization's mission. Good health is a key part of the system's view of work as advocated by this chapter (Badiru 2010, 2014; Schulte et al., 2015).

REFERENCES

Agustiady, T., and A. B. Badiru (2012). *Sustainability: Utilizing Lean Six Sigma Techniques.* Taylor & Francis CRC Press, Boca Raton, FL.

Badiru, A. B. (2008). *Triple C Model of Project Management.* Taylor & Francis CRC Press, Boca Raton, FL.

Badiru, A. B. (2010). Half-Life of Learning Curves for Information Technology Project Management, *International Journal of IT Project Management* 1(3): 28–45.

Badiru, A. B. (2012). *Project Management: Systems, Principles, and Applications.* Taylor & Francis CRC Press, Boca Raton, FL.

Badiru, A. B. (2014). Quality Insights: The DEJI Model for Quality Design, Evaluation, Justification, and Integration, *International Journal of Quality Engineering and Technology* 4(4): 369–378.

Badiru, A. B. (2019). *Systems Engineering Models: Theory, Methods, and Applications.* Taylor & Francis/CRC Press, Boca Raton, FL.

Badiru, A. B. (2023a). Soft Skills in Industrial Engineering and Management. In: Bopaya Bidanda (Ed.), *Maynard's Industrial & Systems Engineering Handbook.* (6th ed.). McGraw-Hill, New York, NY, pp. 1383–1400.

Badiru, A. B. (2023b). *Systems Engineering Using DEJI Systems Model: Design, Evaluation, Justification, and Integration with Case Studies and Applications.* Taylor & Francis CRC Press, Boca Raton, FL.

Badiru, I. A. (2016), "Comments about Work Management," Interview of an Auto Industry Senior Engineer about corporate views of work design, Beavercreek, OH, October 29, 2016.

Schill, A. L. (2016), "Advancing Well-being Through Total Worker Health," Keynote Address, 17th Annual 2016 Pilot Research Project (PRP) Symposium, University of Cincinnati, Cincinnati, OH, October 13, 2016.

Schill, A. L., and L. C. Chosewood (2013). The NIOSH Total Worker Health Program: An Overview, *Journal of the Occupational Environmental Medicine* 55(12 suppl): S8–S11.

Schulte, P. A., R. J. Guerin, A. L. Schill, A. Bhattacharya, T. R. Cunningham, S. P. Pandalai, D. Eggerth, and C. M. Stephenson (2015). Considerations for Incorporating 'Well-Being' in Public Policy for Workers and Workplaces, *American Journal of Public Health* 105(8): e31–e44.

Troxler, J. W., and L. Blank (1989). A Comprehensive Methodology for Manufacturing System Evaluation and Comparison, *Journal of Manufacturing Systems* 8(3): 176–183.

11 The Big Picture of Engineering Design

GRAND INTRODUCTION

Like a capstone, this chapter contains a collection of papers, thoughts, ideas, and guidelines comprising the big picture of engineering design with respect to various topics related to the theme of this book: "Industrial Engineering in Systems Design: Guidelines, Practical Examples, Tools, and Techniques".

As a reference book, this chapter may contain materials previously presented in the earlier chapters. This is by design, either to re-emphasize the topics covered or to ensure that readers catch the topics, in case the book is read by picking chapters of interest here and there.

With varying levels of development and background research, some may be up to contemporary standards and recency while others are more from the legacy realm. The chapter is a sequel to the works of Peacock (2019, 2020, 2021) on Human Systems Integration. The purpose of the chapter is to communicate some ideas on both Big Picture and focused industrial engineering topics and offers some approaches and tools that may be of interest to the practitioner and stimulate interest regarding the purpose, methods, and scope of industrial and systems engineering. Four subject areas receive particular attention: the aging workforce, Manufacturing, Transportation, and Education.

MEASUREMENT, MODELS, AND THE AGING WORKFORCE

This chapter addresses some of the technical challenges of measurement and modeling of the aging process and some design opportunities based on experience in the automobile industry. Measurement of aging is addressed by assessment of records for particular well-defined activities, such as athletic events, rather than sample averages. Modeling employs regression methods to describe the aging process and additional effects of sub-system deterioration. The design process uses a case study in which some two hundred features, related to physical, cognitive, and psychosocial ACCESS to transportation, were addressed. The data for this study are publicly available from the World Masters Athletes website http://www.world-masters-athletics.org/ and analysis of contemporary vehicles that contain many of the features identified during a two-year study of the needs of older drivers (The GM ACCESS Car).

MEASUREMENT AND ANALYSIS

Methods for measurement and analysis of aging come in different forms with different levels of reliability, validity, and generalizability. There is no shortage of anecdotal evidence of great feats of old people. At the other end of the spectrum, there are voluminous statistics related to morbidity and mortality. In between, the literature

DOI: 10.1201/9781003328445-11

contains many reports of surveys and formal performance studies of cohorts of older people. The discussions hover between emphases of older human capabilities and their limitations.

MODELING

Models of aging typically show an accelerating performance deterioration curve, much like the Weibull equipment wear-out curves. Indeed, this is a good model of the human reliability process as affected by aging and associated component deterioration (cancer, arthritis, obesity, and neurological deterioration). As with physical systems, human systems are sometimes subject to significant acute incidents (fractures, strokes, and heart attacks) that are followed by recovery at varying rates. Parametric performance studies of physical, sensory, and cognitive capabilities commonly make use of the familiar Normal Distribution for comparison purposes. A shortcoming of such studies is that the results may be simple artifacts of the study sample. A second shortcoming is that for one reason or another, the data may not be Normally distributed; rather it may be more appropriate to use distributions that have skewness (positive or negative shape parameters), such as the Gamma or Gumbel. The major challenge to modeling is to distinguish between aging per se and the associated sub-system failures and deterioration due to injury, illness, or simple disuse.

MEDICINE

The overt purpose of (geriatric) medicine is to assess, treat, and manage those illnesses and acute incidents that are associated with aging but which are not necessarily a result of aging per se. Typically geriatrics deals with the confluence of multiple, often functionally inter-related disorders – such as osteoarthritis and obesity or diabetes and amputation. The late Charles Sheffield wrote plausible science fiction related to aging and concluded that even an infinite amount of money, spent on medical and engineering interventions, could not prolong life indefinitely, although he did articulate the vision of cryogenics. Contemporary medicine for aging populations is dominated by pharmaceutical interventions and organ replacement and is responsible for the prolonging of life through intervention in disease processes. However, physical, mental, and social activities in the preventive mode and rehabilitation in the reactive mode remain the essential components, genetics aside, of longevity for most individuals.

DESIGN

The design approach to aging has two facets – engineering and organizational. The engineering philosophy is to replace (automation) or assist human functions. In the case of transportation systems, we address ease of entry and egress, accessibility of controls and seatbelts, and the design of information interfaces, both with the vehicle itself and the outside world, to accommodate the limited sensory, perceptual, and cognitive functions of elderly people. Driver communication systems now extend to external support as well as the traditional within vehicle displays.

In the broader context, engineering also addresses supplementation devices such as eyeglasses and hearing aids, powered wheelchairs, and kitchen aids. The proliferation of remote control devices may be a boon for the arthritic but a barrier for the cognitively challenged. Self-driving cars are a technical possibility but an organizational nightmare.

ORGANIZATIONAL DESIGN

Organizational design addresses the psychosocial, operational, and economic limitations of older people. Social security and Medicare are a start and prescription drug subsidies are essential if older people are to reap the benefits of the research that their tax dollars subsidized. The private sector also sees an organizational opportunity through the construction of comprehensive habitats away from the challenges of roads with 18 wheelers. On a smaller scale, there are local "agent-broker" transportation systems for the elderly, and organizations, such as AARP, aggressively sponsor political and functional interventions on behalf of the elderly. At the other end of the spectrum, the World and National Senior Games Associations sponsor athletic competitions among older age cohorts. These competitions cover most of the events of the Olympic Games but fall short of some of the recently popular "Extreme Games". Community organizations, which prefer "senior" to "old", sponsor many physical and social activities at the local level. In other cultures, the extended family fulfills many of these psychological, social, and economic support roles.

ERGONOMICS ANALYSIS

So what then is the role of ergonomics in aging? We carry out measurement and analysis, develop models, and prescribe interventions for individuals and populations. There is merit in formalizing the measurement process, much like developmental psychologists assess progress at the other end of life. Broad-based test batteries, such as the "available motions inventory", have merit as an indicator of functional age. Similar functional, sensory, and cognitive tests, such as tests of situation awareness and workload, may also be adapted for older cohorts.

MODELING

Models of aging lack a "gold (or silver?) standard". However, the World Masters Athletes (http://www.world-masters-athletics.org) publishes statistics on a wide variety of track, field, and road running events in five-year age groups between 40 and 100. Many athletic events use these records to create a ratio with actual performance measures as a form of handicapping. Such a broad-based physical performance battery, which involves strength, stamina, and skill, offers a "pure" model of aging with minimal contamination by illness or lack of motivation. This concept may be usefully applied to other aspects of human functional performance, wherever standardizing task demands can be achieved (chess, scrabble?). The variability associated with such records may be modeled by the Extreme Value (Gumbel) distribution, which reflects their inherent skewness. Regression models may be applied to the general aging variables together with elements related to particular system short-comings, due to disease or other individual bodily system functional deficiency.

INTERVENTIONS

Ergonomics intervention may be offered at the population and personal levels. Such interventions may involve the design of hardware, information systems (often software), medication (the usual prerogative of medical "ergonomists"), humanware (selection, training, rehabilitation, and assignment), and organizational design. At the population level, the design basis must comprehend the age effect per se, very wide variability, and bodily system functional deviations. At the individual level, a rating in comparison with the record will indicate the amount of accommodation (handicap) needed in the design.

HUMAN PERFORMANCE DATA

These concepts will be demonstrated using published data from the World Masters Athletes Organization (formerly WAVA – World Association of Veteran Athletes), experience from a major automobile industry project related to the design of transportation systems for elderly users and in the physical rehabilitation of elderly patients.

A DESIGN CASE STUDY

The General Motors ACCESS car program involved studies by many universities, a series of customer clinics with a wide variety of vehicles, and the design and evaluation of some 200 features. The features were classified into physical, informational, and psycho-socio-economic. The entry egress features of interest were related to the design of door openings, including step-over and head clearance, seat height and profile, obstructions, and the availability of support devices, such as grab handles, seat backs, instrument panels, and steering wheels. Similar consideration was given to stowage, for large and small articles, both within the vehicle interior and the trunk. Seat design and seat comfort faced the inevitable discussion of the differences between "comfort" and "discomfort"; on balance, however, the older drivers preferred and performed better with firmer seats with less contouring. Access to restraint systems was also a concern with the placement of the seatbelts and the fastening mechanisms. Also, the particular fragility/vulnerability of older people to air bags was and remains a concern.

At the informational level, there was a polarization between the demand of many, usually younger customers for features of all varieties to enhance their driving and peripheral experience, and the requirement of older people for simplicity. Much of this debate revolved around access to a hierarchy of features and interface design. The vehicle system information displays pursued the classical sequence of: tell the driver about the existence of a deviation from normal, indicate the seriousness of the deviation, indicate the source, communicate what should be done to resolve the problem, and indicate the urgency. The vehicle operation information display evidence agreed on the visual needs of older people for size and contrast but bounced back and forth among the merits of status ("idiot lights"), analog, digital, and representational displays. The special needs of older drivers were addressed by the development of an emergency communication system that at the press of a (large) button connected the driver with assistance for medical, vehicle, navigational, and security assistance. This system was a prototype for the commonly available driver communication facilities that involve GPS and agent-broker systems.

The agent-broker concept was also applied to the psycho-social-economic aspects of transportation. Ownership of a vehicle has many challenges – purchase, licensing, insurance, driving, maintenance, repair, disposal, etc. The provision of mass, small group, and personal transport for elderly people required a high-level systems approach that started with the analysis of journey type. One outcome of this analysis resulted in the development of a neighborhood car concept that required attention to vehicle design, organizational access, and journey management. The design implementation of this analysis resulted in the confinement of small electric or hybrid vehicles to largely self-contained neighborhoods, isolated from the 18 wheelers, frantic commuters, and distracted teenagers. The more recent development of widely accessible ride-share systems (Uber, Lyft, etc.) can address the various requirements of older people.

BOATS, TRAINS, AND PLANES – SYSTEMS AND PROCESS ENGINEERING

This chapter addresses some opportunities for human factors to become involved early in the design cycle. Brief descriptions are given of five case studies in which different human factors approaches and systems engineering tools were applied successfully. The case studies include a mass transit railway, supertankers, cars, car manufacturing, and space vehicles. All design processes start with requirements – the customer wants the device or service to fulfill some function. Of course, there are conditions associated with the requirements that are articulated in criteria, such as quickly, safely, and inexpensively. In all cases, the measurement and communication of these and other human factors benefit from the design of a common communication currency. The chapter employs frequent but consistent use of "jargon" – requirements, specifications, verbs, adjectives, etc., and examples. The reader is encouraged to develop alternative examples to investigate the concepts.

When we discuss human factors, it is common to describe a machine, a user, an interface, and an environment in the broadest sense of the words. In systems design, we focus on the interfaces with the equipment, tools, environments, and organizational structures with due regard to user (and misuser) capabilities and limitations. Human factors usually faces complex "human-machine systems". For example, an operator in the Space Shuttle may be controlling a robotic arm on the end of which is perched another astronaut. A third, tethered astronaut is outside assisting with the placement of a multimillion-dollar piece of hardware on a space telescope. The environment is characterized by minimal gravity; zero air pressure; and alternating hot, cold, light, and dark. The interface has to deal with the control of six or seven degrees-of-freedom "robotic arm" using quite limited camera views and communication facilities while the whole world is watching. Whatever the context the rules are always the same – we must design the tools and tasks so that the users can be successful and safe.

Visions and Missions

All good ideas start with visions. "Beam me up Scotty". Visions are free, but when we translate a vision into a mission, we need funding. We also need specific mission requirements that can be measured so that we may know whether the mission has been accomplished. Missions usually have constraints – "put a man on the moon before the end of this decade" or "put a group of people on Mars". Next, come the details – the specifications of how, what, when, and how much. There may also be a "where" thrown in, although usually the "whys" are beyond question. Most complex missions have a way of costing more than the first estimates because as the plans unfold, problems arise due to a lack of information or opportunistic subcontractors holding the main organization to ransom. The more successful missions have clear requirements and accurate specifications very early in the design process.

In the 1970s, Hong Kong decided to solve its traffic problem – five million people all wanting to go to work at the same time. Rickshaws and old buses could not handle the demand, so a mission to build a mass transit railway was born. The mission had requirements – move one million people from A to B quickly, safely, comfortably, and cheaply. And "oh by the way expect to move two million people from A to B to C in a few years' time". The plans involved digging a channel up the main street, putting in a tube, and then returning the main street to its former condition a couple of years later. In the 1980s, the oil crisis directed attention to reserves in Arctic Canada but it was deemed to be unwise to have a ship full of LNG collide with an iceberg or run aground. After considerable human factors, economic, and engineering analysis, the plans proved to be untenable. In the 1980s, General Motors was feeling the heat from overseas competition and so developed a mission to improve the quality and productivity of their automobiles and reduce the costs. They embraced many concepts of systems engineering and learned all about "the voice of the customer". They found that there are many external and internal customers and they learned the language of systems engineering. By 1990, the U.S. unions decided that "ergonomics" would help to reduce the effects of the increasing levels of repetition needed to improve productivity. Work-related cumulative trauma disorders began to increase to epidemic proportions. The plans included the creation of the General Motors Manufacturing Ergonomics Laboratory and the development of a plant-based "reactive ergonomics program".

The mission to Mars turned out to be more formidable than the mission to the moon in the 1960s, despite the enormous advances in technology. It may be tolerable to lose a robotic mission but a manned mission had to have more guarantees. So, NASA refocused its attention to Low Earth Orbit and the International Space Station in a hope to answer some of the questions associated with long-duration manned space flight. Many people still have a vision of going to Mars and many have laid out elaborate plans but at the present time, there is no mission. All of these case studies confirmed the importance of clear operational definitions in human factors and systems engineering. Unfortunately, the evolution of this subject area has resulted in ambiguities. The next few pages outline some definitions that may help to improve the reliability of communication as human factors engineering interacts with the system design process.

SYSTEMS AND PROCESSES

A key concept lies in the definition of what is a "system" and what does a "system" do? A convenient operational description of a "system" is any hardware, software, and naturally occurring or human entity that, by themselves, have no functions. When two or more systems interact, in a physical and organizational context, to achieve an objective, then this interaction is called a "process". Usually, the objective or outcome of a "process" is a change in the characteristics of one or more contributing systems. In the case of human factors, one of the contributing "systems" is a human "system". For example, a person may have the characteristic of "being at home". Only when he/she interacts with a car and a roadway does he change this location characteristic to "being at work". During this journey, the human subsystem may interact with other systems – such as a coffee cup, a cell phone, a frosty road, and another human-vehicle system to engage in a process that results in an accident – characterized by a change in the shape of the vehicle and the owner's wallet.

Design processes create various human, hardware, information, and organizational systems with the purpose of producing a new entity or service. There are multiple purposes for such processes. First, the product must meet with customer expectations – this in its broadest sense is called product quality. The next process objective is to be efficient or productive; i.e., it must achieve its quality objectives with minimal use of consumable resources (systems), such as people, money, materials, or time. This last resource "time" often stands out as a key aspect of process design. The customer would like the elapsed time between his want being expressed and fulfilled to be as short as possible. "Time to market" is a key objective of most product design processes. One way of achieving this objective is through the practice of "concurrent engineering" in which phases of the process are implemented in parallel rather than in sequence so that, e.g., the demands of manufacturing can be addressed during the product design phase. These process objectives are of particular interest to the eventual paying customer, management, and shareholders. However, the unique nature of the human system elements is that they may have their own agendas and objectives. For example, employees would like to maximize their own salaries and minimize the risk of accidents, both of which may conflict with other process purposes – such as productivity. A more detailed look at the design process identifies multiple overlapping stages. The term "concurrent" is somewhat optimistic in practice.

Given the vision of putting a man on Mars, there are distinct but interdependent phases that must be addressed. The first phase is a function identification – launching, navigating, and eating – each with its own purposes that are characterized by "quality", "productivity", "safety", etc. Next comes the realization of these functions through the design and construction of the appropriate hardware, software, "humanware", and "organizationware". The process integration phase focuses on interactions, interdependencies, and interfaces. Of course, the advantages of concurrent engineering are particularly evident here as, e.g., the human and hardware systems must be compatible. The penultimate phase of operations design really addresses the time element. In the Mars mission example, it is critical that various supplies (food, water, oxygen, shelter, etc.) would be on the planet before the humans arrive. Another good example is to be found in automobile production – it is one thing to design and build a car, but to produce 1000 cars a day presents altogether new operational challenges, not the least of which is "just-in-time" materials delivery. Finally, there is operations implementation, which has its own local objectives and its contribution to the next mission through "lessons learned" (feedback).

THE GRAMMAR OF DESIGN

The grammar of design offers a discipline for communication that increases the effectiveness and efficiency of the design process. The first concept is that "processes have requirements" and that requirements relate to the adverbs associated with the process functions (verbs). The process "verb" may be "transporting" a vehicle and human systems in some context or environment. The purposes or objectives of "transporting" may include speed and safety. They will certainly include "quality" – the payload should arrive at the correct destination. Thus, "transporting" may be measured in terms of how "quickly", safely", and "accurately" – adverbs. It is important to emphasize that quantification of these adverbs is important if the process requirements are to be reliably assessed.

The achievement of these process requirements will depend on the characteristics of the contributing systems. In the above example, if speed were emphasized, then a vehicle with a big engine and a driver with a heavy foot would assure the desired objective. Again, for precise system design, it is necessary to quantify the adjectives – "big" and heavy" associated with the system nouns – engine and foot. Otherwise, the engineer cannot design the system and the human factors engineer cannot evaluate the quantitative relationships between the system specifications and the process requirements. Give the engineer a number!

VERIFICATION AND VALIDATION

Once the system is built (or modeled) and the process is implemented (or modeled), then the human factors engineer is faced with the important task of evaluation. This consists of two sub-processes – verification and validation. If the system characteristics have been specified precisely and quantitatively, then verification of adherence to these specifications is simply a matter of measurement. On the other hand, the evaluation of process requirements implies the process of validation, which in turn implies the performance of the interacting systems in a real-world context. The key challenge to validation is the inevitable presence of the user, context, and temporal variability. A precisely specified and constructed car may not "perform" adequately with an inebriated driver in thick fog. Development of the validation process begs the question of "humanware" design – who are the expected user and possible misuser? Validation also may exclude certain contextual conditions, such as inebriation, fog, ice, or 100 mph. The contribution of human factors engineering lies in a clear description of usage requirements, user capabilities and limitations, design specifications, and evaluation conditions.

PERFORMANCE, BEHAVIOR, AND PREFERENCE

Human factors measures of process requirements may be classified at three levels – performance, behavior, and preference. Performance can usually be measured in terms of time and accuracy, given the context. For example, running a mile on level ground will differ from the time taken to run the same distance up hill. The accuracy (quantitative deviation from the objective) needed to thread a needle is different from that needed to park a car. Behavior relates to how a task is performed. In cricket, it is possible for the bowler to achieve his objective by swinging or spinning. Behaviors can be categorized and counted. Using the cricket example again, three bouncers get the bowler suspended; in baseball, one beamer gets the pitcher mugged. However, these "bench-clearing brawls" (ungentlemanly behaviors) can be modified (despite preaching of Skinner) by negative feedback – fines. The most elusive measure associated with humans is that of preference. Preferences may be stated and counted but may not affect behavior or performance. However, extraction of the "voice of the customer" or the mechanisms of "usability studies" often resort to the assessment of preferences or subjective judgments of differences. We must be ever vigilant that we do not put too much store in observations elicited by improper application of one of our most widespread techniques – psychophysics. Unfortunately, this is often the only technique we have available.

Working in space involves many processes, such as staying in one place, moving, eating, and assembling. The more complex activity of assembling involves the interaction among multiple human systems, components, robotic arms, and communications facilities. The context of microgravity, the vacuum of space, radiation, and very high cost presents unique constraints on the process requirements. The interacting tasks of controlling the robotic arm while perching exemplify the challenges, especially when the higher level task of assembling an expensive component may take all day. The adverbs related to this task include carefully, slowly, and comfortably. Slowly can be and is defined precisely, and carefully is described in terms of deviation from a prescribed, tight trajectory. Comfortably is one of those unfortunate human factors challenges that defy reliable quantification, although it is likely that uncomfortable perching may create a distraction that in turn may compromise carefully.

REQUIREMENTS AND SPECIFICATIONS

These process objectives and requirements can only be realized through the precise specification of the contributing system characteristics. What kind of restraint design results in comfortable perching? What kind of joystick design contributes to the activity of careful control of a heavy payload (crew colleagues, components, and tools)? What kind of organization of multiple pairs of eyes and brains is conducive to reliable communicating? How much light should be provided, given that the daylight only lasts 45 minutes up there? How much oxygen should be provided for the strenuous tasks of fighting the resistance of a pressurized space suit? How long should the workday be? Give the engineer a number!

The development of process requirements and design specifications is not simple. It is rarely possible to simply translate empirical data into a number that can be applied to reliable validation. At a simple level, if asked why the height of a door opening is 7 feet, we may waffle about percentiles and allowances for shoes and hats. Similarly, we may look at accident statistics on the freeway and determine that 100 mph (160 kph) is acceptable for 95% of journeys. Or we may state unequivocally that 15 minutes of arc is the design specification for font height on a computer screen. If we don't give the engineer a number, he can't design or verify. But we all know that the number includes a policy overlay and will usually be modulated by domain experience. Consequently, it is essential that requirements and specifications be developed by consensus, with management (or the law) imposing policy, our scientists providing the logic and the data, and the engineers – the eventual users of the standard – providing the domain experience. Of course, all standards (requirements and specifications) should be subject to iterative evaluation and an effective technical memory should lead us to convergence. Unfortunately, those policy makers often change their minds when faced with tradeoffs.

THE HONG KONG MASS TRANSIT RAILWAY

The main performance requirements of this transportation process were to maximize the safe throughput of passengers, given the constraints of size, speed, and the need to show a profit. Throughput is constrained by spatial capacity and passenger behavior, which in turn is affected by spatial arrangements. The seat design and grab rail specifications were based on anthropometry and human behavior. The approaches used to generate the anthropometry and behavior evidence involved the human factors literature, surveys, analysis, and evaluation of performance and behavior in physical mock-ups. The use of adjustable physical mock-ups of both the passenger compartment and the operator's cab proved to be very instructive. In fact, there were substantial discrepancies between the simple application of anthropometric accommodation principles (5th and 95th percentiles) and the actual behaviors of representative samples of subjects in the physical mock-ups. An ironical twist in the seating systems development, based on emphatic input from the Hong Kong Fire Department, resulted in the adoption of flat stainless steel bench seats rather than the scalloped aluminum that was first proposed. This resulted in an adaptive rather than constrained seating arrangement. The vertical poles and horizontal grab rails were positioned to allow optimum accessibility, stability, and motion, given the wide range of anthropometric characteristics of the expected user population. A horizontal bar reachable by a 5th percentile female would hit a 95th percentile male on the chin! Compromise!

The operator's seat design followed a task analysis that indicated that the operator would have to get in and out of the train every 90 seconds as he checked that the platform was clear prior to starting the train. This resulted in the design of a seat that could be folded back (allowing easy egress in an emergency) or in the down position for longer, between station, transits. The seat also had a padded front edge to accommodate the preferred lean sitting posture. In these examples, human factors was applied in a somewhat ad hoc way in the very early design stages. The principal "tool" was a physical mock-up evaluation.

LIQUEFIED NATURAL GAS TRANSPORT

In the early 1980s, serious consideration was given to the exploitation of the vast oil and gas reserves in Arctic Canada. The two transportation options were pipelines and large double-hulled ice-breaking tankers. Given the cargo and the context, there were substantial safety concerns – a collision or a grounding could result in a cloud of escaped gas descending on a town and then exploding. The preferred analytic approach was to use fault tree analysis – both for the mechanical and electrical systems and the human systems. It should be noted that, unlike military vessels, commercial vessels are designed to be operated with very small crews – thus reducing the human redundancy in case of error. The approach to the assessment of human error was based on the human reliability assessment techniques developed for the nuclear power plant industry at Sandia National Laboratories. A massive (paper) fault tree was developed and an assessment of the performance-shaping factors indicated that an incapacitated crewmember would be a likely cause of catastrophic failure. Six years later, the Exxon Valdez confirmed these findings. Fortunately, the LNG project was abandoned for a combination of environmental, engineering, economic, and human factors reasons.

CAR AND TRUCK DESIGN

Car and truck design is a fashion business. Some quip "function before form", whereas others say "form before function". This can be translated into preference before the performance. There are, however, basic functional requirements of capacity, operating, maintaining, etc., before aesthetics takes over. Human factors contributions in product design cover the full spectrum of customer needs – from basic physical issues, through sensory and information processing, to their requirements for alternative features and styling. Different vehicle types attract different customers and have different uses. Much of this evidence is elicited early in the design process through competitive review, clinics, and more precise laboratory investigations, involving simulators of various levels of sophistication. As the design process progresses, various iterations of prototypes are assessed using modeling, checklists, and "drives" on closed courses and the open road. The formal process is iterative and involves concept evaluation, selection, and refinement, through processes of analysis, testing, and, eventually, board review.

Quality Function Deployment is a technique that has been applied widely in the automotive industry to translate the voice of the customer into design specifications and on down through the manufacturing, production, and distribution processes. Early uses of the technique resulted in very large and unwieldy matrices that became increasingly less than useful. However, the principles are sound and lend themselves well to the discipline of requirements and specifications development. Unfortunately, the user (customer) does not always adhere to these grammatical rules of design. The dutiful customer should ask for a vehicle that "goes fast", "is easy to maintain", "enhances his social image", and "protects him in the event of an accident". Instead, the customer may stipulate engineering nouns and adjectives, such as 300 horsepower, maintenance-free, red vehicle with side air bags. This lack of discipline (customers stating specifications rather than requirements) occurs among the many internal and external customers.

An example of QFD in product design would be to address the operation or driving of the vehicle. One adverb might be top speed and the range of top speeds of competitive vehicles might be available from market research. The engineer would recognize that top speed would be accomplished by, among other things, engine size, which would be described in engineering units of cu ins. The adverb "fast" (speed) is of course affected by more factors than engine size – there is the mass of the car, the gearing ratio, the aerodynamics, and the type of fuel, etc. Similarly, the "safety" requirement might conflict with the "fast" requirement. Thus, the task of the human factors engineer becomes more complex in the optimization of often conflicting requirements.

The human factors engineer is a surrogate for the end user. He or she should identify the populations of interest on the dimensions of interest. He should communicate clearly with the design engineer by relating the associations between levels of design specifications (independent variables) and performance outcomes (dependent variables). The relationships are affected by human variability, which can only be reduced by curtailing the "expected user population" by selection or training. There will be many occasions where the relationships are affected by the prevailing conditions and interactions with other variables; there may also be multiple, sometimes conflicting outcomes. Eventually, the designer will have to settle for a single value on each dimension, unless he can design an adjustable feature. The task of the human factors engineer is to communicate the acceptable ranges for each independent variable, given the percent accommodation policy, and to articulate the likely sources of interaction.

DESIGN FOR MANUFACTURING AND ASSEMBLY

Contemporary vehicles have many more features and components, especially on the engines, than they had a few decades ago. However, the drivers have not changed in stature and so it is not possible to increase the height of the hood. Thus, more things have to be compressed into a smaller space, which produces challenges for packaging, assembling, and maintenance. Contemporary "design for assembly" approaches use mock-ups and computer models to assess these manufacturing challenges. There are also certain well-defined, though not always feasible, ground rules for design – such as layered assembly and upward and outward facing fasteners. Given the best possible design, with manufacturability in mind, the ergonomist is next faced with materials delivery and presentation, tools, workplaces, and task content. This last challenge of "line balance" attempts to maximize the utilization of every second of the assembly operation. The ergonomist looks for ways of increasing physical (and mental) job variety through team structure, job enlargement, and rotation, but may be constrained by seniority agreements and quality concerns.

One fundamental challenge of manufacturing ergonomics lies in the difficulties associated with measuring people in their working environment. These conditions do not lend themselves to the rigorous demands of the experimental laboratory for accuracy, precision, reliability, and even sometimes validity. The sample size is also usually restricted. Consequently, the thrust of manufacturing ergonomics should be the assessment of the workplace using population data while allowing the eventual individual operator(s) to fine-tune the arrangements to suit their particular needs. Systems approaches to workplace and task evaluation using various levels of analytic tools should therefore limit themselves to population data while leaving room for some flexibility.

Manufacturing ergonomics assessments are applied at all stages of the manu-facturing system design and implementation process. These assessments take the form of computer modeling, "wall reviews", "prototype reviews", and "slow build" reviews in which each motion is evaluated in great detail. This up-front assessment leads to much improved designs of production systems. The practicing manufacturing ergonomist has made available a wide variety of analysis tools that range from checklists through integrated analysis methods to digital simulations.

Ultimately, any design, design change, or operational intervention will be based on a risk-cost benefit assessment. The solution may be a change in the component (the product engineer's responsibility), a change in the tool or workplace (the manufacturing engineer's responsibility), a change in the amount of work in the job cycle or the line rate (the industrial engineer's responsibility), a change in who does the job (the supervisor's responsibility with due regard to seniority), or the method by which the job is performed (the operator's or trainer's responsibility).

OPPORTUNITIES FOR INTERVENTION

The application of systems engineering and human factors in car and truck manu-facturing addresses the opportunities for change that are presented during each of the product, manufacturing process, production, and operations phases. By way of example, rather than use a specialized manufacturing process, one can consider the baggage handling processes at an airport. The first design phase is the product – an item of baggage. There are restrictions on shape, size, weight, materials, and content. The handling process design includes consideration of all sub-processes that occur between the parking lot and the aircraft's hold and back to the parking lot. These sub-processes include mechanical handling and information processing devices, human handling and information processing activities, the design of interfaces, and due consideration of the environment. The production system design element takes the problem from the handling of a single item to that of millions of items a year. It requires the coordinated activities of sufficient handling devices, sufficient information processing capacity, sufficient people (with appropriate training), and sufficient numbers of interfaces. Finally, operations management involves a full complement of baggage handlers, maintainers, customer service agents, second-level problem solvers, and managers all being appropriately selected, trained, and assigned to achieve a desired level of customer service. As the overall process moves toward the operations, there will be increasingly greater levels of scrutiny. Hopefully, the baggage and handling systems design issues have been dealt with early in the design process. However, unusual, seasonal demands to handle golf clubs, skis, bicycles, and fish may expose the shortcomings in design accommodation.

WALLS AND THE ERGOCOP

One way of modeling each phase of the process is through the development of physi-cal or computer-based "walls" that contain an array of standardized details of the systems and processes as they mature toward the implementation. These "walls" provide the media for multidisciplinary teams to comprehensively evaluate each stage of the process so that late developments don't interfere with the critical path and late changes don't result in excessive costs.

Manufacturing ergonomics has its own elements on the "walls". Assuming that the product design issues have been addressed on an earlier "wall", the manufacturing ergonomics wall will contain questions that address workplace design issues of fit, reach, targets, and task content. At the production level, the physical and temporal aspects of an operation will be amalgamated to assure an acceptable job cycle workload. Later, the operations wall will address individual job and team assignment questions. Finally, the operations output wall will document quality, productivity, and health and safety issues associated with each operation.

The practice of manufacturing ergonomics provides important lessons for many other practice domains. The traditional practice of imprecise process requirements and unrealistic design specifications leads to inappropriate designs to be addressed by the "ergo cop". Eventually, battles ensue in the "review boards" often result in requests for waivers and either a loss of face or an inflated ego of the ergonomist. This process is both inefficient for the company and unhealthy for the profession. The ergonomist should participate with engineering, management, and the operations/ user community in the establishment of clear performance requirements and sufficiently precise design specifications and design implementation. In this way, there will be no surprises at the board (performance) reviews.

Space Vehicles

The design of space vehicles differs from high-volume manufacturing in product cost and product life cycle. The environmental challenges, power requirements, and human interactions are unique. The complexity and remoteness of the operations lead to massive information management challenges and costs. The space program is deliberately very visible – the whole world is watching. Finally, because of these things, there is relatively limited opportunity for the program to capture sufficient "lessons learned". Much of the evidence that cannot be based on analysis must be based on small samples of empirical evidence. Human factors specialists become acutely aware of the challenges of human variability, given the relatively small number of experienced astronauts. Manned exploration missions, e.g., to Mars, present even greater challenges of evidence from robotic missions.

Over the past two decades, NASA has developed extensive statements regarding the human factors issues of manned space flight. These statements are in addition to the extensive medical requirements. The NASA Standard 3000 – the Man-Systems Integration Standard (MSIS) – is a compilation of evidence from both the profession of human factors (and other sources) and domain knowledge. Military standards such as Mil-Std. 1472 were particularly influential in MSIS development. The basic MSIS standard has been adapted to program-specific statements for Space Transportations System (Shuttle) and the International Space Station. These basic and derived standards have, like the programs to which they refer, been subject to hostile attacks in the requests for waivers from engineers, programs, and contractors. As the space program matures, working groups, tiger teams, and review boards all contribute evidence on which the next generation of standards will be based.

A general challenge to human factors is exemplified by the NASA review processes. Almost all human experiences are dogged by individual, contextual, and temporal variability. For example, an ideal thermal environment is affected by activity, clothing, individual acclimatization/tolerance, and duration of exposure. The requirements for strenuous exercise are different from those of reading. The challenge to a human factors review panel is to address all the, possibly interacting, criteria in coming up with a decision that the engineer can design. Clearly, adjustability is required, but how much adjustability? Further complications arise because of constraints on design or change. For example, a particular intervention may be too costly or not feasible in the time scale of the overall project. The final complication of the review process is that the judges (usually experienced managers) overlay their own experience/prejudice on the decision. The task of the human factors engineer is to apply his/her own principles to the conditions surrounding the review process. It is the responsibility of human factors engineers to be "user friendly" in their own practice. The response: "come back in a year when I have done a comprehensive study" is only occasionally warranted.

The human factors community at NASA makes extensive use of digital modeling in the design, evaluation, and real-time mission support phases. The primary contractors – Boeing – made extensive use of anthropomorphic modeling during the early design phases of the International Space Station. Currently, the Interior Volume Control Working Group uses models of the ISS interior, together with anthropomorphic models to evaluate additions and changes, such as sleep quarters, the galley, exercise equipment, and protruding racks that interfere spatially and temporally with routine and emergency activities. The application of digital human modeling in the evaluation of the conditions of work in an assembly task was shown to be more precise, faster, and far less expensive than the alternative of a full-blown trial in the Neutral Buoyancy Laboratory. Modeling was particularly useful in the iterative design and analysis cycle of the crew quarters rack that provides facilities for sleep, computer workstation storage, and privacy.

Lighting models, using the Lawrence Berkeley Laboratories "Radiance" software, are critical to operations, given the changes from extreme brightness to complete darkness every 45 minutes. Differing viewing points for crewmembers and cameras, shadows, and glare compounds the difficulties. Just-in-time modeling and prediction of lighting conditions are invaluable to many Extra Vehicular Activity operations. Exterior robotic operations in the rapidly changing day/night cycles make use of both human vision and camera vision for both training and real-time activities. These models are particularly useful to aid decision-making when contingencies change the timeline and hence the lighting conditions for particular activities.

Although there are very few crewmembers, the multiple demands on their time and the many resource constraints, such as equipment, power, materials, and lighting, make activity scheduling a very difficult challenge. The difficulty is compounded by sparse and imprecise evidence regarding the duration of human activities in the microgravity environment and the ever-present challenge of human variability. The problem is being addressed by enhanced data collection approaches and a range of complex and simple scheduling models. The crew work day on the International Space Station is broken up into three main categories – work (including scheduled science investigations, assembly, maintenance, planning, and communications), sustaining activities (sleep, exercise, eating, and personal time), and responding to contingencies (such as caution and warning signals). The considerable spatial restrictions of the ISS complicate work activities through stowage constraints and spatio-temporal interference. The personal preferences of individual crew members in highly congested conditions sometimes result in excessive time spent in finding tools and materials. The crew workday is categorized in detail; however, the variability of times of activities within categories is not well understood. Consequently, steps are being taken to collect better data and develop simulations of activities on a daily, weekly, and mission basis. These models show not only the occasions when the schedule is overbooked but also how different priorities of activities can be used to accommodate this overbooking – such as sacrificing sleep or personal time.

GLOBAL INTEGRATION

Human Factors and Systems Engineering are essential to effective and efficient design. However, all designs of processes and contributing systems are complicated by change and human, and situational and temporal variability. A major thrust over the past decade has been attention to common processes of both the design activity itself and the resulting product and manufacturing processes. Unfortunately, times and best practices change, so the processes must be flexible to assess and accommodate these changes. These challenges are particularly evident in international operations, where economies of common processes often conflict with different national practices that have been established over many years. Of particular value in the human factors area is the establishment of a common communication currency that enables comparisons to be made between widely differing alternatives and conditions and which resolves the ubiquitous importance weighting problems.

Global integration efforts are often the source of conflict between the efficiencies of common processes and the perception of what are best practices. "Not invented here" is often an underlying motive. The challenges of competition in the automotive industry, coupled with the explosion of computing and telecommunication facilities have combined to fuel the fires of globalization. Manufacturing organizations seek out high-quality, but lower cost labor markets. Also it is not efficient to have an engineering design center in every country – why develop essentially the same product separately in multiple markets? But engineers worldwide are conservative and resistant to imposed change. In these cases, there is no substitute (other than dictatorship) for extensive face-to-face interactions among the design teams, including the human factors engineers.

The International Space Station faces similar challenges. The program has very important political underpinnings, the costs are extremely high, and national identities need to be clear. And there are other constraints – only three crew members at a time, relatively few modules, and only occasional opportunities to visit. The management of such a program is not limited to the handful of astronauts and cosmonauts; there are very large support staff in operations, engineering, medicine, and science management. The ISS is at once a miracle of systems and safety engineering and at the same time a management nightmare.

CHANGES

The mechanisms of dealing with change are described in different organizations as request for waivers, change requests, or engineering change orders. In many cases, these requests are appropriate, albeit due sometimes to poor planning or unclear requirements and specifications statements earlier in the process. Often, however, they are seen as frivolous – made only to accommodate the failure of a supplier to be able to deliver on earlier agreements. Where waivers are processed on an individual basis, they may not comprehend the implications on other aspects of the process or the trickle-down effects on other subsystems. Cost is a common reason for the change and the systems engineer and the human factors engineer must work with the managerial accounting community to establish a common basis for the rational processing of waiver requests.

Human factors cannot be practiced without engineering – the people who design the systems – and management – the people who make the policies. Sometimes policies are imposed from elsewhere – through technical standards, government regulations, or labor agreements. Some human factors practitioners feel that it is their duty to convey policy, especially where engineering and management do not have the appropriate information to decide on policy. On occasion, human factors specialists may substitute dogma for policy. The notorious 5th percentile is a prime example. It is a useful concept, often with good rationale, but it is widely misunderstood and often inappropriately applied.

A major problem is that someone who represents the 5th percentile on one measure is unlikely to hold that relative position on another; furthermore, when accommodation is based on multiple dimensions, then it may be difficult to define who or what is a 5th percentile. Monte Carlo simulation methods may be applied to somewhat relieve this problem. Another difficulty is that the implications of a design decision may be more or less important. Thus, in the case of a highly sensitive design decision, it may be appropriate to accommodate the 1st percentile. On other occasions, design for the average may be an adequate approach, given that the dimension in question is not related to an important outcome. An example of a highly sensitive design decision would be the walking speed of old people crossing a busy road.

OUTCOME AND DESIGN SCALES – COMMON CURRENCIES

Human factors engineers are made aware of the processes and theoretical underpinnings of scaling methods from their earliest training in the statistical methods applied in the broad context of human variability. Scaling systems abound – percentages and Yes/No are separated by Lickert-type scales of varying degrees of resolution. A prerequisite of any scale is the establishment of anchors for both endpoints and intermediate thresholds. The fuzzy classification reflects reality but is often a practical inconvenience. Appropriate resolution is always needed. The zero to ten scale is probably the most universally familiar one and has stood the test of time. It usually has adequate resolution and can easily be linked to a response categorization.

Given this ten-point scale, it is relatively easy to visualize a nonlinear mapping function that covers the full range of outcomes from ideal to unacceptable. Examples include lighting, noise, temperature, spatial, and force scales of design specification ranges. It is also possible to comprehend single and complex variables, although, for engineering design purposes, the evaluation will ultimately have to identify individual variables for change. Where the relationships are not monotonic, as in the choice of an optimal temperature, it is convenient to use two scales – one for hot, the other for cold. There may also be multiple outcomes – some of which might be conflicting. For example, a spatial scale related to controls, such as vehicle pedals, may have movement time and inadvertent actuation conflicts. Such conflicts point toward the importance of consensus processes in the establishment of mapping statements and cut-offs. The reality of human variability is such that a single mapping function will never be precisely "right". Again the inclusion of human population accommodation policy in the consensus decision is essential to assure "buy-in" of all concerned as the design process develops.

Consensus

The development of human factors design standards is best pursued through a consensus process, using the common currency described earlier. The credo that standards should be data driven is over simplistic. Policy, scientific logic, technical feasibility, and experience must all contribute to the establishment of a standard. It is also essential that representatives of the customers – internal or external – who will have to apply or be affected by the standard should be involved. In this way, upfront agreement in both the principles and the values related to the standard will help to assure "buy-in" and less demand for waivers. Human factors standards are also iterative in that they should be verified, validated, and evaluated as the project or program evolves. A clear example lies in the establishment of speed limits for different road conditions.

Design trade-offs should be made with a full global view of all the relevant information, preferably with a common communication currency. The choice of cut-off points on individual variables can be used to assign explicit weightings. A broad range would imply wide tolerance (less importance) and a narrow range of greater importance. The common currency outcome prediction scale facilitates the process of amalgamation. At the simplest level, a count of the number of variables in each of the outcome ranges produces an index or profile that reflects the general nature of the problem. The count can also be used as a decision aid – e.g., decision policies could be "no reds" or not more than "5 yellows".

Addition (as opposed to counting) is rarely justified, although this is the preferred method for some checklists. However, the case for multiplication in the amalgamation process may be justified where interactions are likely. Such situations are best handled by two-dimensional matrices using the same common currency described here. The special case of interactions between basic variables and time may also be approached in this way. This big picture decision aid can be used to indicate before and after change situations, comparisons between alternatives, or progress of a project through the design process.

The common currency scale can be viewed as an estimate of the "probability of failure on a single transaction". For example, the probability of "failure to accommodate" of a 12-inch wide seat would be of the order of 0.7. Similarly, the probability of failure of a 2mm high font, given an elderly reader population could be 0.99. At the other extreme, the probability of failure of a 24-inch diameter escape hatch might be 0.05.

DECISIONS

The decision process must also involve the benefit of the transaction, the number of transactions per unit time, the number of people affected, and the various costs of failure. To complete the assessment, an evaluation must be made of the costs, benefits, and probabilities of alternative designs and outcomes. Where possible, objective evidence (including data) should contribute to the probability and costs/benefits and exposure estimations. Where individual, personal decisions are made, subjective probability and cost estimates may be sufficient – this is of course the basis of most naturalistic decision-making.

The hypothetical example of our choices of transport to work – car, bus, or tank – can be used to illustrate the quantum decision process. Assuming a decision horizon – a day, year, or project lifetime – all measures can be reduced to base ten arithmetic. In the case of choice of transportation mode based on cost and safety, the analysis shows that we should ride the bus. However, if we are the president of a country in political turmoil, then we may wish to revise our probability, exposure, cost, and outcome estimates and at least buy an armored car. This quantum arithmetic approach is appropriate for a cursory analysis and exploration of the sensitivity of the different elements of the decision to changes. Greater resolution may be obtained by including multipliers and decimal components while still adhering to the basic decision logic. However, in this case, it may be appropriate to use some computational aid.

CONCLUSIONS

It is important to reiterate that it is the role of management or governments to communicate policy, given human factors evidence of outcome likelihood and effect. The role of consensus in the establishment of design standards was also addressed. In this respect, it is important to address the realities of false consensus and the sometimes inappropriate or overly weighted influence of experts. The opinion of experts should always be weighted heavily but only in the area of their expertise. A better approach is to use independent and interdependent "voting" processes, with sufficient allowance for discussion as the standard or decision scenario is developed.

The design of boats, trains, cars, and space vehicles as most other human factors opportunities necessarily involves teams of one kind or another. Generally, the teams consist of an exhaustive and exclusive "set" of people with technical and domain knowledge. Individual team members may have discrepant objectives. Team dynamics create challenges to both the effectiveness (accuracy) and efficiency (speed) of standard, risk, and design decisions. Greater effectiveness and efficiency can be achieved through the application of common currency and clear visual aids to comparison, trade-off, and decision-making.

Even the decision processes common in contemporary design projects face the challenges of over/under reliance on expertise. On occasion, the efforts to substitute process for expertise may also be counterproductive. Overly enthusiastic attention to "process" can be cumbersome, where simple experience may be sufficient. One hundred years ago, craftsmen built outstanding automobiles, slowly. Nowadays, processes result in the high-volume production of automobiles. Hospitals used to be run by medical experts but now the processes of HMOs have diverted the purpose to profit rather than a caring motive. Soccer is essentially a game of experts, it has experts that are bound by processes. Music was once the realm of experts, now it is relegated to simplistic marketing processes. We went to the moon on the backs of experts; processes will get us to Mars.

Human factors is alive and well early in many design processes. The appropriate place for human factors is as a branch of engineering, not as an after-the-event "ergocop" in safety or consumer protection functions. Human factors is rightly a component of systems engineering, it can contribute important knowledge and tools both to the designs themselves and to the design processes. One important contribution is the establishment of a common currency for communication of human factors and implications of design.

Why did the chicken cross the road? To get to the other side! Or so the old joke goes. Why did the self-drive car crash? Because it was too late to see the chicken and it did not know that this was a common chicken-crossing area. It was late at night, raining and slippery and the tires were worn and the driver/monitor was in a hurry to get home, under stimulated, drunk, on pain killers, fatigued, complacent, inexperienced, or maybe asleep. Furthermore, the vehicle was last year's model without the updated sensor technology or software, and anyway, the driver was baffled by the interfaces with this new technology. A Fault Tree Analysis would put all these conditions together and conclude that the chicken or driver should have stayed at home.

Another model, SHEL (Edwards, 1966), of system behavior and system failure would describe interfaces, interactions, interruptions, interferences, interdependencies and integration within and among people (and chickens), technology, processes, and contexts. An expansion of the SHEL model adds energy, information, space, time, and purpose. That said, the transition to self-driving cars, supported by ever-increasing and improved technology, will take a long time. This transition period will inevitably involve interactions between self-driving cars and conventional vehicles. The only constants will be human and situational variability. The $64,000 (or greater) question is when, probabilistically and politically, the failure rate of self-driving cars will be demonstrated to be lower than the failure rate of conventional vehicles. The lawmakers will have a field day developing constraints and conditions for this newfangled technology, including segregation and driver/monitor training and certification.

Now systems designers would realize that the many and various stakeholders in this complex system – drivers, passengers, chickens or other pedestrians, road builders, traffic managers, emergency services, vehicle designers, etc., might have different and perhaps conflicting priorities related to system (and process) design purposes and outcomes. The E4S4 model summarizes these (arguably exhaustive and largely mutually exclusive) expectancies and outcomes: Effectiveness, Efficiency, Ease of use, Esthetic appeal, Safety, Security, Satisfaction, and Sustainability, which may include both reliability and resilience. For example, the driver may have wanted to get to his destination (effectiveness), quickly (efficiency) and safely while riding in his fancy new automated car (esthetic appeal). The emergency services would be out of business if the system didn't fail from time to time. And the vehicle manufacturers would soon go out of business (sustainability) if there were too many failures that they couldn't blame on the drivers. It remains to be seen when the lawyers will focus on the designers rather than the drivers/monitors of these nonsegregated vehicles and routes.

A key to diverse behaviors of humans and animals is the innate and acquired ability to pay attention, perceive, think, and anticipate, and to prepare for any one of various actions ahead of time and then (re)act in a timely manner. In other words, we learn from experience or, better still, learn by being taught, having had a similar transferable experience, reading a book, searching the internet, or simply thinking about it. The processes of anticipation, attention, and perception presuppose a more or less well-developed mental model (framework) that is used to explore and simulate information regarding the sources and effects of various external or internal inputs over time, various model parameters, and various outcomes. Sometimes the desirability of a particular outcome outweighs the risk of failure, with catastrophic consequences. Sometimes there are just too many factors related to the decision (to cross or wait) that the poor chicken becomes overwhelmed and just "goes for it". The Bayesian, (Bayes, 1763), community has much to say on this problem, although the human capacity for forgetting makes them generally sub-Bayesian. Another problem for the chicken, people with chicken brains, and even ordinary people is the word "various". The earlier described complex systems model – people, technology, procedures, contexts and their interfaces, interactions, interruptions, interferences, interdependencies, and integration – is incredibly complicated by variability within and among all these components. Furthermore, these situations are replicated all over the world (places), all through the day and night (times), and involve millions of people and transactions. No wonder people have accidents from time to time or are blamed for causing accidents, especially those playing chicken with the traffic. As suggested earlier, this probabilistic problem will be interpreted politically.

The matter of interest here is the viability of self-driving cars (primary technology) without the concomitant design, integration, and otherwise management of all the supporting technology (including roadways), contexts (weather, traffic, and pedestrians), procedures (drive on the right, keep a safe spacing, obey traffic signs and signals, etc.), and the selection, training, and assignment of all the primary, secondary, and tertiary stakeholders. The solutions are technical, economic, political, legal, and probabilistic. Segregation is or is not an option? Are self-driving cars safer statistically than conventionally driven cars? The lawyers are happy to assign blame and sue errant drivers. But who will they sue when the automated system fails, albeit less frequently? The driver/monitor, the traffic management, the designer of automation, the company that designed the vehicles, or the traffic managers that permit the integration of automated vehicles. Of course, the lawyers will find a way and the self-driving car manufacturers have deep pockets, albeit full of equally cunning lawyers.

So we are back to probability and statistics, perhaps even Bayesian models if the lawyers can estimate the likelihood of a particular event given the existence of other (also probabilistic) evidence. The juries may not have studied Bayesian convergence and so may revert to simpler statements of likelihood and the assignment of blame when the system fails, depending of course on who suffered the unwanted outcome and who is lined up to defend the failed system components. The applied psychologist or human factors/ergonomics practitioner or researcher may again fall into the trap of reductionist research and analysis. The astute safety practitioner will apply the 5 Whys, and end up with 5 to the nth power explanations. Meanwhile, in the absence of latter-day Luddites, technology will develop until the automatic driving machine brains have learned, not only to learn from past experience and selectively take in (perceive) relevant and timely information but also to anticipate and make (probabilistically) sound decisions, just like or even better than drivers. Unfortunately, chickens will continue to cross the road out of habit and get to the other side, until that unfeeling automatic car decides to sacrifice the chicken as braking hard is more dangerous (to the car occupants rather than to the chicken).

This was a very brief discussion of human cognition, attention, behavior, and performance; the challenges of automation and chickens would not be complete without a handful of case studies to explore the underlying theories:

1. The first situation is a crossroads consisting of two to three lane divided highways with sidewalks on both sides. It is 6 o'clock in the evening and going dark. The roads are full of tired and impatient drivers and pedestrians on their way home from work. It is raining cats and dogs and beginning to freeze. Note that 50% of the cars and large trucks have various generations of self-drive technology. The traffic lights have failed and a lone policeman has stepped into the breach. The automated vehicles do not comprehend the subtleties of the policeman's glances, nods, and hand waving. Half the automated vehicles resort to manual control. The other half makes a mess of the intersection, blocking the way for the automated emergency vehicle. Meanwhile, the early nightly news drone hovers to capture the details of the carnage.

2. The next situation has a genius 17-year-old riding in his parent's automated car. Now, this 17-year-old has spent 90% of his waking time playing video games, which are filled with exciting system failures, and has become immune to the realities of mass and acceleration. He (auto) pilots his way onto the freeway and joins the fast track. He decides to take over from the automation, with disastrous consequences.

3. A third scenario involves grandma, who uses her newfangled golf cart-sized technology to go shopping, just along the street. At the same time, the automated garbage collection truck is performing its start-stop regime in the neighborhood and cannot tell the difference between grandma in her private pod and the garbage bin.

These examples are offered to explore the limitations of technology and its applications, especially where it interacts with people of different abilities and motives, in different contexts.

This chapter is about natural laws – aging and gravity – and man-made laws – taxes and "ergonomic" dogma. They all apply to everybody. We are not very good at these challenges but valid ergonomics intervention can help to analyze the problems and lighten the load.

AGING WORKFORCE

The general problem of aging is that many human systems and sub-systems show a functional decline and, to make matters worse, these systems often interact to compound the resultant performance problems. Yates (200X) carried out substantial research on individual biological system decline rates. Over the 40–70 years, our musculoskeletal, endocrine, cardiovascular systems, etc., decline at an average (linear) rate of about 2%–3% per year, the exact rate being somewhat influenced by nurture. Peacock (2003) reported on the use of athletic records to delineate the boundaries of aging, and noted that when the age range is extended to 100, a non-linear decline is observed. Nurture is also a big player in tooth decay and a massive industry has grown to prop up or replace our pearly whites. Joint replacement is now commonplace, as also are hearts, blood vessels, lungs, and skin. The declines in our sensory systems are legendary and eye and ear doctors apply very sophisticated technology to counter this deterioration. Our cognitive systems – attention, perception, memory, decision-making, and learning – have no such prosthetic solutions (as yet), although smartphones offer a plethora of (sometimes) easily accessible, situationally pertinent information to reduce our mental workload. The Internet has an enormous fund of information, much more than any individual will ever use or need. However, technology aimed at the general (above average) public can often bring its own barriers and hazards to the less gifted often older users. Information on the Internet is not always easy to access or use, and sometimes it may even be "fake". The confluence of physical and cognitive decline often gives rise to a decline in our social interaction capabilities that are offset somewhat by combinations of community and family interventions. The only reliable recourse is to exercise our bodies and minds if we wish to reduce the rate of inevitable decline. Unfortunately, disease processes associated with aging can hasten system deterioration and hence compromise performance capability.

One general response to the [gradual] decline associated with aging is to [gradually or abruptly] change our behaviors. We reduce our performance goals and are happy when we continue to achieve success. We give up running because our knees hurt and we ask our grandchildren to help us thread our sewing needles. Our roles as parents and protectors are reversed as we age.

Less than 50% of the U.S. population develop and submit their own tax returns. Furthermore, a majority of tax returns, even those completed by specialists, have errors of one kind or another (Peacock and Karwowski, 1993). These authors conducted a study in which the format of both the tax forms and the instructions were rationalized according to general ergonomics principles – larger spaces for entries and rearranged instructions for easier search and links to entries. These alternatives were presented to subjects with varying levels of experience – from novices to experienced tax accountants. Completion times and errors were recorded for returns of various complexities. One finding was that the effects of form and instruction (interface) changes were minimal among the experienced tax preparers. The implication is of course that it is the cognitive challenge of the tax laws that is the main culprit and that improvements in interface design can only go so far in improving performance among less practiced individuals.

The dual purposes of display design are to extract and communicate situationally pertinent system data and, where possible, to present abstractions, combinations, and derivations of the raw data in order to reduce the cognitive load. These so-called composite displays can enhance attention, reduce memory load, and guide decision-making. For example, in aviation displays, one can automatically derive density altitude by combining pressure and temperature information. Fuel consumption displays can be converted to predict vehicle range, given assumptions about future driving behavior. In the tax return situation, the use of a computer program can prevent calculation errors, particularly those errors that are induced by frequent page turning and cross-referencing during the hectic hours of April 14. The computer can also check the validity of raw data entries by flagging unlikely values, such as an extra zero here or there.

But even with all these packaged computer programs for tax return preparation, the portion of tax returns completed by individuals has not increased substantially and the number of errors remains high. The culprit is still the ever-increasing complexity and loop holes in the tax laws that are accompanied by the inevitable cry for tax law reform (the cognitive challenge). Aviation displays have progressed from the traditional "six pack" to an integrated glass cockpit that provides a more holistic view of the many variables associated with flying in the context of unforgiving gravity, weather, and traffic congestion. As with tax returns, the opportunity for error is a function of complexity but, in ·flying, the outcome may hurt more than your pocketbook.

The time-honored process of measuring human performance and its modifiers, such as the physical and cognitive effects of aging and the relationship between performance and income tax form design, is to obtain a random bunch of subjects and assign them to the different categories identified as independent variables of interest. Then run the experiment, analyze the data, and make wild generalizations and extrapolations about the differences and associations that are attributable to cohorts or formats of interest. Research-trained ergonomists are sometimes cautious about the last part, but many ergonomics practitioners may forget all about experimental design, assumptions of normality, randomization, personal biases, etc., and they often choose convenient subject samples of insufficient size. Their conclusions may not be "true". Common causes of untrue conclusions are to be found in the concepts of construct, sample, and contextual validity. Generalizations about aging are mired in a mess of individual and situational variability. The tax laws continue to outpace most of the user population and the skies are becoming as congested as the roads. Complexity rules!

It is well known (that's a dangerous statement) that aging effects can be offset by exercise. Sages may point out that it is not the aging process per se that is affected, rather it is the effect of concomitant factors, such as disease or disuse/abuse atrophy that allows performance levels to be above (or below) the observed "population averages" for that age cohort. Examples include the ability to walk upstairs (which is affected by one's history of exercise and eating) or hear conversations in a busy restaurant (following a career in a stamping plant). Filling out the EZ tax form, when you no longer have dependents, mortgages, second jobs, second homes, and multiple investments, indicates a natural adaptation to the cognitive demands. Hardly a day goes by without an advertisement in the mailbox offering a "free" lunch in return for long-term tax and money management advice. The aviation industry is swamped by great gadgetry that addresses many of the separate challenges of flying (autopilot, ADSB, synthetic vision, highways in the sky, etc.), but cannot deal with situational complexity. Fortunately, one manufacturer now offers a parachute with its smaller products. I suggest that we should invest in parachute technology for the commercial and private Unmanned Aerial Systems of the near future.

AGING, GRAVITY, TAXES, AND ERGONOMIC DOGMA

Aging, gravity, and taxes are inevitable hurdles for all people. Ergonomic dogma should not add to the challenges. Ergonomic dogma suggests that we should reduce the physical and cognitive demands faced by older people (of all ages) by design to assure both immediate and continued success. On the other hand, the conventional wisdom and research data of geriatricians indicate that longer term success will be assured by continuing exposure to physical and cognitive demands, albeit with the aid of "parachutes" – that is to rely on an individual's ability to learn from experience.

The most basic of "physical ergonomic inventions" – the chair, the wheel, and the engine – all have their place. They are highlights of contemporary society that are accompanied by a "parachute" – medicine that remedies the human effects of abuses of these devices. Unfortunately, the long-term effects on the environment have no equivalent to parachutes. Or at least none that appears to be generally acceptable. The chair, the wheel, and the engine are less used in primitive societies but unfortunately, their success, as measured by longevity, is hampered by more fundamental challenges – diseases, which are preventable and curable by the appropriate deployment of modern medical knowledge. This deployment can be helped by ergonomic analysis of the process. The iatrogenic effects of contemporary ergonomics – disuse atrophy in a general sense – are due to ergonomists paying lip service to issues of short-term comfort, convenience, and productivity without due regard to long term.

This is not about the Hippocratic Oath (first do no harm); rather it is about the hypocrisy of engineers, scientists, and operations managers in their attempts to patch the interfaces between their designs and the end users. These patches are euphemistically called facilitators and cover a spectrum of devices, such as labels, warnings, instructions, procedures, and a plethora of other so-called job aids. Modern computing technology has now made available the rich medium of augmented reality to facilitate the activities of the human operator.

A facilitator is a temporary device to supplement the user's knowledge or current situation awareness to ensure an accurate and timely transaction.

For example, a label on a switch that says on/off obviates the need for the operator to know which way to move the switch; after using the switch a few times the user no longer needs the label. Warning signs on the edge of the Grand Canyon caution the adventuresome tourist to "Be careful". Labels on food and drugs present a frequent challenge to manufacturers who have to strike a delicate balance between advertising their products and avoiding the consequences of misuse. An augmented reality example is the yellow line on the computer screen to mark the 10 yards needed for a first down in football and the chains which are the (10) yardstick. In this example, the facilitator is used regularly, because the human judgment of the 10 yards is unreliable. On the International Space Station, the crew members have to perform many complex system and payload tasks, and respond to contingencies that may escalate into emergencies. The complexity of these tasks is addressed by extensive training and the even more extensive use of facilitators. The usage and utility of facilitators (such as procedures) depend on the degree and currency of training at the knowledge, rule, or skill level (Rasmussen, 1987 – *Cognitive Control and Human Error Mechanisms* in *New Technology and Human Error* edited by J. Rasmussen, K. Duncan, and J. Leplat). At Christmas time, parents are often faced with the dreaded messages "some assembly required" and "read me first". Mothers do as they are advised; fathers, who know about these things, put the instructions on one side.

The key, as with medicine, to good facilitator design is to "first do no harm". Harm in this context includes inaccurate, cumbersome, and untimely advice. However, this raises another common human factors engineering dilemma – the speed-accuracy trade-off. Commonly instruction and procedure designers err on the side of accuracy, by expanding detail and sacrificing speed. Conversely, label and warning designers often favor speed (or space conservation) in their use of acronyms and symbology, and may sacrifice accuracy, and, because of the subsequent need for recovery, ultimately they may also sacrifice time. Perhaps the greatest mistake made by designers is the pseudo-sacred cow of standardization. There are two fundamental reasons why standardization is not universally useful in facilitator design. First, not all problems are alike, and second, not all users are alike. Trying to fit a medical emergency procedure into the same format as a hardware maintenance or assembly procedure begs the question of the level of training and knowledge of the user. Also, the obsession of many organizations with technical jargon and acronyms often overestimates the knowledge of the nonspecialist user. A prime example of this issue is with the international crews on board the ISS, whose language differences impair the effectiveness of acronyms. On the other hand, certain conventions and forms of standardization have their place – a prime example being the familiar "Windows" interface.

THE HUMAN FACTORS APPROACH

The human factors engineering community approaches this challenge of the bridge between engineering or operational designs and the user in a number of ways. The first level is through the design of the permanent displays and controls that are normally essential for successfully carrying out a transaction or procedure. The second step is to carry out a task analysis that addresses the sequences of information and actions that comprise the transaction, including the human and system contributions. The third step is to assess the conditions and context of the transaction(s). The fourth step is to address the knowledge and skills of the intended user – it should be noted here that users' capabilities change with practice and may fluctuate with the conditions under which a particular transaction takes place. The final analytical step is to evaluate the expected use and possible misuse of the interface or procedure. This step should involve formal "usability testing", in which users of varying degrees of knowledge, skill, and experience perform representative tasks, under a variety of conditions and contexts with a candidate set of facilitators. Facilitators should first be designed with due regard to the expected users, possible misuses, and likely conditions, and should then be subject to formal evaluation.

These analytical activities lead to the design of facilitators to fill the knowledge and skill gaps. Because conditions and users vary, it is necessary to include flexibility in facilitator design while remembering the important rule: "first do no harm". In the context of facilitator (e.g., procedure) design, "harm" includes both inappropriate action as a result of the design, with various possibilities of outcome from minor inconvenience to catastrophe, and unnecessary delays in the transaction caused by the verbatim following of the procedure. It should be noted that errors in the use of facilitators (procedures) commonly cost recovery time, which, in turn, may lead to catastrophe in a time constrained condition, lack of productivity, or simply frustration in "being treated like a child". Unfortunately, these possible outcomes are affected in their likelihood by the particular user, conditions, and contexts, thus making evaluation a probabilistic process at best.

Some Cases

In case the reader is skeptical about this discussion, he should consider the NW255 accident at Detroit Metro Airport in 1987:

> About 2046 eastern daylight time on August 16, 1987, Northwest Airlines, Inc., flight 255 crashed shortly after taking off from runway 3 center at the Detroit Metropolitan Wayne County Airport, Romulus, Michigan. Flight 255, a McDonnell Douglas DC-9-82, U.S. Registry N312RC, was a regularly scheduled passenger flight and was en route to Phoenix, Arizona, with 149 passengers and 6 crewmembers.
>
> Of the persons on board flight 255, 148 passengers and 6 crewmembers were killed; 1 passenger, a 4-year-old child, was injured seriously. On the ground, two persons were killed, one person was injured seriously, and four persons suffered minor injuries.
>
> The National Transportation Safety Board determines that the probable cause of the accident was the flightcrew's failure to use the taxi checklist to ensure that the flaps and slats were extended for takeoff. Contributing to the accident was the absence of electrical power to the airplane takeoff warning system which thus did not warn the flightcrew that the airplane was not configured properly for takeoff. The reason for the absence of electrical power could not be determined.

In this instance, the use of a checklist – a facilitator – was complicated by the crewmembers' considerable familiarity with the procedure, a failed warning system (another facilitator), and the unforgiving context.

Closer to the interest of space human factors are the comments from an astronaut.

> The biggest issue with all of the Payload experiments from a Human Factors perspective is that the procedures are too lengthy and are not clear (they read more like an airplane checklist as opposed to furniture assembly). The crewmember thinks they need

to use fewer words and more line diagrams/drawings. They perceived this as a common problem for all payload procedures during training and on-orbit. The procedures typically resulted in very slow and inefficient operations (ex. was 20 pages telling how to remove a screw). They mentioned that they did comment about this in other briefings also. They stated that someone should take ownership of this issue and coordinate with all payload developers to simplify their procedures.

A lot of time and productivity is lost when the crew has to follow airplane checklist-style procedures (e.g., Medical, Payloads, and IFM.) This style of procedure writing increases the chance of errors. Simple, clear, line drawings with a few words are best for efficient procedures.

But the biggest issue with Payloads is that too many acronyms in a multi-lingual environment will cause confusion among the international crewmembers. It was not a problem of inaccurate/incomplete acronyms, but that there were "way too many" acronyms used for the hardware. They suggested that it should be stressed to the engineers to not use acronyms if at all possible and to spell out a name that describes what the hardware/system is and "stick with it".

Previous Work

Human Performance Engineering: A Guide for System Designers presented an extensive discussion of facilitators, including their role in supporting human performance, selection of facilitators, the design of instructions (procedures), and other performance aids. Facilitator designers who heed Bailey's advice and rules of thumb will go a long way toward achieving acceptable facilitators but it is clear that many designers do not consider this advice or use their own flawed common sense to supplement their designs with facilitators. There are many other examples of "good" labels, warnings, and instructions, often stimulated by "failure to warn" lawsuits. An extensive bibliography of research in this area is contained in Miller, Lehto, and Frantz (1990) – *Instructions and Warnings – An Annotated Bibliography.* The American National Standards Institution (ANSI) and the International Standards Office (ISO) both offer extensive guidance on instructions, warnings, and labeling, usually in domain-specific contexts.

http://www.ansi.org/

http://www.iso.ch/iso/en/ISOOnline.openerpage

NASA Standard 3000 (6.4.3.3.4) states simply that "Warning labels shall be provided where potentials are hazardous to crewmembers".

Despite this extensive activity, facilitators (labels, warnings, instructions, and procedures) continue to be a major source of system failure, inefficiencies, and personal frustration in many domains.

Research, Measurement, and Analysis

Although there are many "rules of thumb" for facilitator design, there is no robust theory of the pervasive role that they play in everyday life and particularly in technical areas. Such a theory must address the level of complexity, outcome/recovery importance, and human and contextual variability, including learning and the speed-accuracy trade-off. Measurement of the use and utility of facilitators must consider the human capabilities to learn and forget. Analysis of facilitators requires tools that can be used both by researchers and designers to evaluate the effectiveness of alternative ways of aiding human performance.

Another research opportunity lies in understanding the relative roles and interactions among the system designs, human learning, and facilitators. For example, facilitators can become crutches that interfere with knowledge or skill acquisition and result in human performance shortcomings in the absence or other deficiencies of the facilitator. Reliance on contemporary vehicle navigation technology is a case in point.

MODELING AND DEVELOPMENT

A simple model of a facilitator places it between the varying demands of a system design and the varying capabilities of the human operator, with the purpose of assuring that the human capabilities will not fall short of system demands. Such a model, although conceptually meaningful, fails to capture the specific form and content of a facilitator, given the environment of complexity and variability.

REQUIREMENTS AND DESIGN

The literature cited earlier presents many considerations, guidelines, and standards for the design of facilitators but these are rarely supported by clear system performance requirements. Consequently, many facilitators are designed as an afterthought and refined iteratively following untoward incidents or user comments. Most facilitators are static in nature and do not address the changing needs of users as they become more familiar with system behavior. There is a need, therefore, to develop performance requirements and detailed design specifications for the facilitator to assure safe and productive human performance.

IMPLEMENTATION AND EVALUATION

All system design processes should require the integrated design of facilitators along with the formal assessment of expected human performance and performance variability. The changing role of facilitators should be evaluated as the system is validated in an analog or operational context. Formal guidelines and evaluation protocols are needed for the implementation of facilitators.

INTEGRATION

Facilitator design needs to be formalized and integrated into the system design process. This integration requires answers to the research, modeling, design, and evaluation questions described in the previous paragraphs.

OPPORTUNITIES FOR FOCUS

There are many opportunities in which facilitators play a key role in system effectiveness and efficiency. The first area is in the management of its complex activities and its notorious reliance on acronyms, despite the often observed speed accuracy trade-off. As in aeronautics, space operations rely heavily on procedures and eso-teric jargon for vehicle operations. Assembly and maintenance of space vehicles is engulfed in labels, acronyms, instructions, and procedures, all of which contribute to the possibility of catastrophic system failure, inefficiencies, and inconvenience. Space vehicle payloads are notorious for their use of cumbersome and inconsistent procedures. Finally, the medical treatment procedures, full of jargon in English and Russian, and supported by minimal training, are a time bomb waiting to explode.

MOMENTS

Newton's findings were pretty important, but as these examples show there are com-plications that have to be dealt with other mechanical principles. For the tall number 8 to put the impudent number 9 on his back, he had to impart a turning moment to the small particle by aiming the stiff arm some distance from the small particle's center of mass. Many sports injuries are caused by a combination of Newton's laws and the mechanical advantages offered by moment arms. Some, like skiing into a tree or heading a soccer ball are direct examples of Newton's principles. Games played with racquets and bats and feet and hands make great use of moments. How can a pitcher throw a ball at 100 mph? How can a golfer hit a drive over 300 yards? Long before Newton – a few hundred years BC to be exact – Archimedes was relaxing in his bath-tub, thinking about how to move the earth. Being an empiricist, he suggested getting a very long piece of wood, anchoring it under the earth and over the moon, and then getting his buddy Phidippedes to run along the lever until his weight was sufficient to move the earth. Archimedes discovered that a moment is calculated by multiplying the force by the length of the lever arm but would Phidippedes have been weightless when he got to the end of the lever?

FRICTION

Another mechanical principle that was demonstrated by the needle incident is that of friction, which is helpful in stopping things. The amount of frictional resistance to motion depends on two things. First, there are the characteristics of the two surfaces. The coefficient of friction takes on a value between 0 and 1 or a little more if we introduce adhesion into the discussion. Also, there are the normal forces (N) applied to the objects in question. The limiting static frictional force is designated as F. Now the ice has a pretty low value – the coefficient of friction between metal ice skates and ice is about 0.04, whereas that between two pieces of wood is about 0.5. If we rub two pieces of aluminum together, we may see coefficients as high as 1.5. I suspect that the frictional resistance to a needle penetrating soft tissues is somewhat higher! Slips and falls, because of low friction surfaces, cause very many accidents. Trips and falls, on the other hand, are caused by large moments.

FORK TRUCKS

So how long does it take to stop a forklift truck and how far will the truck have trav-eled before it stops? Well, first of all, the driver has to be able to recognize that it would be wise to stop. This decision is hampered somewhat by high loads or broad "masts" while the driver is going forward, and the reliance on peripheral vision while driving backward. This visual part of the story is sometimes amplified by poor factory layout and lighting, and the fact that most fork truck drivers have high seniority (age) and often deteriorating eyesight (driving the trucks is a desirable job when compared with working on the line). The next part of the reaction process is moving a heavy boot to the right pedal (unless a rocker pedal is installed). Then Newton and friction and a few other mechanical and fluid principles take over. The floor may be composed of oil-soaked wooden blocks – a coefficient of friction between 0.1 and 0.3 between this floor and the solid rubber wheel. But fork trucks, especially loaded ones, have a big N – the normal force that is a sum of the mass of the vehicle and load, and the effect of gravity. Now, we come to the important issue – the brakes. When the driver puts his foot on the brake pedal, brake fluid is compressed in a cylinder that in turn pushes a high friction pad or shoe onto a shiny metal brake drum, which is directly connected to the wheels. The effectiveness of this truck-stopping system is related to the surface area of the contact between the brake pads and the drum and the amount of normal force that can be generated by the hydraulic braking system. The matter of rolling resistance complicates the discussion somewhat.

There are three more factors that compound this problem of stopping a fork truck. First, there are the mass and speed of the truck. Of course, there are speed limits and governors but those are for the other guys. We now have to move into part 2 of Hibbler's book – *Engineering Mechanics – Dynamics*. So, this big vehicle is trucking along at 8–10 miles per hour and we encounter linear momentum, which is defined as the mass (m) of the vehicle times its velocity (v). We also have to consider impulse, impact, deformation, elasticity, and restitution – mechanical and financial! Now the brakes really have to work – assuming the vehicle doesn't skid on the slippery floor – they have to overcome this combined effect of mass and velocity. And the law of conservation of momentum suggests that the small mass and velocity of the unwary pedestrian who just stepped into the isle way is not going to help much – it's back to number 9 versus number 8 on the rugby field only more so. The second factor is when the floor slopes and this mess of mass are moving with increasing momentum downhill. The third factor is more insidious and controversial. Suppose instead of carrying the load on the forks, the truck is towing a train of trolleys, all loaded with heavy components. So, the normal force over the fork truck's wheels is reduced, thus reducing the frictional effect, while the mass of the material in the trolleys is increasing the momentum, and the brakes of the fork truck were not designed to deal with this double whammy. The result is that the truck will take a little longer to stop and travel a little further in the process, which is just fine, providing it does not meet Newton standing in the way with a small mass (Peacock, 2020).

NEWTON THE ASTRONAUT

Sir Isaac Newton, who lived between 1642 and 1727, was an astronaut. He must have been because his laws of motion actually work when there is not much gravity and not much friction due to the drag of the atmosphere. Up there about 240 miles above the earth's surface, the velocities and the masses are quite big. The International

Space Station is trucking along at about 17,500 miles per hour and the driver of the Space Shuttle or Soyuz is trying to catch up and dock the two vehicles together. Now, this is all right if the driver can see where he is going and has practiced a bit in the simulators back at the Johnson Space Center but when Progress comes along, the cosmonaut in charge of docking, like our fork truck driver, may not have a very good view. Another space job is to stick a remote-controlled multi-jointed arm out and catch a satellite that weighs quite a few tons on earth and will do funny things if it gets an off-center Newtonian nudge. Remember the number 8 who caught the number 9 some distance from his center of mass? Once one of these satellites is spinning it takes a well-trained rodeo star to stop it – that's why they put the Johnson Space Center down in Texas.

Just recently we were asked to advise on the location of handles on large pieces of equipment that had to be manipulated in and around the International Space Station. Just imagine a Newtonian nudge with a good lever arm setting a large rack careering around the confined interior of an ISS module. Now, these astronauts are quick learners and it is a joy to see how they maneuver large objects around the ISS interior with a little push of a foot here and a little pull of a spare hand there. The trick is to get all the handles and restraints in the right places to allow maximum accessibility, stability, and mobility – this is where biomechanical analyses and models come into play (and intelligent astronauts).

REFERENCES

Peacock, B. (2003). Running, Self-Published.
Peacock, B. (2019). *Human Systems Integration*, Self-published manuscript, Fernandina Beach, FL.
Peacock, B. (2020). *How Ergonomics Works*, Self-published manuscript, Fernandina Beach, FL.
Peacock, B. (2021). *Ergonomics Tools and Applications*, Self-published manuscript, Fernandina Beach, FL.
Peacock, B., and W. Karwowski (1993). *Automotive Ergonomics*. Taylor and Francis, Boca Raton, FL.

12 Measurements and Design Criteria

INTRODUCTION

We are surrounded by variability, but this chapter will be limited to the topic of human variability. The empirical observation of variability can, fortunately, be described by probability theory and statistics. These mathematical tools allow us not only to describe human variability but also to provide the mechanism for making decisions about people and for designing the world in which people exist. Coincidentally, people with no knowledge of these useful tools also deal intuitively with human variability and exploit it. When we speak to a large audience, we use a sound level such that all can hear – "can those in the back hear me?" Conversely, in a crowded room, we lower our voices so that only those in the immediate circle can gain the advantage of our insights. Exploiters of variability include the gamblers, the game players, and the politicians. When these people harness the power of probability theory, they get an advantage over those who limit their observations to subjective probability. We – human factors engineers and ergonomists – come armed with these tools – that is why we are useful. Those "Johnny-come-lately" who limit their observations in a reactive way, solely to the "voice of the customer", are doomed to an eternity of feedback control or tampering (Peacock 2019, 2020, 2021).

PROBABILITY

Variability comes in all sorts of shapes and sizes. Evans, Hastings, and Peacock (2000) describe some 40 statistical distributions, many of which can be brought to bear on the analysis of human variability. One distribution – the Normal Distribution or "bell curve" – stands out as the most widely applied tool. Many people blame Gauss for developing this theory, but the real culprit was the father of probability – a Frenchman named de Moivre, in 1773. Gauss simply used the ideas to explain how the universe works.

CORRELATION

The example given above – the heights of people – is an unfortunate example that almost everyone knows about. They know that a 5th percentile person is taller than 5% of the population and go on to extrapolate that they are also stronger, have better eyesight, and are more intelligent than 5% of the population. What gobbledygook! But don't laugh – the clothing industry has been making such assumptions for years about the various shapes and sizes of people. And they sometimes get away with the assumption. Why? Because many human variables are correlated – i.e., taller people are usually heavier and have longer legs than smaller people. The problem with this correlation thing is that we need another pot full of statistics to describe it, again

DOI: 10.1201/9781003328445-12

based on that pervasive Normal Distribution. Suffice it to say that correlation may be perfect and positive, perfect and negative, and all shades in between. When we measure people, stature (height) is highly correlated with leg length but less correlated with back length, girth or weight or strength or stamina or memory or vision or age, or political party. You get the idea.

We can measure people on thousands of dimensions that are more or less positively or negatively correlated. These similarities and differences have genetic and environmental causes. What is particularly striking about genetic factors is that we have more similarities than differences as a race. We are not like apes or aphids or apples. Most people can sleep on a standard size bed, eat a standard size big burger, read the standard font on a newspaper, drive a standard car, or carry a standard size suitcase. One size fits all? Well, it depends on how precise you want to be or how important it is that the glass slipper fits only one princess. Given our genetic underpinnings, opportunity and practice allow us to adapt to our environment. Expertise is determined by dedicated practice. Unfortunately, age and disease fight our attempts to exaggerate our differences. But to get back to the central thesis of this article, subjective and objective measurements of people can be made on thousands of dimensions and, given a sufficient resolution of our measures, we can articulate substantial variability on each dimension. Furthermore, these measures are not necessarily correlated.

Say we are a 95th percentile height, 70th percentile weight, 80th percentile strength, 90th percentile stamina, 30th percentile dexterity, 50th percentile sociability, 40th percentile at swimming, 10th percentile at singing, and so on. What is the probability that we will find someone else just like us? Fat chance. But with all this genetic stuff pushing through, we are likely to be more like our parents on many dimensions than our overseas pen pals.

By now, we have established pretty convincingly that people are not alike. We have explained why these differences occur and how probability and statistics can be used to describe the variability. No two persons are alike ergo nobody can be Normal.

DESIGN

Now to the challenges of design to accommodate this variability. The challenge hit me between the eyes (almost literally) when I contributed the ergonomics input to the design of the Hong Kong Mass Transit Railway. Assuming that two million people have ridden the train each day over the past 20 years, I feel that I have contributed substantially to the comfort, convenience, and safety of a lot of people – I leave it to the reader to do the arithmetic. One challenge was a decision regarding the height of a horizontal grab rail to be used by passengers as they moved up and down the train. Of course, we should take the comfortable upward reach of the 5th percentile Chinese female. (Note that 98% of the Hong Kong population is Chinese and their stature, due to genetic and dietary factors, is less than Westerners or Northern Chinese.) So, I arrived at a number that hit a 50th percentile male on the chin or thereabouts. I discovered the ergonomics challenge of conflicting criteria! Compromise! The reader is invited to travel to Hong Kong to assess the utility of the final design decision.

The reader is referred back to the diagram earlier in this chapter, specifically to the value of X. Suppose our job is to design the instructions related to income tax forms. We hope that most citizens will be able to read and understand them. Question: citizens must be members of the human race, can tax lawyers be citizens? So, the well-intentioned advisor to the IRS persuades the government to make a law that the instructions must be written at a 8th-grade reading level – we will assume that this X represents the 5th percentile tax-paying adult. This is a pretty difficult task because the cognitive content of the tax laws is determined by these tax lawyers and the instructions only represent the interface. So, the result is that 50% of the tax-paying public makes use of a professional service. Great interface? Don't blame the interface – it is the system designed by tax lawyers themselves that discriminates.

CONFLICTING CRITERIA

Now for an example of conflicting criteria, suppose we work as an ergonomist for one of those parcel distribution companies. Our job is to answer the question: how many (few) young, strong, students does it take to move N parcels, with an average weight of P pounds from A to B in an ideally designed "ergonomic" (I hate that word) workplace, in order to fulfill the company's promise of next day delivery? One more thing: without hurting these young, strong students. Fortunately, we have a wealth of industrial engineering knowledge and the NIOSH lift equation. If we pay our workers enough money, they will work very hard and we can set the value of X – the work rate – very high. Another way of looking at it is that we can call x the materials handling capability of the population and we can set the X selection screen to say the 75th percentile parcel handling capability – young strong. Looking at it from the safety perspective, we know that as the demands on the body increase, there will be two outcomes – first, we will get a training effect – the bodies will become stronger but not younger. The second effect of increasing the demands is that a greater proportion of the population will break. The NIOSH Lift equation lift index is a surrogate for the probability of breaking. (I don't want to get into a lengthy discussion of how good a surrogate it is – it's probably as good as a test of reading level in the tax form example.) So, back to the design question. We have done our best at work place design; we have controlled the mean parcel weight by economic and physical restrictions; we have selected our worker population perhaps with the help of the union; we are left with the design dimension (X) of parcels per person per hour over a ten-hour day. Our conflicting criteria are productivity and safety – both having probabilistic underpinnings. Set X high and we get great productivity and a high incidence of injuries and vice versa. What are we going to do? Well, we are not alone – this question has been debated in the highest circles in the land and was resolved at least temporarily by a five to four vote in the U.S. Supreme Court. Productivity rules OK! Ergonomics is junk science. (But by the way, don't hurt anyone.)

ERGONOMICS, VARIABILITY, AND POLICY

These two examples – tax instructions and parcel handling rate – highlight the fundamental separation of human factors engineering and policy. The human factors engineer can measure human capabilities and limitations on thousands of dimensions. It is up to the policy makers – management, government, and union negotiators, etc., to draw the lines in the sand. It is just like traffic speed limits and blood alcohol levels. We are working on a probabilistic continuum that is based on human characteristics, behavior, and performance measures. So, where does this leave our sacred 5th percentile female or even our 95% confidence limits? These are policy concepts; they are human factors dogma; they are convenient; they are useful; they may be used as the basis of rules; rules are meant to be broken, laws are meant to be interpreted, or negotiated. Human factors engineers are good at measuring, they should also help management and the legislature with the evidence of human variability that is the basis of policy. Nobody is Normal. People vary. Ergonomists by whatever name, have job security because of it. But human variability is our Achilles heel.

RISK AND BENEFIT

The safety and risk community uses data to assess the probabilities and consequences of system failure, particularly with those systems that involve human beings as controllers or simply as participants in hazardous operations. For example, when we drive to work, we rarely think about the fact that some 40,000 people a year die on the roads of the United States. We are more likely to think about our contribution to and benefit from the Gross National Product, which usually necessitates us driving to or for work. We may recognize that drivers who are incapacitated by alcohol or drugs cause some 20,000 fatalities on the roads. But even that sobering figure and the severe penalties that are in place to address DUI do not stop a very large number of people from using their cars to provide access to recreation involving alcohol nor does it prevent many people from using cold and allergy medicine or more powerful prescription drugs that may interfere with driving or driving when they are tired or in very hazardous conditions. At a more spectacular level, the data tell us that 2 out of 113 Space Shuttles have failed catastrophically, which, if you wish to put 95% confidence limits on the probability of failure, the relevant statistic of the order of 1 in 30; of course, those responsible for the space program will be quick to point out that these statistics are invalid because the Shuttle is not a member of a large homogenous fleet for which component failure rates are known and that significant changes have been made to the engineering and managerial issues related to known failure causes – O rings, debris, and tiles. The astronauts are still enthusiastic to fly because, in their minds and those of the Space Program management and many of the U.S. population, the benefits of human space flight outweigh the enormous risks involved.

The business and operations research communities have long been interested in cost-benefit analysis. Their examples and mathematical models show us that decisions under uncertainty often involve multiple variables and multiple criteria, which have to be weighted and optimized in the process of business decisions. Psychologists who investigate human decisions and choices also point out that human beings (e.g., managers, engineers, doctors, juries, generals) rarely reach the lofty ideals embodied in Bayes Theorem) in that they rarely use all the data that are available. Even sophisticated consumers, like you and I, are usually sub-Bayesian when it comes to buying the more expensive model of car, computer, cell phone, or camera. In the laboratory, cognitive psychologists can easily demonstrate the fallibilities of operational memory when the subjects are faced with time or other psychological/social stress.

Strayer and Drews (2004) quantified the decrement in human driving performance as affected by the secondary task of having a cell phone conversation. They produced numbers like an 18% increase in driver response delays and a doubling of rear-end collisions (in a driving simulator). They also pointed out that the effect on younger drivers is to make them perform as if they were old. Like most good human factors researchers, these authors reported averages and standard errors – the chances are that those with average responses would have survived, which leaves a whole bunch of accidents waiting to happen in the right-hand tail of what is probably a positively skewed response time distribution. These issues of driver distraction have been around for decades. Years ago car radios were the target of the safety advocates. Since the late 1980s till today, the car companies have been conducting extensive research on driver distraction with in-vehicle navigation systems, mobile phones, and head-up displays. Many of these studies showed that, although the driver's visual and motor resources may be aided by such things as hands-free operation and better (often bigger) display features, the issue of cognitive capture remains the premier challenge. During the GM ACCESS car study, I remember one old lady taking almost a minute to fathom out how to tune the radio while she was traveling at 45 mph. I was also very active in the assessment of the physical and content characteristics of head-up displays, which in the simulator sometimes had similar cognitive capture difficulties. On another occasion, I had the opportunity to evaluate whether it would be safe to drive a car equipped with a video player that projected into a small mirror next to the rear view mirror. I remember watching "Top Gun" and that, although I was attuned to time sharing between the movie and the road ahead, the extended headway between my car and the one in front was typical of many (but not all) cell phone users on our roads today.

This brings me to my main point. It was possible for me to drive without incident while watching a movie. Also, the vast majority of cell phone use in cars do not result in accidents, even under heavy traffic conditions. This is due to what used to be called "spare mental capacity", with a bit of probability thrown in. Speaking on the phone, even in a car, is perceived by some eighty million cell phone users in this country as being a reasonable thing to do and most of the time they survive the experience and feel that the conversation, like drinking (a little or a lot?), was of a level of benefit that greatly outweighed the risks. Meanwhile, our profession offers (even with excellent data) the simple statistic that response times are increased by 18%.

In the early 1980s and throughout the 1990s, I was involved in manufacturing and office ergonomics. Work-related musculoskeletal disorders, including carpal tunnel syndrome, rose to national prominence in epidemic proportions, fueled by a contentious debate between labor and business and between the two main political parties. Early attempts to push the problem under the table were soundly defeated when it was realized that on many occasions these disorders were real, even though varying in their levels of pain, debilitation, and reporting. However, nobody caught carpal tunnel syndrome by assembling one car a fortnight, cutting up one turkey a year, taking one geriatric patient to the toilet once a day, or entering one page of data an hour into a computer. Meanwhile, management and one of our parent professions – industrial engineering – were pressured by the competitive economic climate to improve productivity. So, now automobile assemblers may face one car every 40 seconds, typists one page a minute, turkey carvers one bird every few seconds, and nursing homes are like an ice hockey rink.

Students of human factors/ergonomics will also realize that this mess of force, posture, repetition, and politics was complicated by the ever-present problem of human variability. Just like cell phone users and accidents, not all assemblers or disassemblers caught carpal tunnel syndrome. The benefits of work and competitively priced products greatly outweigh the prevalence of work-related musculoskeletal disorders, to managers, consumers, shareholders, unions, and workers, unless you are one of the unfortunate ones. The reader is also invited to consider the repetitive work involved in building the pyramids and the Burmese railroad as described by the movies: "The Ten Commandments" and "The Bridge over the River Kwai". Closer to home there are the self-inflicted stresses commonly experienced by athletes, for whom the reward is worth the pain.

PRODUCTION WITH PROTECTION

We may debate the gross accident, fatality, and injury statistics. We can also pick the low-hanging fruit – drunk drivers and drivers who cause accidents by cell phone use should be punished, but the cell phone will not go away. Similarly, the demand for cheap meat and cars, affordable extended care, and computer-centered work will not go away, so we will continue to face the physical and political problems of work-related musculoskeletal disorders, at least until the automation utopia arrives. Human factors engineers/ergonomists will always find themselves wandering in the middle of a mess of variability. Some of our professions veer toward the side of protection. They say, "Ban cell phones from cars" and "Slow down the line rate". Meanwhile, those that pay our salaries – the consumers and managers (academics aside) say, "If you want to get at my cell phone, you'll have to go through me first" or "Productivity provides profit, which makes us all happy" (except of course for those relative few who die in car accidents or can't sleep at night because of carpal tunnel syndrome). Now, the academics may be politically neutral and simply report prevalence, associations, and sometimes solutions. Or they may consider that our profession is a humane profession largely driven toward health and safety. It wouldn't be the first time that academics showed their social conscience but some academics work on nuclear weapons and other killing machines. Is science and policy really an oxymoron? A policy cannot do without science but science alone does not dictate policy.

It makes no sense to discuss risk outside the context of benefit. Space flight is risky, but who wouldn't wish to walk on Mars? Cell phones in cars are risky; they are very useful when you have a flat tire or someone runs into you. When you are lost, forgot to pick up the children from daycare, are late for an appointment, or can't wait to relay the latest office gossip, who wouldn't be tempted to a brief conversation while driving? The mental workload puts you in a dilemma Also, repetitive work is useful, productivity is good and if you can find a more efficient way than the production line (or work cell), then you will be a hero; I'll not hold my breath!

Somehow the human factors engineer/ergonomist must assess the triple whammy of costs, benefits, and human variability and provide the customer with usable advice, not just statistics and rhetoric. You may argue that Strayer and Drews' number of 18% is both statistically and operationally significant. But just look around you while driving along the highway. The NIOSH/BLS findings on work-related musculoskeletal disorders are pretty convincing and so is the trade deficit. As in the forensic branch of our profession, we must give both sides of the aisle unbiased science and go further – give them a number that accepts that risk and benefit go hand in hand

ERGONOMICS DECISION ANALYSIS (EDAN)

> A missing element in many ergonomics processes is a formal tool to support design or intervention decisions that are often made on only a limited assessment of all the implications.

The concept of risk is widely developed in the safety world. In the UK, criteria have been established to define regions of acceptability, tolerability, and unacceptability in terms of the number of fatalities per year. Negligible risk is considered to be at a level of 1 in ten million. The borderline into the tolerable region starts at 1 in one million and into the unacceptable region at 1 in 10,000. A distinction is made between the risk to the general public and to workers employed in a hazardous industry, in which case the borderline between tolerability and unacceptability is 1 in 1000. The tolerability region is further defined as that region where the benefits may be judged to outweigh the risks. The risk calculator takes into account outcomes of lesser severity than fatalities and includes quantitative concepts of exposure in the denominator of the equation.

In the United States, the National Association of Manufacturers has strongly supported the Regulatory Improvement Act of 1998, which aims at laws (particularly environmental laws) that "are based on sound science and focus on the comparatively higher risks to society". The recommended process includes:

1. Scientifically sound risk analysis
2. Risk-based prioritization
3. Benefit cost analysis
4. Flexible efficient and cost-effective risk management
5. Public participation in all phases of the process

A fundamental issue in the assessment of risk is uncertainty. Many hazard sources are innocuous at low exposure levels but lethal at high levels. Such sources include physical energy such as heat, sound, light or force, and exposure to chemical and biological stressors. In many cases, there are direct trade-offs among the risk of harm to people, equipment, or the environment and the benefits to society or a subsection of society. An intuitively clear example is traffic speed – there are benefits of completing a journey quickly but the risk and severity of accidents increase with speed. A second issue with risk on a particular dimension is that sometimes the effects (benefits or risks) are not monotonic in relation to hazard intensity. Sometimes too little is as harmful as too much, as in the case of heat. The same problem occurs with physical work. Biological systems thrive on (mechanical) stress, without it bodily systems atrophy as commonly seen after bed rest or space flight. It is generally understood that underuse is a primary risk source for ill health. On the other hand, overuse is a common risk for athletes and production line workers.

A second complication of risk analysis is the inherent variation among people in their response to hazardous sources. People are also capable of adapting to increasing levels of stress, given a gradual increase in exposure. These sources of variability confound the issue of risk analysis. They beg the question *what proportion of the population should be accommodated or protected?* Clearly, the answer to this question must factor in an assessment of the possible severity of the outcome and the acceptability of selection strategies to remove vulnerable individuals from exposure. In the case of driving, the selection is accepted at both ends of the age spectrum, although not always without a challenge. In an unionized industry, seniority-based job choice is a prime method of self-selection. In the area of recreation, self-selection is the primary method of hazard avoidance and, in general employment, given a healthy economy, personal choice of job is widely used, although here again there may be a trade-off between personal benefit (wages) and risk.

An appropriate decision tool must comprehend all the relevant aspects, including both the positive and negative outcomes and the costs of analyses and alternative designs or interventions. It must also address the short- and long-term implications. The decisions must also be made in light of the broader relevance of the situation, such as the cost of a design or intervention alternative when compared to the value of the functional process. Again, a traffic control example is pertinent where the design alternatives for an intersection may range from a yield sign to an overpass, depending on traffic density.

The proposed model is comprehensive and accommodates the errors that accompany many outcome and cost estimates. It addresses the nonlinearities of human judgment by the application of fundamental psychophysical principles. An elementary mathematical process involving exponents is used to allow the model to be applied manually. Alternatively, more complex model elements can be applied through a spreadsheet. Units are chosen to facilitate the estimation and meaningfulness of the quantities involved in the model.

The decision model elements and ranges have been chosen for convenience but expansion is possible, should the situation warrant it. Although it is understood that there are nonfinancial elements of outcomes, it is convenient for the purposes of this model to reduce outcome assessment to a dollar equivalent of the daily availability of employees or systems. The purpose of this assumption is to obtain common currency measures for the comparison of outcomes and design interventions and address the wide-ranging variability of the severity of outcomes. The time unit of a day is used because of its universality.

The data capture process requires the estimation of the expected value of each element by an individual analyst or, preferably, by a consensus of experts. The latter approach is more likely to produce a more stable estimate. Another benefit is that consensus can be used to account for multiple sources of costs, benefits, and other factors affecting system performance. Given these expected values, high and low estimates are identified as those values above and below the expected value, respectively.

SCENARIO ANALYSIS

The ergonomics researcher, professional, or practitioner is often faced with complex situations and has to decide on an approach to the problem including consideration of:

1. What are the important problems? What are the purposes of the investigation?
   a. Are there problems with system effectiveness?
   b. Are there efficiency, productivity, or cost implications?
   c. Are there health and safety problems? What are the probabilities of "human error or failure"? What are the implications or consequences of system failure? Are the effects acute or long term?
   d. Are there issues of human preference? Opinion?
   e. Are there individual and social issues involved?
2. What data or evidence are needed?
   a. Does accurate, reliable, and valid data/evidence already exist?
   b. Are there any existing design guidelines or standards that are not being applied?
   c. Can data be obtained by simple observation or questioning of those involved?
   d. Should a formal survey or observational approach be used?
   e. Can physical mock-ups, modeling, or simulation be used?
   f. Can the investigation be carried out in the laboratory? In a simulator, or in the field?
   g. Is a formal controlled experiment needed? What about sample selection and size? What about experimental design? What about confounding? What about subject selection?
   h. Is there a need for a multi-pronged approach to the collection of evidence?
   i. How long will it take to collect the data? How much will data collection cost?
   j. What equipment and methods should be used?
   k. Are there any human subject and institutional review board issues with your proposed approach?
3. Which measurement and analysis tools should be used?
   a. What tools should be used to collect, process, and analyze the data?
   b. Are the approaches validated and calibrated?
   c. Is the significance of differences or associations important?
4. What designs or changes are indicated?
   a. Do the suggested designs or changes include hardware, software, or operations?
   b. Are user selection and training needed?
   c. Will facilitators help? – Instructions, warnings, labels, checklists, procedures, tutorials, augmented reality?
   d. Will the changes require careful implementation, persuasion, and monitoring?

5. How should the changes be justified and communicated?
   a. What effects will the changes be expected to have?
   b. How long will it be before the changes have a detectable/useful effect?
6. How should the effects of human variability be assessed and communicated?
   a. To whom should the results be presented?
   b. How should the results of the investigation be communicated? – Presentation, report, process requirements, system design specifications, guidelines, and standards.
   c. How should graphs, tables, diagrams, photographs, statistics, verbal arguments, and demonstrations be used?
   d. Are the results of the investigation suitable for publication in academic, professional, or trade literature? Are the results proprietary? Could the results be used in court cases?
7. How should the effects of the changes be evaluated?
   a. Is there a plan for monitoring the implementation and effects of the changes?
   b. Will these evaluations be specific to the particular situation or will they have general applications?

The reader is invited to address each of these aspects of an investigation with regard to the following scenarios?

1. The observations of air traffic control and the communications from the pilot of a small airplane indicated that he was acting irrationally shortly before the plane crashed. The pilot was relatively inexperienced and was not trained to fly at night or over open water using his instruments. You have been asked to comment on the possibility that hypoxia was a leading factor in the accident. Describe how you would prepare for your presentation to the board of inquiry, which may include members not familiar with physiology.

2. You are SCUBA diving at a depth of about 90 feet in relatively warm water in the Caribbean. You notice that a relatively inexperienced pair of divers appear to be in difficulties and are repeatedly looking at their pressure and depth gauges and dive computer. Given that you have a PADI instructor's certification, how would you assess the possible causes of the problem and what information would you seek and what advice/instructions would you give? What are the implications of alternative responses to the problem? How would you evaluate alternative equipment designs and training to prevent such incidents?

3. You are a newly hired ergonomist in a large gas and oil company that produces, transports, stores, processes, and distributes products under pressure. You have been given the task of reviewing the instrumentation used throughout the processes in order to reduce the probability of human error by the process controllers. How would you use your knowledge of the gas laws (Charles Law, Boyles Law, etc.) to recommend changes in display design? What help would you seek from experienced chemical engineers?

4. You have been asked by a university vice president for facilities to advise on the seating design for a new basketball stadium donated by a rich alumnus. How would you address the challenge of human variability in your recommendations? What criteria would you use in the evaluation of alternative designs? What recommendations would you give and how would you justify them?

5. You have been asked to design a seat for a new fork truck, which is to be used to move materials in and out of trucks parked at the loading bay, to temporary storage areas, and then to sites of operations throughout a large manufacturing plant. How would you address this task and evaluate the appropriateness of alternative designs?

6. You have been engaged by a commuter airline to investigate the numerous complaints about seats. How would you approach your task? An approach proposed by the CEO is to provide adjustable seating and to charge a premium for people who take up more space. How would you implement this approach?

7. A major league baseball club is planning to address the challenge of soaring salaries by opening its batting cages for the general public to pay to participate in a new game – hitting baseballs for distance with different implements and different ball speeds – from 0 to 100 mph. What factors would affect the distances achieved? How would you design an investigation to evaluate the effects of equipment design, ball speed, and individual differences?

8. A major league football club is interested in alternative biomechanical strategies for their defensive linemen. They have engaged you to develop a computer-based model of the locations, directions, and sizes of the forces involved, including issues of friction but excluding holding and tripping. How would you go about this task?

9. OSHA has engaged you as a consultant to evaluate and modify the NIOSH lift equation for application to single lifts. Their objective is to account for all possible factors associated with acute injuries to people in the warehousing and construction industries. What biomechanical factors would you consider important? How would you develop a model of injury likelihood? How would you validate this model empirically?

10. You have been engaged to investigate an epidemic of bad backs in a large data processing facility. The union leadership has offered the opinion that all the desks and chairs should be replaced by adjustable units, based on a numerical fitting matrix. The union membership would rather just have the adjustable chairs. Management is prepared to pay for either new chairs or new desks, but not both. How would you approach this problem? How would you plan an evaluation of the effects of the changes? How would you justify your findings to the scientific community?

11. In a nonunion precision, labor-intensive manufacturing facility is being pressured to reduce costs and increase productivity while maintaining its historic high-quality standards. The management approach is to introduce a team structure along typical Japanese lines. What would you expect to be the long-term outcome of this situation? How would you monitor the effects of the changes?

12. A colleague with a business school background has asked you to explain the Hawthorn effect to a professional seminar for accountants. You may expect that they had all taken an organizational behavior class. How would you communicate to your audience that the causal factors for individual and group behavior may not necessarily be related to their organizational design strategies?

13. When you get into a rental car, you have various expectancies regarding the locations and direction of operation of the secondary controls such as climate control, entertainment systems, seats, and windows. These expectancies are based on basic spatial relationships and your experience with other vehicles and control operations. How would you address the designers of future cars, the marketing departments, and their managers regarding the importance, or otherwise of standardized control locations and operations? What kind of arguments against your ideas do you expect to face? How would you deal with these arguments?

14. When you want to increase the flow of water from a tap, you turn it counterclockwise. When you want to increase the volume of your radio, you turn it clockwise. How would you investigate and design an electronic interface for the control of the temperature of a shower (assuming that you could be sure that you would not electrocute the subjects during your usability trials)?

15. One difficult job for an astronaut is the control of a six-degree-of-freedom robotic arm. This involves the use of two controls – one for pitch, yaw, and roll and the other for the three translation axes. The astronaut also makes use of multiple cameras, with similar controls, but which, because some are mounted on the arm and some are on fixed stations, provide different views of the scene at different times of an operation. How would you apply your knowledge of expectancy and compatibility to the design of controls and training methods?

16. Foundry workers are required to clean out hot furnaces. Time is of the essence as the costs of shutting down and starting up the furnaces in terms of productivity are very high. Consequently, management would like to get the maintenance operators into the furnaces as quickly as possible after shutdown. Describe your approach to data collection, analysis, and intervention, including personal protective equipment and administrative controls. How would you explain heat stress to the workers and managers?

17. You have been engaged by the Army to acclimatize the infantry for very physically demanding work in a desert war zone. How would you assess the effects of alternative and complementary intervention approaches?

18. You have been engaged by a local school board in Texas to investigate the dangers of starting aggressive football training on 1st July. How would you assess the physiological effects of alternative training, monitoring, and intervention strategies?

19. You have been asked by the government to develop easy-to-use rules for manual materials handling in large hardware stores that can be easily implemented assessed and understood, similar to traffic control rules. How would you research the background of this task? How would you quantify, implement, enforce, and evaluate the rules?

20. The biggest challenge to ergonomists is human variability. Individuals vary on many dimensions and most tasks involve many human attributes. How can you justify simple 5th percentile type rules? What other forms of rules could be developed and applied so that the majority of the population could be productive and safe in their work?

21. Many ergonomics situations involve trade-offs between productivity and safety. These issues are notoriously contentious. How would you develop and implement a rule-based approach to the challenges of baggage handling at airports or in nursing homes?

22. You have been engaged by the Department of Homeland Security to design the jobs of inspectors who have to screen thousands of people a day with the challenge of identifying a few terrorists. The Department is increasingly sensitive to the problems of false accusations. Given that all physical avenues (lighting, image enhancement, training, etc.) to improved detection have been exhausted, what organizational strategies could be applied to improve inspector performance?

23. Car driving is becoming increasingly complex with fast speeds, traffic congestion, complex vehicle features, and increasing numbers of nondriving tasks led by the ubiquitous cell phone. Given that driver performance is related to distraction levels but that the effects of distraction are situationally specific and often of little importance, how would you use laboratory, simulator, and field methods to develop guidelines for driver attentional management?

24. You have been asked by a major wine-producing company to plan an evaluation of wine-tasting methods in a task that compares home-produced wines with imported wines. The plan is to have large groups of volunteers assemble on Friday evenings for three-hour wine-tasting sessions. Given that wine can be assessed by sight, smell, and taste, you are expected to distinguish between the roles of these senses and the possible interaction effects. You are also required to produce and evaluate criteria for measuring individual differences in wine-tasting behavior and performance. Finally, you are expected to assess the likely order effect in performance and time into session, assuming of course that designated drivers accompany all participants.

25. People come in all shapes and sizes and clothes manufacturers address these challenges with discrete sizing systems, sometimes with a little bit of adjustability. Fine tuning is carried out on the sewing machine and if all else fails sartorial license rules. Given the advent of whole body scanning techniques, there is a wealth of new information that could be brought to bear on clothing design. Devise a field fitting method for men's suits using whole body scanning while you wait for sizing and adjustability.

26. Common experience demonstrates that individuals familiar to the observer can be recognized under limited lighting and exposure duration conditions, especially if movement information is added to shape and size. Identify a set of features that are likely contributors to this shape recognition task and develop a laboratory experiment and computer-based method to test your hypotheses.

27. Sheldon developed an easy to understand method of somatotyping that has become less popular in modern times. Devise an experiment involving both anthropometric measurement and subjective perception to evaluate Sheldon's model.

28. Contemporary advertisers highlight key features and associations of products, such as cars, to inflate customers' general impression of the product. A challenge to such processes is that in reality, the products do not differ a great deal objectively on these dimensions. Devise a field experiment to investigate the halo effect in car purchase behavior.

29. Multiple choice or "objective" tests are used widely in education and professional certification examinations. However, many educators still consider that essay-type or problem-solving answers tell more about the student's analytic and communication capabilities. Devise an experiment to test the reliability of both forms of examination.

30. The interview has been shown on many occasions to be an unreliable method of personnel selection. However, the vast majority of managers still place considerable weight on their own judgment capabilities during interviews. Devise an experiment to evaluate the reliability or otherwise of individual and panel interview processes.

31. The methods of psychophysics are generally applied to assess the abilities of individuals to make judgments on single physical dimensions such as size, weight, or color. The results of these experiments are usually reported in terms of "just noticeable differences" that are detected on 50% of trials. Devise an experiment to establish the relationship between a physical dimension and a standard.

32. You have been asked to advise a retail chain on the perception by customers of package size and perceived value. How would you apply psychophysical methods to this task?

33. Psychophysically derived differences do not always estimate operationally significant differences. You have been asked by a fast food retailer to address the perceived hamburger size issue, with a view to cost cutting on the amount of meat used. Develop a field experiment to detect customer judgment of hamburger size.

34. Use a psychophysical experiment to assess the perception of speed – baseballs, cars, trains, airplanes, hand movement, running, walking, etc.

35. Carry out a simple experiment to measure human errors (and human variability) in the recall of digit and letter strings of different lengths. Repeat the experiment with mixed lists and grouped sequences, including meaningful and nonsense syllables.

36. Write down at least 50 codes associated with a person or product description. Investigate how these codes are processed.

37. You have been asked by the State Department of Transportation to revolutionize the way in which people, vehicles, and licenses are coded. How would you investigate the current problems of errors and productivity? What new designs would you suggest? How would you implement and evaluate the new designs?

38. Everybody knows about learning curves but very few people actually quantify their own learning curves. Devise an easy to use process to measure and predict changes in performance over time in booking a flight on the web, knitting, running or walking one mile, throwing darts, reading, etc.

39. You have been hired by an automobile manufacturing company to investigate how many cycles it takes to get a new car line moving at a rate of one car an hour to 60 cars an hour. How would you go about this task? What about quality?

40. Murphy's law states that if anything can be done to cause a catastrophe, then someone will do that thing. Research the web for the Darwin Awards and explain why these priceless individuals did obey Murphy's Law.

41. You have been engaged by the Office of Homeland Efficiency to investigate why airline flights are held up because of bizarre human behavior. How would you collect appropriate data on these relatively infrequent events? What changes would you suggest recognizing that you must not interfere unduly with the obedient majority?

42. You have been engaged by a cell phone company to investigate all possible failure modes of the system that are due to inappropriate human behavior. How would you investigate these problems?

43. You have decided to go into the wall and ceiling papering business, as you believe that there are a lot of surfaces out there waiting to help you make your fortune. You have estimated that one casual helper can paper three $20 \times 15 \times 10$ rooms (3000 ft^2) in an 8-hour day and that you should charge by the square foot ($0.10 plus materials) and not by the hour. Unfortunately, on day one you only manage to finish one room. How would you investigate the physiological causes of this low productivity? How would you revise your estimating procedure in order to make $1,000,000 per year, always assuming that you stayed competitive?

44. You wish to describe and demonstrate local muscle fatigue and recovery to a class of unruly high school students, noting that different muscle types fatigue at different rates. Devise a demonstration that involves the whole class and results in data that can be used for a peer-reviewed publication. How would you account for individual differences? What methods would you use? What experimental design would you employ? How would you address the order effect?

45. You have been engaged to investigate why the productivity of some employees in a parcel handling center stays constant over the 12-hour shifts while that of others declines significantly. How would you investigate the effects of general physiological fatigue? What changes in work design would you make given that no new equipment could be purchased?

MEASUREMENT

There are jobs, hard jobs, and jobs your fathers (and forefathers) used to do. All work has its rewards and much work has significant costs that sometimes outweigh the rewards. Some work is hard on bodies, some on minds, and some on souls. Some work provides enormous personal satisfaction, with minimal tangible gain. Other work is boring but lucrative. The relationship between these different rewards and costs of work is not simple – some people, who "work smart, not hard" may get significant financial rewards, and others who work physically hard may pay exorbitant costs but have the satisfaction of substantial personal achievement. A challenge to students of work is to provide a common currency for the description of the context, content, and outcome of work, so that valid comparisons may be made.

THE MULTIPLE PURPOSES OF WORK

There are four principal categories of work purposes or outcomes – effectiveness, efficiency, safety, and satisfaction. Effectiveness of quality addresses the expectancies of the customer for the product or service. Efficiency or productivity relates to the consumption of resources, such as time, money, space, equipment, and materials. Safety addresses the prevention of acute or cumulative harm to the people or other resources associated with the carrying out of work. Satisfaction is a uniquely individual outcome or work – one man's meat is another man's poison. Additional purposes will include: ease of use and esthetic appeal, security, and sustainability (reliability and resilience).

COLLABORATION, SLAVERY, AND EMPLOYMENT

From the very earliest of times, people have formed organizations or collaborative groups to expand the effectiveness, efficiency, safety, and satisfaction associated with work. When these groups are formed, there is usually a division of duties and leaders emerge to guide the strategic management of the group. The leaders may be appointed (perhaps self-appointed), selected, or elected and sometimes these personnel processes are subject to bias. This bias may not necessarily lead to bad leaders. Indeed, many slaves may be treated very well and have considerable autonomy, except for their opportunity to choose their master.

The line between employment and slavery is very fuzzy in reality although the two conditions of work are distinct politically. In the early days of transportation, it was common for groups of slaves to row boats across the sea or move large blocks of stone for the construction of pyramids, castles, churches, and factories and be encouraged by physical methods. These slaves typically came from a separate ethnic, geographic, or cultural community. In the recent past, slave labor became a tradable commodity and shiploads of Africans came to America; to this day employee slaves are forced by their leaders and circumstances to give up their practical rights to choose what they do in the form of work. Golden handcuffs or the promise of better times ahead may be as effective as the whip.

When someone joins the military or the college football team, they are typically directed by appointed or selected leaders called sergeants or coaches. When they fail to perform quality work, they may be subject to physical or emotional abuse with the intention of making them stronger. In times of war, if a soldier, for very good reasons of personal survival, chooses to opt-out, he may be shot. During practice, the deficient footballer may simply be given the symbolic public humiliation of ten pushups or a run around the field. Human deficiencies in the real world of war and sporting competition may be rewarded by a career-ending public humiliation and untold personal trauma.

It is common in business and industry to use the expression: "I work for" or "who do you work for?" The operational meaning of these expressions is simply a reflection of the appointment or selection of a leader and the symbolic placement of his name in a box on an organization chart. Sometimes appointments and selection of leaders are focused on a subset of the purposes of work – such as productivity – and sometimes these leaders may choose a managerial style that is incompatible with the wishes of the subordinate, especially where that subordinate was not a party to the appointment or selection process. The boss also has a difficult task, because his boss or customers may have expectations that diverge from those of the employee. As in war and football, rule number one in business and industry is that "the boss is always right" and rule number two is that "when the boss is wrong, go to rule number one". The penalty for not understanding these important rules may be career shortening or a more subtle modification of the conditions of work.

In government, labor organizations, and academia, it is common for leaders to be elected by those who they lead. The campaigns leading up to such elections commonly involve promises and quid pro quo. Another characteristic of elections is that the electorate may not be aware of all the facts about the candidates or the implication of decisions made by the successful candidate, once elected. Also, elections are rarely unanimous so that once again as in war, football, industry, and business, those that are being led may not always be happy with their elected leaders. They do have a chance at the next election or even at a recall election but once again there will usually be a minority of unhappy workers. You can't please all the people all the time, and the context and content of work will always be constrained.

Hertzberg addressed this issue at the individual worker level. He argued strongly that it was the intrinsic content of work that motivated the worker and that the context simply contributed to dissatisfaction. Can a slave be happy picking cotton? Can an automobile assembly worker be happy attaching an exhaust pipe, overhead, to a car every 30 seconds? Can a quarterback be happy if his receivers can't catch or his protectors don't protect? Can a soldier be happy if the war is being lost? Can the voter be happy?

HUMAN VARIABILITY

Work and play are good indicators of individual differences. In professional sports, these differences are documented in great detail and the influence of intangibles, halo effects, and pitchfork effects become dominated by objectivity. Play can be measured by wins and losses and by salaries. So can work, but there is enough uncertainty left to create work for students of work and play for the foreseeable future. The outcome of work is measured by effectiveness, efficiency, safety, satisfaction, and their derivations. The human input to work may be physical, informational or motivational, and humans vary enormously on these dimensions.

DEMOGRAPHICS OF WORK

Life is divided into three parts – learning, working, and smelling the roses. Ideally, these activities correlate with the aging process but the borders are so fuzzy that they may overlap considerably.

The concepts presented thus far can be drawn together using the familiar grammatical construct of a sentence:

- I want to operate my car.
- Some people want to drive their cars to work in less than half an hour while listening to their voicemail.
- Qualified drivers want to drive their racecars quickly and safely around the wet, winding racetrack.
- Well-trained astronauts want to capture a large satellite with a robotic arm.

These sentences articulate process requirements with varying degrees of specificity. Designers of the processes – driving, capturing – need more information to satisfy their customers. The first task is to identify the customers and their perhaps differing requirements. The end user – I, the driver, the astronaut – may not be the only customer. Other customers include trainers, maintenance engineers, managers, legislators, and the general public and they may have differing, perhaps conflicting, requirements. Sometimes the customer may be an individual that has a tailor-made [space] suit. On other occasions, the end user may be one of a large population of users for minimally adjustable hardware. (If the glove doesn't fit, try the next size up and if that's too big, tough!)

A requirement must be articulated as an adverb that may be evaluated, assessed, or validated by objective or subjective methods. The adverbs quickly and safely can be assessed subjectively; and for this assessment to be useful, it should reflect the consensus of all the customers. One way of improving the reliability of requirement assessment is to provide verbal or numerical anchors to the assessment statements. For example, the requirement "quickly" could be quantified by a speed or a time to cover a fixed distance, under controlled environmental conditions. The "safely" requirement is more problematical and may be quantified in a safe/not safe scale, by articulating a continuum of possible outcomes – such as mission failure, crewmember fatality, and subsystem damage – or by adding probabilistic statements that can only be assessed reliably in the light of experience with the operational system, or through simulation. Parenthetically, the best available estimate of the probability of failure of the Space Shuttle mission process is 2/113; however, this only predicts an expected value, and where confidence limits are placed around this point estimate, using the Binomial, Poisson, or Normal distributions the 95% confidence level is of the order of an unacceptable 1/30. Such historical estimates are always suspected as learning occurs in most human-managed processes and possible failure modes are eliminated in the light of experience, thus, changing the system design and process reliability. In the car driving context, the biggest process reliability change would be to eliminate drunk driving that contributes to almost half of the 40,000 fatalities a year in the United States, but on balance, driving is a very reliable process and most drivers who are "under the influence" don't have accidents and most drivers who drive faster than the speed limit don't get caught.

RISK ASSESSMENT

The technology of risk assessment has progressed over the past few decades and there exist various standard processes for linking outcomes and likelihoods. These methods typically use nonlinear probability scales and ordinal outcome or severity scales, which are sometimes converted to a common currency, such as dollars. A fundamental shortcoming of these risk assessment approaches is that they do not usually address the trade-offs that must be made with positive outcomes – benefits. Where a common currency approach is adopted, it is possible to develop key ratios that relate to costs and benefits. A common metric in space flight engineering is equivalent system mass (ESM); this also fails to comprehend trade-offs between costs and benefits and is therefore an insufficient decision tool. More sophisticated analytical processes are essential if we are to comprehend how space flight trade-off decisions are made. Another challenge is related to the costs of the development of countermeasures as well as the countermeasures themselves. For example, the development of a planetary surface suit that is both protective and offers good mobility and where there may be acceptable trade-offs between protection and mobility, may have very high development costs and conflicting operational advantages and disadvantages. Very few car buyers purchase Hummers in order to increase their personal safety but the evidence is clear that vehicle mass is a major contributor to accident outcomes). The issue of the trade-off between "production and protection" is also very apparent in high-volume manufacturing and materials handling processes. The shareholders and management want "productivity", whereas the union fight's for "protection" and this trade-off has been escalated to the highest circles in the country with the debate about ergonomics standards.

THE DESIGN PROCESS

The engineer cannot do his job effectively without requirements that can be validated. In other words, the requirements statements must contain verbs and their associated adverbs that define process performance and the conditions or tests under which this performance is to be evaluated. For example, a sports car may be expected to go from 0 to 60 in 6 seconds. A suited astronaut may be expected to travel 100 meters over a planetary surface in 5 minutes, with a heavy load of equipment.

Given these validatable operational or process verbs and quantitative adverbs, together with contextual information, the engineer and operations designers are in a position to start addressing the systems that may be needed to satisfy these requirements. Designers create things – nouns – and can only design them with quantitative information – adjectives. For example, from 0 to 60 in 6 seconds may be achieved with a big heavy car with a big powerful engine or with a small, aerodynamic car with a small, efficient power train. The planetary surface astronaut's task may be achieved with a rucksack or a golf cart. Once the engineer has the process requirements, he/she can then explore the systems (nouns) and their characteristics (adjectives) to develop concepts that may satisfy the requirements.

Unfortunately, the design process does not always work in this tidy way. The customer may not articulate clear requirements but may seek to specify design options and impose requirements after the fact. For example, the customer may ask for a small, aerodynamic sports car and may be disappointed when his luggage doesn't fit in the trunk. An exploration program manager may specify a lunar rather than an orbital launch platform. Conversely, the engineer sometimes seeks requirements that fit his predetermined design specifications, much like the health care specialist with a limited set of interventions may seek diagnostic information to justify those actions. Such conservatism is sometimes justified as the system design characteristics may be well evaluated. The challenge occurs when the system is expected to meet new requirements. "If your only tool is a hammer, very soon everything begins to look like a nail". These possibly unfair references often occur because of a shortage of research and development funding, but in the long run, the new challenges of long-duration space travel will require new technologies.

QUALITY FUNCTION DEPLOYMENT

An adaptation of Quality Function Deployment can provide the discipline of separating systems, processes, nouns, verbs, adjectives, and adverbs. Quality Function Deployment employs a series of matrices that transfer information from market research through product design, manufacturing, and production processes to sales and the rest of a product life cycle, including maintenance and recycling. The vertical axes of the matrices contain information about customer requirements and their quantitative adverbs, often obtained by benchmarking tests. For example, a space suit user may expect good shoulder mobility and the adverb may require this to match unsuited shoulder girdle function (an impossible task with current hard upper torso technology). A vehicle maintenance function may involve visual, hand, and tool access and the quantitative adverb may expect spark plug change in 5 minutes – a process performance standard derived from comparison with other similar vehicles.

The horizontal axes contain descriptions of the systems (nouns) and their quantitative adjectives. These are system design specifications. For example, maintenance access may require a cone with a minimum diameter of 20 centimeters. Radiation protection may require a material thickness of x millimeters. Eventually, the engineer must design the system with these quantitative values. Give the engineer a number. Unfortunately, no single number will ever be "correct", at least where human subsystems are concerned. A tradition in engineering has been the inclusion of tolerances in specifications – a range of values around a point that is acceptable. Commonly, the engineer may assume that if he or she stays within the upper bound of the tolerance range, then the implications in terms of performance will be acceptable. Unfortunately, tolerances have a way of "stacking up" and although all subsystem designs may be within tolerance, the total system may fail. For example, a space suit may have sets of different-sized modules that accommodate a range of expected crew member segment sizes, but because of human body size and shape complexity, including imperfect correlation of segment sizes and changes due to microgravity, the performance of an EVA activity may be compromised.

An alternative to traditional "tolerances" is the use of loss functions. This involves the identification of a target value, which will be ideal and not interact adversely with collaborating subsystems, and a nonlinear function that "penalizes" deviations. As the system design develops, these penalties are amalgamated and a total system score is calculated. The decision process for system acceptance is based on a policy statement regarding the total system score and identification of those subsystem deviations where the greatest impact may be made regarding process performance. For example, a vehicle interior may specify loss functions for headroom, shoulder room, knee room, and eye height as well as many other parameters. In the final assessment of the perception of interior spaciousness or performance in a standard entry – egress test – the design compromise will optimize the amalgamation of these multiple loss functions.

Formal testing of the relationship between the individual (or sets) of system adjectives (independent variables) and process performance outcomes or adverbs (dependent variables) is the very basis of human factors engineering and its regression or analysis of variance tools. A shortcoming of this reductionist approach is that experimental management of many interacting and concomitant variables is often prohibitive, because of system complexity. An interesting alternative approach is described by the paradoxical statement that "if a nonconforming system passes a [process] performance test then the system can be considered to conform." Or to use a familiar truism: "the proof of the pudding is in the eating".

HUMAN FACTORS IN DESIGN

This paradox envelops the relationship between human factors engineering and their designer and user customers. When human factors enter the design process late with usability tests of the total operational process, it is often too late or too costly to rectify fundamental system design shortcomings. For example, ergonomics intervention in automobile manufacturing may influence workplace design, tool selection, and task content but cannot change the main design problem of inaccessibility of a particular component. The same is true of the maintenance of space hardware; if a suit is to be maintained on a remote planetary surface, there will be very different challenges from those encountered in a well-equipped workshop on earth. Conversely, when human factors is involved in the life cycle requirements planning early in the design process, it is more likely that a comprehensive set of performance requirements will lead to a corresponding set of system design loss functions and the sequence of evaluations as the design matures.

System design specifications can be verified and process performance requirements can be validated in an appropriate context. These important design evaluation processes are effective only if reliable testing processes accompany requirements and specifications. A generic phase of the design process can be described by analogy with the familiar educational process. The first component is the articulation of performance requirements – will the existing students have obtained knowledge that fits them for their next course or phase of their careers? The proof of the pudding is in the starting salaries of graduates or better still, the final examination should include an evaluation (validation) of performance in analogous situations. Curriculum or course design specifications flow from the outcome requirements. If the outcome requires problem-solving capability, then the course curriculum should specify practice in problem-solving. Verification of the curriculum, like verification of system design specifications, should be straightforward if the specifications have been articulated clearly and reliable tests have been planned and implemented. All too often classes are designed based on historical specifications, rather than customer requirements. The limitation of this analogy is that the educational and design processes are extremely complex and involve many subsystems, including teachers (engineers), classroom facilities (design facilities), students (internal customers), and employers (external customers). However, the discipline of process performance requirements (verbs and adverbs) first followed by system design specifications (nouns and adjectives) will assure a more satisfactory outcome.

The root of the design and education challenge lies in human variability and adaptability. Students may succeed despite their professors; vehicle customers may tolerate poor quality if the styling is exceptional, astronauts may succeed in their tasks despite design shortcomings. Conversely, unprepared students may fail despite good professors and facilities; poor drivers may fail despite well-engineered systems; and astronauts (or their support entourage) may fail if their training, experience, or readiness to perform are insufficient to meet novel or emergency situations. Examples of the former performance successes, despite subsystem failures, are found in the Apollo 13 and Skylab solar array incidents. Evidence of the latter failures was observed in the Progress collision and the Soyuz/Salyut tragedy.

Design for human variability may be addressed in several ways. The obvious way is to reduce the [human] variability by meticulous attention to "humanware design" – selection, training, assignment, and performance monitoring. Historically, NASA has had great success in this respect although performance monitoring has always been a bone of contention among crewmembers who are reluctant to publicize their shortcomings. An analogous process in professional sports does not suffer from this shortage of evidence. The sports pages are full of the most detailed performance statistics of these highly talented, selected, trained, and paid athletes. At the other end of the design spectrum, consumer product design, including automobiles and their usage contexts, must accommodate a wide variety of minimally talented, marginally selected, inadequately trained, and rarely monitored users. Only catastrophic failures are documented and the usually forgiving context allows recovery from gross human error and minimal monitoring of inappropriate behaviors.

Design for highly trained and talented human operators is easier and more forgiving than for the broad population of consumers. But this can lead to complacency and overreliance on the human operator to accommodate for design shortcomings. It should be noted that even the best operators suffer from human fallibilities, such as inattention, fatigue, overload, and debilitation. Picture a good driver finding his way through a strange city, in the fog, on icy roads, to an important meeting deadline. Translate this into an astronaut, debilitated by a long interplanetary journey (or EVA), wearing a cumbersome suit, and finding his or her way to a safe haven with limited consumables. The focus of system design must acknowledge expected use and foreseeable misuse. An automobile must be designed to protect an inebriated driver in the event of a high-speed collision. Space hardware must be validated in similarly challenging contexts.

Automobile design has a considerable advantage of an enormous amount of data. Space exploration is relatively data poor. Consequently, space system design must take advantage of contemporary modeling, simulation, and analog facilities. These facilities, to be predictive of human performance in space, must address human shortcomings as well as their successes. It is one thing to winter over in the Antarctic and suffers from frostbite, or run out of air in NEEMO and have your buddy lend you his spare regulator. It is altogether different with a minimally redundant crew on their way to a distant planet when the doctor gets a toothache, a solar flare erupts, or a piece of software misbehaves. The Advanced Integration Matrix (AIM) program aims to answer the challenges of expected use and foreseeable misuse with a comprehensive suite of digital and analog simulations and an extensive repertoire of what if questions, with particular reference to the many sources of human performance variability.

SIMPLE RULES FOR PROCESS AND SYSTEM DESIGN

- Differentiate between process requirements and system specifications.
- Develop tests for specification verification and requirements validation.
- Develop a comprehensive picture of system design interactions and process performance outcomes.
- Develop digital simulations of mission process performance and carry out sensitivity analyses of hardware, software, humanware, and organization-ware subsystem design ranges.
- Use contemporary tools such as Failure Mode and Effects Analysis, Fault Tree Analysis, Human Reliability Analysis, Quality Function Deployment, and Discrete Event Simulation to evaluate expected use and foreseeable misuse.
- Comprehend human variability on all dimensions, including physical, sensory, cognitive, psycho-social, and affective.
- Design in redundancy and forgiveness – make space travel as safe as driving to work.

THE MYTH OF USER CENTERED DESIGN

User-centered design is not a myth – all design is user-centered – but human factors engineers do not own the practice unless of course all designers are considered to be human factors engineers. The following discussion contains an explanation of this paradox together with an articulation of the boundaries of our profession.

All animals take steps to protect themselves from their physical and social environment and extend this defense into offense by physical and operations designs that give them an advantage over their competitors, sometimes through collaboration. Darwin explained the motivation and many of the mechanisms of this competition among and within species. The annual Darwin awards describe some ingenious but failed efforts (by *homo sapiens?*) to gain some advantage, either through engineering or operations design.

The purpose of engineering and operations design is to extend human capabilities; even some small-brained primates have been observed to make use of tools, such as nutcrackers. The process of design is predictive and should cater for expected use and foreseeable misuse. Use involves sometimes-conflicting criteria, such as effectiveness (quality), efficiency (resource utilization), safety, health, and satisfaction. Human use sometimes extends satisfaction to the lofty, but nebulous, heights of pleasure. The phases of this process include a mental model of how the engineered device or process will work and how it may fail, together with an assessment of the possible positive or negative outcomes. This stage models the behavior and performance of the resultant design in its context. The second design stage is to create some representative or physical (or electronic) model of the eventual system, commonly in the form of a drawing and some specifications, a computer model or a physical mock-up. In many instances, an analog will contribute information to the design process. The next stage (unless your designer took a concurrent engineering class (another myth)) is to design the manufacturing process. Most of us could design or at least visualize a functional pyramid but not comprehend the challenges of getting a bunch of reluctant Hebrews to build it. Even if we could design a car and the tools to form the metal and fasten it together, it is left to the operations designer to make this manufacturing process efficiently produce 1000 similar units a day, all with high quality. After all this design has taken place, we have to manage the use of the product or process. This involves making sure that some foolish or incompetent user does not misuse the product and that the product continues to perform reliably over its designed lifetime. Finally, we have to bury the worn-out thing and its batteries in a landfill or REDUCE, REUSE, and RECYCLE.

Where does user-centered design fit in? For a start, all people are users and by definition will choose and operate some device, such as a chair, car, or computer to enhance their capabilities or pleasure. *The challenge is complexity.* Although most people can visualize the operational use of a particular product, the actual design, manufacture, production, or maintenance of most products (including hardware, software, and organizationware) is beyond the capabilities of most people *(Norman – The Design of Every Day Things)*. In fact, with the exception of a diminishing number of craft industries, most products are actually designed by teams of people, including inventors, market research, conceptual design, engineering design, manufacturing design, production design, production operations management, marketing, distribution, sales, service and maintenance, and operations management. Each of these teams turns specifications into products and hands over to the next phase, unless, of course, they have heard about concurrent engineering. Concurrent engineering is simply a way of adding constraints to the design process as early as possible. Human factors engineers join in this gig to use their knowledge and tools to anticipate expected use and foreseeable misuse. Human factors engineers, like cost analysts, industrial hygienists, lawyers, doctors, and personnel people, are rarely responsible for any one of these design activities but they do have a lot to contribute to each stage, depending on how people interface with the process.

In fact, in large organizations, human factors engineers may not directly help the person responsible for a phase rather they may play the role of advising the advisers. For example, in manufacturing operations, it is the line supervisor who actually manages the operations personnel and he or she, in turn, is advised by the manufacturing engineer, the safety specialist, the quality guy, and the industrial engineer. The human factors engineer is usually at least one step removed. In product design, the conceptual designer, marketing specialist, and the various engineers and managers collaborate on a product development team, and the human factors specialist has to be very polite when he is invited to the table, often at the behest of the company lawyer. And even then, the contribution of human factors may be only to add color to the warning label that informs the eventual user that the use of the product may be acutely life-shortening.

This dismal picture paints the human factors engineer as a cosmetic afterthought and puts our aspirations of being in charge of user-centered design in perspective. Unfortunately, some models of the human factors process place the human factors engineer as a purveyor of ambiguous requirements, using the excuse of human variability, followed much later in the process as an ergocop with an exaggerated view of self-importance as he or she interprets the vague requirements and signs off on waivers when told to do so by a wise manager. Too often this clumsy process is due to the lack of knowledge of the human factors engineer of the domain in question. In defense of the HFE, the acquisition of domain knowledge is not always easy, even after grounding in the local jargon and acronyms. This lack of domain knowledge of the human factors engineer can lead to a lack of credibility, both for the individual and the profession. The excuse that "you may understand the domain, but I understand how people behave and perform" is not good enough.

The challenge to the human factors engineer is, of course, human and situational variability. It may be easy to design car, navigation system, entertainment system, and cell phone interfaces from first principles, focus groups, and usability trials and estimate the average or even 5th percentile performance levels of the user population. But the reality is that all these come together in the infamous single channel that is narrowed by grandma's age, grandad's tipple, or some unforgiving traffic light. Even without human factors engineers, designers will design systems that suit many people most of the time, occasionally with the help of legislation, a product liability suit, or simply product failure in the marketplace. Human factors engineers must learn to deal with their Achilles heel – variability, and not hide behind it in the hope that Lady Luck will be on their side.

There is a better way. Human Factors Engineers should adhere to the following rules:

1. Learn about all of the Human Factors – body, mind, and soul; behavior, performance; and preference
2. Learn about statistics and investigation design – confounding and significance
3. Learn about your domain or find a friendly expert to run interference for you
4. Know your place as a supporter of the design process, not as an ergocop
5. Recognize that there are many "users" in the design and operations process

PROCESS AND TRANSACTION TIMES

The fundamental measures in ergonomics are space, mass, energy, information, affect, money, and time. A "transaction" adds a time dimension to the mix of energy, information, and affect, and provides the opportunity and basic unit for measurement and analysis. A transaction may be lifting a suitcase, checking the time, or buying a loaf of bread. Alternatively, it may be a day at work or a journey across the country. A transaction can be described, measured, and repeated. It will always have a time dimension and one or more spatial, force, information, or affect dimensions; it may also cost or produce money. Human participation in a transaction may result in fatigue, learning, or enjoyment. A more elaborate analysis of the outcome of a complex transaction includes effectiveness (quality), efficiency (use of resources such as energy, time, people, materials, or money), enjoyment, and satisfaction for one or more participants (the affective dimension); transactions may be easy or difficult to conduct on energy or information dimensions. Transactions may have desirable or adverse safety and health outcomes. Finally, transactions may be repeated indefinitely under compatible contexts and may be resilient or otherwise to extreme contextual demands. The observable and measurable transaction, including its variable human, technological, operational, and contextual participants and its various outcomes, is central to the practice of ergonomics, in research, simulation, analysis, design, and evaluation.

COMPLEXITY, COLLABORATION AND COMPETITION

Transactions will usually involve one or more people, simple or sophisticated technology, and a plethora of physical and operational contexts that cannot be changed but must be considered in transaction analysis. The outcome of a transaction may be success or failure, in the eyes of one or more of the people involved. Outcomes may be discrete and clear or based on subjective assessment. The operational context of commercial aviation involves schedules and communication among various participators regarding the use of resources, such as airspace and runways. Other participants may represent the airlines, customs, and immigration officers. In competitive field sports, there are opposing teams participating in a zero-sum game – if one team wins, the other loses. Various basic transactions are combined to constitute a game or series of games. In soccer, the transactions involve dribbling, passing, tackling, and shooting, which may lead to successful or unsuccessful outcomes, depending on which side you are on. Despite these varying levels of complexity, transactions can be described, observed, measured, and repeated many times.

PHYSICAL TRANSACTIONS

Physics deals with combinations of mass, space, and time to describe energy: 1 Joule = 1 kg·m^2/s^2. Energy is described in different ways, such as potential energy, kinetic energy, mechanical energy, and thermal energy all of which obey a law of conservation that suggests that energy is not lost but it may be transferred, change its form, or stored. Other important physics considerations are that we are surrounded by gravity; that Force = mass times acceleration, and that Newton offered a perceptive description of why things move or stay in place. People eat and move, and if these are not balanced they store energy as excess fat; furthermore, with training, and the right choice of parents, people greatly improve their capabilities at transferring food into mechanical energy, plus "waste" energy in the form of heat.

A basic physical transaction is "the step", a milestone in a child's development. This involves a contraction of the plantar flexion muscles of the ankle and knee and hip extensors to raise the body against the downward pull of gravity; the body then falls forward and the downward motion is arrested by the other heel striking the floor. The physiological mechanisms of this transaction include controlled muscle contraction, which requires energy plus tactile, proprioceptive, vestibular, and perhaps visual and auditory feedback. Over time the basic mechanisms become reflexive and automated and many variations on the basic transaction (the step) are developed to enable walking, running, stair climbing, side stepping, dancing, and gymnastics. This progression is the result of maturation and practice as the person develops larger, more varied, and more complex transaction chunks, which are observable, measurable, and repeatable.

INFORMATION TRANSACTIONS

Information science describes a "bit" of information as $\log_2 n$, where n is the number of (equiprobable) choices; greater mathematical complexities are needed where items of information are not equiprobable, which is usually the case. Shannon's theory of communication describes the following elements of Information transfer: Source – transmitter (encoder) – medium – receiver (decoder) – destination, to which may be added, as with energy, the concept of storage or memory. Another important aspect of information transfer is feedback from the receiver to the sender to confirm accurate and timely transmission, although the one-way broadcast mode may sometimes be chosen, sometimes with delayed feedback. The transfer of information is measured in bits per second and some communication media have greater capacity than others. Over time larger and larger chunks of information are created to produce meaningful messages. In psychology, information manipulation and transfer are called cognition, which has numerous components, including perception, prediction, decision-making, and control. Unlike energy, information may be lost during the transfer process or clouded in noise – information from other, usually unwanted sources. The familiar process of forgetting involves a failure to retrieve once-stored information. Also, multiple sources of information, when combined before, during, or after a transfer process may give rise to new larger and more complex chunks of information; this, as with walking, is the process of learning. A metaphor is a book that consists of words, sentences, paragraphs, and chapters that are bound and labeled to produce a story.

In applied human factors, the communication of these more or less complex chunks of information is characterized as transactions. The transaction may be measured in terms of time and the accuracy or otherwise of the outcome, plus some form of immediate or delayed feedback. We switch on our computer and a login screen appears; we enter our credentials and our home page appears; we enter a URL or click on an icon and our e-mail or a web page appears; we examine the contents, type a message, and press return to complete our transaction sequence. This transaction takes a variable amount of time and the outcome success or otherwise will be fed back to us at some future time. In reality, there are millions of such transactions that are measurable and repeatable with variable times and outcomes. The intent of human factors is generally to minimize transaction time, ensure transaction accuracy, and guarantee a successful transaction outcome. This simple intent is complicated by variability among the people involved, the vagaries of the technology, and the influence of other external factors that cannot be changed but must be guarded against.

Consider the simple challenge of ordering a meal at a restaurant. The customer looks at a menu and makes a selection following suggestions by a smiling waiter who scribbles down the order and reads it back for confirmation, then passes it on to the, perhaps less happy, short-order cook; a short time later the still smiling waiter brings the, hopefully correct, meal to your table, followed sometime later by the bill. There are, of course, quite a few variations on this transaction format. Now, in the interest of accuracy and productivity, we develop an automated ordering and payment system, with many colorful pictures of sumptuous meals and prices according to size and embellishments; these pictures are surrounded in emoticons aimed at replacing the waiters' smile. Whether automated or with the help of a waiter, this communication transaction will take a variable length of time and may generate errors at various stages. The eventual bill, thanks to the addition of automation may contain a dozen or more information items for the sake of posterity. The transaction sequence is repeated many times, with or without automation, throughout the world in many contexts; it is observable, measurable, and repeatable.

Consider next you are driving a car in a stream of traffic. The car in front brakes hard. You see the LED center high-mounted stop light flash and automatically move your foot off the accelerator and across to the brake; you may be a foot lifter or a foot rocker if you leave your heel in one place. Hopefully, you hit the brake hard enough and stop short of the car in front. But the driver of the car behind you is talking on his cell phone and has cruise control activated. Bang! You are shunted into the car in front and then bang again as the fourth car pushes the third car into you again. The keys to this sequence of overlapping transactions are reaction time and response time. Your reaction time was aided by the short rise time LEDs being in your forward field of view rather than the slower incandescent traditional brake lights. You may also have gained a head start on this transaction by noticing the slowing of vehicles further up the line thus enabling anticipation. The driver behind you was distracted and the cruise control continued to operate until his foot left its resting place and moved over to the brake, causing a further extension of his transaction time. This sequence of events can be described as an overlapping series of transactions, each of which is observable, measurable, and repeatable. Changes in the parameters of these transactions, such as reaction times and vehicle speed and spacing, may affect the outcomes. The debate is whether adaptive cruise control is superior to attentive drivers in avoiding unwanted outcomes on urban freeways.

A third observable, measurable, and repeatable transaction involves landing an airplane, with or without the aid of automation and in variable wind conditions. This transaction is constrained by time, vertical and horizontal distance, and speed. The attentive pilot will look at his instruments and out of the window, and if the changing scene suggests that he is "low and slow", he must quickly add power and "go around" to make another approach and landing. In the case of Asiana Flight 214, the pilots "over relied on automated systems that they did not fully understand". This confusion led to a delay in the decision to manually initiate a "go around", given the airplane's unsafe approach.

Affect

Affect is less easily measured or understood; it deals with likes and dislikes that may be combined with importance or weighting factors. These weighting factors result from past experience and anticipation of future outcomes. Affect may also be vulnerable to environmental and social pressures and fashions. Affect is what drives our emotions and motivations to create opinions and behaviors. Affect also varies over time, sometimes it drifts gradually and sometimes perhaps a minor incident may cause a monumental change; such changes in affect and perception can be articulated by the "halo" or "pitchfork" effects, depending on the direction of the change. Money is a derivation of affect, it intermingles with energy and information, to constrain behaviors, form opinions, drive motivations, and explode emotions. Money makes the world go round, it is the root of all evil and *Geld macht nicht Glücklich, aber es Beruhigt.* From the ergonomics viewpoint, money and affect complicate everything! The analyst who ignores these affective and emotional factors does so at his own peril!

In the case of consumer behavior, the purchaser may assess the physical and informational features of a product but be more influenced by nonfunctional or cosmetic characteristics. Consider the purchase of a car, the selection of a child's toy, or the choice among alternative forms of footwear. In each of these examples, the consumer may allow either fashion or price to dictate the decision, which may be done quickly, on "impulse".

Some Notable Transactions

Drinking a cup of tea and landing an airplane are examples of transactions. There's many a slip twixt cup and lip; spilling a hot cup of tea in your lap may be quite uncomfortable, as too may be a hard landing. These transactions are carried out many times, every day, with a wide variety of people, technologies, and contexts. Both transactions involve energy, information, time, money, and affect. The outcomes of the transactions may be measured on many dimensions, including time, accuracy, and style. Running a marathon certainly takes motivation and time and uses energy, and various outcomes produce a lot of emotion; the motivation to return next year increases after the pain has diminished. Buying a complex product, like a car or computer, on a complex website using a complex mobile phone application also takes time and may or may not have an accurate or pleasing outcome.

These examples represent a spectrum of transactions that are repeated and measurable. They are clouded by variability among the participants, technologies, and contexts. They all take time. They all have measurable outcomes. They are all amenable to ergonomics analysis and design as people are involved in some way in the transaction. The tools of analysis are measurement instruments, probability, and statistics; the tools of design involve compromise and cost constraints and are driven by affect as well as some more tangible considerations of energy and information.

ERGONOMICS PRACTICE

In practice, ergonomics deals with particular cohorts, conditions, and conclusions. The transaction and its outcomes are well defined, at least in statistical terms. The ergonomist may report one thing to his colleagues through the medium of forums and publications and something else to the customer who is responsible for the change. This customer will be thinking of a thousand other things apart from the reductionist ergonomics advice. Design is about compromise. Design is also about the long and short term, including the many repetitions of the transaction of interest, and its variations.

First, a physical ergonomics situation will be explored to explain the compromises in transaction design. Consider a sick patient in a hospital bed who from time to time has to be moved for bathing, sleeping, eating, exercise, or treatment. This movement transaction involves the patient, the nurse, some technology, time pressure, expense, and certainly affect. From the physical viewpoint, the transaction involves moments (force times distance) that will be repeated many times during the nurse's and patient's day. Biomechanical analysis involving large immobile patients and small fragile nurses will conclude that the moments and movements are too large and the transaction is impossible to perform safely. So the ergonomist suggests technology – let a robot grab the old lady and move her up the bed so she can eat her lunch. But the old lady may not like robots, and anyway robots are expensive. Impasse and rapid deterioration! Compromise is needed. The ergonomist then suggests less elaborate technology or administrative controls such as training and method change – including getting two nurses to do the job. But for reasons of mechanics and money, these suggestions too are rejected. More compromise is needed. This doesn't work either. So, the result is that sick patients and motivated nurses all over the world make the best of their situation. Ergonomists need to be both finders (analysts) and fixers (designers); they also need to simulate their suggestions before implementation and objectively evaluate their designs in the real world. The same problem occurs on a vehicle assembly line where well-meaning materials handling aids are bypassed in the interest of convenience and productivity.

A second example, more related to the information dimension, involves a professor and a class of students. Here, we have people, information, technology, money, affect, and time. The transactions are repeated many times for years and years. The solution to this mass transfer of information is technology. First, it was the printing press and books, but these did not make the teacher redundant, because the teacher's job is to motivate and explain and not just to be a knowledge delivery vehicle. Now, the delivery media are the Internet and portable technology, but still, we need the teacher for elucidation and motivation. This contemporary teaching technology manacles the students and teachers to a knowledge conduit (laptop or smartphone?) for long periods of time. This is not conducive to effective learning or good health unless it is punctuated by active participation with the teacher, colleagues in study groups, and with the vast amount of related information on the Internet and in the academic and technical literature, and through empirical investigation. Currently, the explosive development of eLearning tools is being pushed upon rather than pulled by the student community. The transactions include elements of motivation, delivery, discussion, consolidation, and examination at the micro- and macrolevels, according to Blooms' taxonomy of learning stages. In practice, these transactions may be more chaotic than precisely managed. The educational transactions, like chunks in information theory, vary in size and content; they stick together and grow; they need fertilizer and sunlight, and not torrential rain or frost. They need collegiality and collaboration, not inundation and asphyxiation. They need clarity, not obscure metaphors.

Affect – the Kano Model

The Kano model describes different levels of requirements in product design – "must have", "more the better", and "excitement" factors; over time as the product matures, the "excitement" and "more the better" features may become "must have". These requirements are extracted from existing and potential customers by questionnaire, interview, and focus group methods. Although many of the factors may relate to physical and informational dimensions, most have overtones of affect and preference. The earlier application of this Kano approach to product design may be extended to service evaluation, and more recently to job and organization evaluation. A recent study of the employment of elderly Singaporeans indicated, similar in some ways to the Herzberg two-factor theory, that basic requirements of collegiality and job content were must have features, even more so than salary. These investigations using the Kano principles revolved around describable activities or "transactions". The Kano model is a powerful concept because it describes the influence of affect in transaction strategy, behavior, and performance.

EpICTC

EPICTC stands for Effective Performance in Information Complexity and Time Constraints. It is the title of a research project to study why transactions have successful or unsuccessful outcomes. The transactions of interest may include: how does a nurse deal when confronted with a rapidly deteriorating patient? How does a pilot land an airplane when the supporting technology fails? How does a game player or referee decide what to do in a developing dynamic situation? When should a policeman decide to use his weapon? What should a driver do when the car in front stops quickly and the driver in the car behind is on cruise control or using his cell phone? Such transactions occur frequently and have common components – information complexity, outcome stress, time constraints, and various levels of expertise.

These transactions occur for real in the outside world and can be simulated in the laboratory with widely available computer-based virtual reality. Similar transactions are also found in the widely popular video game world. Children of all ages participate in these transactions for entertainment; they fail and succeed and progress to the next level of complexity; they practice to reduce their transaction times. They develop expertise in dealing with information complexity. The key to the success of these games is the provision of immediate feedback regarding the outcome of a transaction. Learning theory is full of material on positive and negative outcomes and feedback. Human transaction performance related to real-world situations may be tested under manipulated conditions of outcome stress, information complexity, time constraints, and expertise.

Transactions may involve one or more people, various levels of sophistication of technology, rules and regulations, and many outcomes. Transactions involving energy, information, and affect often in an unpredictable context. The challenge for the human factors engineer is to define, describe and delimit the transaction, and then develop appropriate measures and methods of measurement, including measures of process – the input variables, contexts, and outcomes. The human factors specialist will also describe the target population and develop reliable samples, in theory, random samples, but in practice usually convenient samples. Given these data and, appropriate statistical analyses the investigator will identify causes and effects and suggest technological or operational interventions.

PREFERENCE, BEHAVIOR, AND PERFORMANCE

The ultimate purpose of ergonomics/human factors is performance – transactions should be carried out effectively, efficiently, safely, and to the satisfaction of all concerned. A transaction is any interaction of a user with an engineered or natural situation in whatever context. Think of a transatlantic flight, jumping out of an airplane, booking a flight on the web, handling your baggage, and deciphering the small print when your baggage has been lost. Performance measures can usually be reduced to measures of time and accuracy – accuracy being a measure of the deviation between the intent and the result; time is usually a reflection of efficiency – the utilization of resources. Sometimes "performance" is indicated by customer satisfaction, perhaps with esthetic underpinnings, such as the "performance" of a hairdresser.

We often have difficulty with performance measurement in usability testing – sometimes this is due to our failure to describe performance in tangible terms. Behavior begets performance. People have different physical, sensory, and cognitive characteristics; consequently, when faced with a given set of circumstances they behave differently. This different behavior may result in satisfactory or unsatisfactory performance. Just look at differences in web page navigation techniques – even among those pages that have passed the usability tests.

Despite all our attempts to study and design for performance and accommodate a wide range of behaviors, we are often faced with the ultimate challenge – individual preference. This fact has been brought home to me in a resounding fashion by my attempts to please the buyers and builders of cars and the users of space vehicles. The customer is always right.

In a fair world, we would define the range of acceptable performance, accommodate the spectrum of possible behaviors, and allow as wide a range of preferences that is compatible with the restrictions of behavior and performance. We should not allow the use of sledgehammers to crack nuts, even if some nuts prefer that behavior. We should not allow our 16-year-old to drive our Corvettes, even if they have passed driver's ed'. We should not design low contrasts into our computer interfaces, just because some (younger) users prefer it that way.

THE RULES OF MEASUREMENT

We apply measurement and analysis so that we can design something. We measure people, the things that we can change (the independent variables), the things that we cannot change (the context of use), and the outcomes of our designs – performance, behavior, and preference (the dependent variables). If we had paid attention to our first few classes in quantitative methods, we would remember that there are some very important rules of measurement. First, we must calibrate our instruments – whether they are rulers, electronic sensing devices, simple checklists, or surveys. We must be confident that our tools don't produce "measurement artifacts". Even simple psychophysical techniques are susceptible to errors imposed by the instructions to the subject. Our measures must be accurate. They must not result in systematic bias, we must minimize the random or residual error and we must avoid blunders. For example, have a sample of measurers measure the joint angles of a sample of subjects? You will observe systematic and random errors and blunders.

The next rules of measurement include resolution and precision. They are not the same. Resolution is the number of decimal places. The distance between my home and my work is 1.2345678 miles, as the Shuttle flies. The last six digits are unlikely to be of any use to anyone. The height of my chair is 16.54321 inches. So what? We should not attempt greater resolution in our measurement than is likely to be useful. Precision works in the opposite direction. The dimensions of my roll-on suitcase are 22 × 14 × 9 and weigh about 70 lbs, more or less depending on how much I have packed. A lack of sufficient precision could cause me considerable inconvenience and embarrassment when I find that the bag will not fit into the overhead bin and I have to pay for excess baggage. When measuring people, our instruments must be sufficiently precise. The questions – how much can you lift? or how fast can you run? – are best supported by precise instrumentation.

Finally, we have reliability and validity. Reliability is about repeated measures, perhaps in different contexts. We may have multiple instruments, multiple observers, multiple subject samples, and multiple occasions. Reliable measures are not affected by these contexts. "Get real", you may be thinking. In practice, reliability (or lack of it) is a major headache for human factors engineers that demands considerable expertise in experimental design. Validity is another challenge that comes with many faces. The bottom line is that the measurement should be predictive of the eventual performance, behavior, or preference in the real world. If the usability test is not predictive, then it is not useful. But validity is a two-edged sword, sometimes we use invalid tests to prove a point that is not important in the long run. For example, we sit the customer in a car in the showroom and ask him to assess seat comfort, without discussing the context of use – commuting off-road or long distance. On some occasions, we have to do our best to simulate the eventual context – it is difficult to study microgravity effects on performance and behavior on earth (except through flying parabolas in a KC135 or riding in a centrifuge).

EXPECTED USE AND FORESEEABLE MISUSE

Engineers design things for intended and expected use. Lawyers thrive on foreseeable and unforeseeable misuse. Ergonomists have the unenviable challenge of pleasing most of the people most of the time, including the users, the engineers, the lawyers, the juries, and the historians. Expected use begs the question of who is the expected user. Will it be average Joe, granddad, or a superhero? Will it be on a train or in the rain, with a spoon or on the moon, with a plant or in a plant? Does foreseeable misuse include the operation of a car at night, when intoxicated or when someone else crashes a red light? Does foreseeable misuse include our less than average Joe using a weight-training machine designed for an NFL lineman? Should grandma be "allowed" to use an ATM, a VCR, a cell phone, a PDA, or the Internet? Should a rock star be allowed to fly into space? Should a handicapped person be allowed to work?

A recently popular term is "universal design". Taken at face value this is an impossible and unrealistic dream. The system design must comprehend the capabilities and limitations of the expected users. This means that some amount of selection, training, and assignment is implicit in the design. In reality, "universal design" implies the accommodation of a greater variety of people in a greater variety of contexts. The challenge to usability testing is to offer reliable and valid evaluations for expected users and foreseeable misusers of all or most shapes, sizes, and abilities in a broad range of contexts. What exactly is meant by "most"? Finally, remember we have to keep both the engineers and the lawyers happy as well as the intended and unintended users.

STATISTICS AND EXPERIMENTAL DESIGN

Fortunately, the well-prepared human factors engineer acquired the knowledge necessary to address all these challenges while he or she was in graduate school. Unfortunately, many conveniently forget the rules of measurement and design because they are too complex, inconvenient, or perceived as being infeasible in the context of their work. Statistics is simple – all it requires is knowledge of probability, agreement on risk levels, and application of various tests to assess probable differences and associations. Experimental design is another kettle of fish. If you don't ask the right questions of a reliable and valid sample of users, no amount of statistical analysis will give you the right answers. This sampling issue is a common shortcoming of human factors investigations.

All human factors investigations (experiments, evaluations, measures – call them what you will) involve the following elements – conditions (independent variables), contexts (uncontrollable variables that must be taken into account in the design so as not to confound the conclusions), and dependent variables – usually measures of performance, behavior, or preference. It is common for human factors investigators to focus on the conditions (independent variables) of interest. This may be a workplace, a work task, a workload, a workpace, or a workpiece. We manipulate the levels of each of these conditions and balance them to allow us to assess interaction effects and to avoid confounding. Confounding is the assignment of cause to the wrong source. For example, if we were to measure driver habitual speed as affected by the type of vehicle and did not control for road and traffic conditions, we may get spurious results.

The biggest challenge is those contextual "uncontrollable but important" variables. The issues of physical context are generally amenable to understanding and produce caveats on the conclusions. Behavior and performance in a simulator will always lag (or exaggerate) reality to some extent. Indeed, the use of simulators to take the user beyond normal safe operations and through failure (and perhaps recovery) may be very beneficial in both training and design. The most important "uncontrollable" variables are the between and within subject factors. Unfortunately, these challenges are ubiquitous and the Achilles' heel of human factors investigators. Even with the best will in the world, it is rare for a human factors investigator to have access to a truly representative (unbiased) sample of expected users and possible misusers. Also, it is impossible to prevent subjects from learning things that affect their behavior and performance (the experimental outcomes – dependent variables). Sample size requirements in reliable "between" subjects design are much greater than the more efficient "within" subjects design where an individual subject may be exposed to multiple experimental variables. Validity is another kettle of fish.

These challenges are attacked by adopting between subject designs – where different subjects or subject samples are assigned to different experimental conditions. The analysis of these "mixed model" or "random factor" designs involves robust comparisons between the main effects and the main effect times subject interaction. But such designs are often inefficient – they require many more subjects. They may also be unintentionally biased because of nonrandom assignment of subjects to conditions. Within subject designs make good use of subjects – while you have the subject there, you might as well use him in all the conditions. You may also use the subject as his own "control" so that inherent subject biases are removed. (Think of the paired sample T-test). But subjects learn and may also become fatigued – which in a general sense has the opposite effect on learning. These between and within (learning) subject effects may often completely obscure the effects of the conditions of interest – the only reliable solution being to have unrealistically large sample sizes.

Because subject and learning effects are ubiquitous, it behooves us to always address the problem in the design of experiments with as much gusto as we address the conditions of interest. If the informed reader will permit the assumption that there are no other systematic interactions between the condition and trial or the condition and subject, then the Latin Square can come to our rescue.

In this basic design, each subject is exposed to each condition and each condition occurs at each location on the learning (trial) curve. The astute observer will note that the conditions are in a fixed sequence (C2 follows C1), which may result in unwanted (but hopefully unimportant) carry-over effects. However, the philosophy behind this design is sound – it accounts for the unwanted confounding of subject and trial effects. Where the main condition effect is complex, such as where there are multiple conditions, some of which may interact, then these factors must be balanced within the main Latin Square. Tombs have been written on the intricacies of experimental design; it is the responsibility of the investigator to ask the question – "are my conclusions accurate, reliable and valid?"

The Voice of the Customer

The foregoing discussion should have convinced the honest ergonomist of the sea of landmines inherent in extracting an accurate, reliable, and valid voice (performance, behavior, and preference) of the customer. Frequently, our sample of customers (users) is biased, our sample sizes are not sufficient, our contexts are invalid, and our predictions are likely to be inaccurate. If you don't get the right answer, ask another customer (focus group). We should give up now while there is still hope! Ergonomists should stop measuring people! But there is hope – we have an ocean of accurate, reliable, valid, and useful evidence in our literature. This evidence was collected by researchers (not only human factors researchers) who obeyed all the rules and whose communications passed the peer review processes. Much of the evidence has stood the test of time and much of it is based on sound scientific logic and principles to support the empirical evidence. Unfortunately, some of our "evidence" is mythical – unsupported by sound theory or data. So, there is job security for our research brethren and there should also be job security for the practitioner community, so long as they obey the rules.

REFERENCES

Peacock, B. (2019). *Human Systems Integration*, Self-published manuscript, Fernandina Beach, FL.

Peacock, B. (2020). *How Ergonomics Works*, Self-published manuscript, Fernandina Beach, FL.

Peacock, B. (2021). *Ergonomics Tools and Applications*, Self-published manuscript, Fernandina Beach, FL.

Evans, M., N. Hastings, and B. Peacock (2000). *Statistical Distributions* (3rd ed.). John Wiley & Sons, New York, NY.

Index

For Product Safety Concerns and Information please contact our EU
representative GPSR@taylorandfrancis.com
Taylor & Francis Verlag GmbH, Kaufingerstraße 24, 80331 München, Germany

www.ingramcontent.com/pod-product-compliance
Lightning Source LLC
Chambersburg PA
CBHW060822170526
45158CB00001B/59